First Do No Harm

RELATIONAL PERSPECTIVES BOOK SERIES
Volume 45

RELATIONAL PERSPECTIVES BOOK SERIES
LEWIS ARON & ADRIENNE HARRIS
Series Editors

The Relational Perspectives Book Series (RPBS) publishes books that grow out of or contribute to the relational tradition in contemporary psychoanalysis. The term "relational psychoanalysis" was first used by Greenberg and Mitchell (1983) to bridge the traditions of interpersonal relations, as developed within interpersonal psychoanalysis and object relations, as developed within contemporary British theory. But, under the seminal work of the late Stephen Mitchell, the term "relational psychoanalysis" grew and began to accrue to itself many other influences and developments. Various tributaries—interpersonal psychoanalysis, object relations theory, self psychology, empirical infancy research, and elements of contemporary Freudian and Kleinian thought—flow into this tradition, which understands relational configurations between self and others, both real and fantasied, as the primary subject of psychoanalytic investigation.

We refer to the relational tradition, rather than to a relational school, to highlight that we are identifying a trend, a tendency within contemporary psychoanalysis, not a more formally organized or coherent school or system of beliefs. Our use of the term "relational" signifies a dimension of theory and practice that has become salient across the wide spectrum of contemporary psychoanalysis. Now under the editorial supervision of Lewis Aron and Adrienne Harris, the Relational Perspectives Book Series originated in 1990 under the editorial eye of the late Stephen A. Mitchell. Mitchell was the most prolific and influential of the originators of the relational tradition. He was committed to dialogue among psychoanalysts and he abhorred the authoritarianism that dictated adherence to a rigid set of beliefs or technical restrictions. He championed open discussion, comparative and integrative approaches, and he promoted new voices across the generations.

Included in the Relational Perspectives Book Series are authors and works that come from within the relational tradition, extend and develop the tradition, as well as works that critique relational approaches or compare and contrast them with alternative points of view. The series includes our most distinguished senior psychoanalysts along with younger contributors who bring fresh vision.

RELATIONAL PERSPECTIVES BOOK SERIES
LEWIS ARON & ADRIENNE HARRIS
Series Editors

Vol. 45
First Do No Harm:
The Paradoxical Encounters of
Psychoanalysis, Warmaking, and Resistance
Adrienne Harris & Steven Botticelli (eds.)

Vol. 44
Good Enough Endings:
Breaks, Interruptions, and Terminations
from Contemporary Relational Perspectives
Jill Salberg (ed.)

Vol. 43
Invasive Objects: Minds Under Siege
Paul Williams

Vol. 42
Sabert Basescu:
Selected Papers on Human Nature and
Psychoanalysis
George Goldstein & Helen Golden (eds.)

Vol. 41
The Hero in the Mirror:
From Fear to Fortitude
Sue Grand

Vol. 40
The Analyst in the Inner City, Second Edition:
Race, Class, and Culture Through a Psychoanalytic Lens
Neil Altman

Vol. 39
Dare to be Human:
A Contemporary Psychoanalytic Journey
Michael Shoshani Rosenbaum

Vol. 38
Repair of the Soul: Metaphors of Transformation
in Jewish Mysticism and Psychoanalysis
Karen E. Starr

Vol. 37
Adolescent Identities:
A Collection of Readings
Deborah Browning (ed.)

Vol. 36
Bodies in Treatment:
The Unspoken Dimension
Frances Sommer Anderson (ed.)

Vol. 35
Comparative-Integrative Psychoanalysis:
A Relational Perspective for the Discipline's Second
Century
Brent Willock

Vol. 34
Relational Psychoanalysis, V. III: New Voices
Melanie Suchet, Adrienne Harris, & Lewis Aron (eds.)

Vol. 33
Creating Bodies:
Eating Disorders as Self-Destructive Survival
Katie Gentile

Vol. 32
Getting From Here to There:
Analytic Love, Analytic Process
Sheldon Bach

Vol. 31
Unconscious Fantasies and the Relational World
Danielle Knafo & Kenneth Feiner

Vol. 30
The Healer's Bent:
Solitude and Dialogue in the Clinical Encounter
James T. McLaughlin

Vol. 29
Child Therapy in the Great Outdoors:
A Relational View
Sebastiano Santostefano

Vol. 28
Relational Psychoanalysis, V. II:
Innovation and Expansion
Lewis Aron & Adrienne Harris (eds.)

Vol. 27
The Designed Self:
Psychoanalysis and Contemporary Identities
Carlo Strenger

Vol. 26
Impossible Training:
A Relational View of Psychoanalytic Education
Emanuel Berman

Vol. 25
Gender as Soft Assembly
Adrienne Harris

Vol. 24
Minding Spirituality
Randall Lehman Sorenson

Vol. 23
September 11: Trauma and Human Bonds
Susan W. Coates, Jane L. Rosenthal, & Daniel S. Schechter (eds.)

RELATIONAL PERSPECTIVES BOOK SERIES
LEWIS ARON & ADRIENNE HARRIS
Series Editors

Vol. 22
Sexuality, Intimacy, Power
Muriel Dimen

Vol. 21
*Looking for Ground: Countertransference
and the Problem of Value in Psychoanalysis*
Peter G. M. Carnochan

Vol. 20
*Relationality:
From Attachment to Intersubjectivity*
Stephen A. Mitchell

Vol. 19
*Who is the Dreamer, Who Dreams the Dream?
A Study of Psychic Presences*
James S. Grotstein

Vol. 18
*Objects of Hope:
Exploring Possibility and Limit in
Psychoanalysis*
Steven H. Cooper

Vol. 17
*The Reproduction of Evil:
A Clinical and Cultural Perspective*
Sue Grand

Vol. 16
*Psychoanalytic Participation:
Action, Interaction, and Integration*
Kenneth A. Frank

Vol. 15
*The Collapse of the Self and Its Therapeutic
Restoration*
Rochelle G. K. Kainer

Vol. 14
*Relational Psychoanalysis:
The Emergence of a Tradition*
Stephen A. Mitchell & Lewis Aron (eds.)

Vol. 13
*Seduction, Surrender, and Transformation:
Emotional Engagement in the Analytic Process*
Karen Maroda

Vol. 12
Relational Perspectives on the Body
Lewis Aron & Frances Sommer Anderson
(eds.)

Vol. 11
*Building Bridges:
Negotiation of Paradox in Psychoanalysis*
Stuart A. Pizer

Vol. 10
Fairbairn, Then and Now
Neil J. Skolnick and David E. Scharff (eds.)

Vol. 9
Influence and Autonomy in Psychoanalysis
Stephen A. Mitchell

Vol. 8
*Unformulated Experience:
From Dissociation to Imagination in Psychoanalysis*
Donnel B. Stern

Vol. 7
*Soul on the Couch:
Spirituality, Religion, and Morality
in Contemporary Psychoanalysis*
Charles Spezzano & Gerald J. Gargiulo (eds.)

Vol. 6
*The Therapist as a Person:
Life Crises, Life Choices, Life Experiences,
and Their Effects on Treatment*
Barbara Gerson (ed.)

Vol. 5
*Holding and Psychoanalysis:
A Relational Perspective*
Joyce A. Slochower

Vol. 4
*A Meeting of Minds:
Mutuality in Psychoanalysis*
Lewis Aron

Vol. 3
*The Analyst in the Inner City:
Race, Class, and Culture through a Psychoanalytic Lens*
Neil Altman

Vol. 2
*Affect in Psychoanalysis:
A Clinical Synthesis*
Charles Spezzano

Vol. 1
*Conversing with Uncertainty:
Practicing Psychotherapy in a Hospital Setting*Rita
Wiley McCleary

First Do No Harm

The Paradoxical Encounters of Psychoanalysis, Warmaking, and Resistance

Edited by
Adrienne Harris • Steven Botticelli

Routledge
Taylor & Francis Group
New York London

Routledge
Taylor & Francis Group
270 Madison Avenue
New York, NY 10016

Routledge
Taylor & Francis Group
27 Church Road
Hove, East Sussex BN3 2FA

© 2010 by Taylor and Francis Group, LLC
Routledge is an imprint of Taylor & Francis Group, an Informa business

Printed in the United States of America on acid-free paper
10 9 8 7 6 5 4 3 2 1

International Standard Book Number: 978-0-415-99648-8 (Hardback) 978-0-415-99649-5 (Paperback)

For permission to photocopy or use material electronically from this work, please access www.copyright.com (http://www.copyright.com/) or contact the Copyright Clearance Center, Inc. (CCC), 222 Rosewood Drive, Danvers, MA 01923, 978-750-8400. CCC is a not-for-profit organization that provides licenses and registration for a variety of users. For organizations that have been granted a photocopy license by the CCC, a separate system of payment has been arranged.

Trademark Notice: Product or corporate names may be trademarks or registered trademarks, and are used only for identification and explanation without intent to infringe.

Library of Congress Cataloging-in-Publication Data

First do no harm : the paradoxical encounters of psychoanalysis, warmaking, and resistance / edited by Adrienne Harris, Steven Botticelli.
    p. cm. -- (The relational perspectives book series ; v. 45)
Includes bibliographical references and index.
ISBN 978-0-415-99648-8 (hbk.) -- ISBN 978-0-415-99649-5 (pbk.) -- ISBN 978-0-203-88519-2 (e-book)
 1. War--Psychological aspects. 2. War victims--Psychology. 3. Psychoanalysis--Moral and ethical aspects. 4. Psychology, Military. I. Harris, Adrienne. II. Botticelli, Steven.

RC550.F56 2010
616.89'17--dc22   2009048677

Visit the Taylor & Francis Web site at
http://www.taylorandfrancis.com

and the Routledge Web site at
http://www.routledgementalhealth.com

For
Lucas Botticelli-Glassman and Jake Tentler
and
in Memory of
Ruth Stein
(1947–2010)

Contents

Acknowledgments xiii
Contributors xv
Editors' introduction xix

PART I
Psychoanalysis and antiwar work: Healing 1

1 Where is the "post" in posttraumatic stress disorder?: First impressions working with Iraq and Afghanistan soldiers 3
   TOM MCGOLDRICK

2 Men learn from history that men learn nothing from history 15
   JEAN-MAX GAUDILLIÈRE

3 The psychoanalytic politics of catastrophe 29
   GHISLAINE BOULANGER

4 Whose truth?: Inevitable tensions in testimony and the search for repair 45
   NINA THOMAS

PART 2
The paradox: Psychology's militarism 65

5 Psychologists defying torture: The challenge and the path ahead 67
   STEPHEN SOLDZ

6 From resistance to *resistance*: A narrative of
  psychoanalytic activism   107
  STEVEN REISNER

7 Torture and the American Psychological
  Association: A one-person play   143
  NEIL ALTMAN

8 Violence in American foreign policy: A psychoanalytic
  approach   153
  FRANK SUMMERS

PART 3
War and militarism deconstructed   **175**

9 Psychoanalysis, vulnerability, and war   177
  ELI ZARETSKY

10 Casus belli   201
   FRANÇOISE DAVOINE

11 Combat speaks: Grief and tragic memory   223
   SUE GRAND

12 War stories   243
   DONALD MOSS

13 Notes on mind control: The malevolent uses of
   emotion as a dark mirror of the therapeutic process   251
   RUTH STEIN

14 The gendering of human rights: Women and
   the Latin American terrorist state   279
   NANCY CARO HOLLANDER

PART 4
Resistance   **303**

15 Living in the plural   305
    EYAL ROZMARIN

16 The politics of identification: Resistance to
    the Israeli occupation of Palestine   327
    STEVEN BOTTICELLI

17 Dread is just memory in the future tense   349
    ADRIENNE HARRIS

18 Resistance to resistance   359
    LYNNE LAYTON

    Index   *377*

Acknowledgments

This book, an edited collection of papers from an international group of writers, represents a deeply important set of issues and problems regarding psychology's and psychoanalysis' relation to militarism and warmaking. For all the authors and ourselves as editors, the issues faced and voiced in this book have been an ongoing part of intellectual and political lives for more than 40 years. So we thank our authors who wrote personal, clinical, intellectual, and political essays, spanning and crossing disciplines and identities. We thank you for your thoughts and ideas, and for taking time from astonishingly busy personal and professional lives to work with us on this project.

We thank friends and colleagues who have supported and provided sources of influence and ideas toward this endeavor. A book comes to life in a community or communities: Robert Sklar, Noah Glassman, Muriel Dimen, Virginia Goldner, Ken Corbett, Ana Daniel, Steven Seligman, and Bruce Reis.

The editors and publishers at Routledge have been enthusiastic supporters of this project. We are very indebted to and grateful for Kristopher Spring, who read, gave feedback, prodded, and encouraged in just the right balance. The problems we are describing in this book are international and we are grateful for the work and capacities of Routledge to see this book is widely read and internationally distributed.

We thank Lewis Aron, Adrienne's coeditor of the Relational Book Series, for his support and steady collegiality. For 30 years, Adrienne and Lew have been colleagues and participants in the relational movement. We like to think that this kind of book, with its mix of theory, practice, and politics, would have pleased Stephen Mitchell, whose vision created this series and so much else.

Thanks to Bruce Reis for his sensitive assistance with editing.

Books are edited and written in the interstices of work and family life. Our families motivated aspects of this project. They are our influences, our support, our guides to the need to work and to play. Adrienne thanks Bob Sklar, Kate Tentler and Justin Tentler. Steve thanks Noah Glassman.

A word about our dedication: We began to think about and plan this book as we were meeting in a series of playdates, starting several years ago, with Jake Tentler, Adrienne's grandson, and Lucas Botticelli-Glassman, Steve Botticelli and Noah Glassman's son. In various parks and playgrounds, watching these two children and others swing, rush about, argue, negotiate, worry over bullies, seek grown-up comfort, have snacks, and play, we talked about and planned this book. We like to think that something about being with these lively children at play guides some underlying hope and dread in this book. The book is made in the hopes that remembering and working through is part of what changes the world. Only part, of course. But an important part.

Another word about our second dedication: On January 2010, our esteemed colleague and contributor to this book, Ruth Stein, died suddenly and so unexpectedly. We add her name to the dedication to honor her brilliant contribution to this book and to so much else in our field.

Contributors

Neil Altman, PhD, is an associate professor at the New York University Postdoctoral Program in Psychotherapy and Psychoanalysis. He is also past president of the Section on Social Responsibility for Division 39 (Psychoanalysis) of the American Psychological Association; and is a founding board member, consultation group leader, and therapist in Fostering Connection, a program that offers pro bono psychotherapy to foster children. A coeditor of *Psychoanalytic Dialogues,* he is the author of *The Analyst in the Inner City: Race, Class, and Culture through a Psychoanalytic Lens* (2nd ed.; Routledge, 2010).

Steven Botticelli, PhD, (editor), is adjunct faculty at City College, City University of New York. A contributing editor to *Studies in Gender and Sexuality*, he maintains a private practice in New York City.

Ghislaine Boulanger, PhD, is a faculty member at the New York University Postdoctoral Program in Psychotherapy and Psychoanalysis, and is the author of *Wounded by Reality: Understanding and Treating Adult Onset Trauma* (The Analytic Press, 2007). She is a founding member of WithholdAPADues, a group of psychologists that withheld their dues and subsequently resigned from the American Psychological Association in protest over the APA's collaboration with Bush-era detention policies. She maintains a private practice in New York City.

Françoise Davoine is a professor at the Ecole des Hautes Etudes en Sciences Sociales in Paris, France. Davoine holds an advanced degree in classics and a doctorate in sociology. Along with Jean-Max Gaudillière, Davoine has worked more than 30 years as a consultant at a public psychiatric hospital as well as in private practice. She is the coauthor (with Gaudillière) of *History Beyond Trauma* (Other Press, 2004).

Jean-Max Gaudillière is a professor at the Ecole des Hautes Etudes en Sciences Sociales in Paris, France. Gaudillière holds an advanced degree in classics and a doctorate in sociology. Along with Françoise Davoine, Gaudillière has worked more than 30 years as a consultant at a public psychiatric hospital as well as in private practice. He is the coauthor (with Davoine) of *History Beyond Trauma* (Other Press, 2004).

Sue Grand, PhD, is a faculty member and supervisor at the New York University Postdoctoral Program in Psychotherapy and Psychoanalysis, where she holds a specialization in trauma as well as couples and families. In addition, she is on faculty at the Mitchell Center for Relational Psychoanalysis, the Manhattan Institute for Psychoanalysis, the National Institute for the Psychotherapies, and the Psychoanalytic Institute of Northern California. An associate editor of *Psychoanalytic Dialogues*, she is the author of *The Reproduction of Evil* (The Analytic Press, 2000) and *The Hero in the Mirror* (Routledge, 2010). She maintains a private practice in New York City and Teaneck, New Jersey.

Adrienne Harris, PhD, (editor), is a clinical associate professor at the New York University Postdoctoral Program in Psychotherapy and Psychoanalysis, as well as an associate editor of *Psychoanalytic Dialogues* and *Studies in Gender and Sexuality*.

Nancy Caro Hollander, PhD, is a faculty member and a member of the Los Angeles Institute and Society for Psychoanalytic Studies. A professor emerita of Latin American history, she travels frequently to Latin America to do research on subjectivity and politics in the context of extreme social situations and economic crises. She is the author of *Love in a Time of Hate: Liberation Psychology in Latin America* (Rutgers University Press, 1997) and coeditor of and contributor to *Psychoanalysis, Class, and Politics: Encounters in the Clinical Setting* (Routledge, 2006). She maintains a private practice in Los Angeles.

Lynne Layton, PhD, is assistant clinical professor of psychology at Harvard Medical School, and a faculty member and supervisor at the Massachusetts Institute for Psychoanalysis. She is the author of *Who's That Girl? Who's That Boy? Clinical Practice Meets Postmodern Gender Theory* (The Analytic Press, 2004); coeditor, with Susan Fairfield and Carolyn Stack, of *Bringing the Plague: Toward a Postmodern Psychoanalysis* (Other Press, 2002); and coeditor, with Nancy Caro Hollander and Susan Gutwill, of *Psychoanalysis, Class and Politics: Encounters in the Clinical Setting* (Routledge, 2006). In addition, she is editor of the journal *Psychoanalysis, Culture & Society* and associate editor of *Studies in Gender and Sexuality*. She maintains a private practice in Brookline, Massachusetts.

Tom McGoldrick, PhD, a candidate at the New York University Postdoctoral Program in Psychotherapy and Psychoanalysis, has worked with combat veterans for the past 20 years as a psychologist with the United States Department of Veterans Affairs Readjustment Counseling Service Center, located in Staten Island, New York. He also maintains a private practice.

Donald Moss, PhD, is on faculty at the New York University Psychoanalytic Institute. The author of *Hating in the First Person Plural: Psychoanalytic Essays on Racism, Homophobia, Misogyny, and Terrorism* (Other Press, 2003) as well as over 50 journal articles, he serves on the editorial boards of *Psychoanalytic Quarterly, Journal of the American Psychoanalytic Association, American Imago*, and *Studies in Gender and Sexuality*. He has maintained a private practice of psychotherapy and psychoanalysis for 30 years.

Steven Reisner, PhD, is an advisor on psychological ethics for Physicians for Human Rights. He is a founding member of the Coalition for an Ethical Psychology, a group dedicated to upholding international standards of human rights in psychological practice and research, and supporting psychologists who work to combat the effects of political violence and oppression internationally. In addition, he is an adjunct professor in clinical psychology, Columbia University, Teachers College; an assistant professor of clinical psychiatry at the New York University Medical School (Psychoanalytic Institute); and senior advisor and faculty member at the International Trauma Studies Program at Columbia University's Mailman School of Public Health. He maintains a private psychoanalytic practice in New York City.

Eyal Rozmarin, PhD, is a clinical psychologist and candidate at the New York University Postdoctoral Program in Psychotherapy and Psychoanalysis. A contributing editor for *Studies in Gender and Sexuality*, he writes on the intersection of psychoanalysis, philosophy, and critical theory. He maintains a private practice in New York City.

Stephen Soldz, PhD, is a professor at the Boston Graduate School of Psychoanalysis; cofounder of the Coalition for an Ethical Psychology, an organization opposing psychologist participation in torture and detainee abuse; and in 2010 became president of Psychologists for Social Responsibility. He is the author of numerous professional papers and many articles on social aspects of psychology, published by such Web sites as CounterPunch, ZNet, OpEdNews, Common Dreams, Scoop, and Alternet, and has participated in numerous radio and television appearances on social issues.

Ruth Stein, PhD, was associate clinical professor in the NYU Postdoctoral Program for Psychotherapy and Psychoanalysis and training analyst at the International Psychoanalytic Association. The author of *For Love of the Father: A Psychoanalytic Study of Religious Terrorism* (Stanford University Press, 2009), she maintained a private practice of psychoanalysis and psychotherapy in New York City until her untimely death in January 2010.

Frank Summers, PhD, ABPP, is a supervising and training analyst at the Chicago Institute for Psychoanalysis and an associate professor of Psychiatry in the Behavioral Sciences at Northwestern University Medical School. President of Psychoanalysts for Social Responsibility, he has worked for the past 5 years to prohibit psychologists from participating in interrogations in detention centers. The author of three books, he maintains a private practice of psychoanalysis and psychoanalytic therapy in Chicago, Illinois.

Nina Thomas, PhD, ABPP, is clinical associate professor, cochair of the relational track, and chair of the Specialization in Trauma and Disaster Studies in the New York University Postdoctoral Program in Psychotherapy and Psychoanalysis. She has written and presented both nationally and internationally on the intersection of the political and the psychoanalytic, particularly as these manifest in individual versus national traumas. She maintains a private practice of psychotherapy and psychoanalysis in New York City and Morristown, New Jersey.

Eli Zaretsky, PhD, is professor of history at the New School for Social Research, and is the author of *Secrets of the Soul: A Social and Cultural History of Psychoanalysis* (Vintage, 2005) and *Capitalism, the Family and Personal Life* (Harper & Row, 1976).

Editors' introduction
Adrienne Harris and Steven Botticelli

This volume takes its title from a most ancient and ethically binding directive to physicians: First, do no harm. We mean to situate this title and this volume of essays in paradox. For over a century, the healing practices that flow from depth psychology and a psychoanalytic theory of unconscious process have been used to offer repair to soldiers and civilians traumatized by war. Psychoanalytic thinking has been crucial to our understanding of the long sequelae of war and war trauma, as they burrow deep into character and collective life. Many of the writers in this collection examine the use of psychoanalytic ideas to explain the impulse for war and destruction, uses that go back to the very beginning of psychoanalysis and to Freud himself.

But what tumbles us into paradox and difficulty is that we must also notice the use of psychology and psychoanalysis to wage war and its most destructive and lethal programs. We must notice with chagrin and horror that psychological theories and concepts underwrite torture. Psychodynamic theories can be tools with which individuals and groups launch attacks on the psychic integrity and on the emotional, mental and physical health, of civilians and soldiers, victims and perpetrators.

There are two interwoven stories here in regard to the destructive and the reconstructive potential for psychoanalytic and depth psychological ideas. There is the potential in the theory for care and damage, as we suggest. There is the potential in analysts or caregivers for transformation from good to bad objects and also there is the perversion of ideas and personnel in the hands of ideologues and forces of domination, which may come in the form of individuals or states (Stein, Chapter 13, this volume). Viewed as technique or instrument, psychoanalysis is morally agnostic. Our examination of this paradox must be undertaken at the social level, the level of state power, and at the individual level. We see the invocation "first, do no harm" as a crucial ethical guide for any mental health worker. We use the phrase with irony, with critique, but also with a moral and ethical imperative, as we explore the recruitment of psychological ideas for damage, domination, and humiliation in circumstances of armed conflict and warmaking, circumstances that may destroy soldier and civilian.

This edited volume was begun in a moment of acute personal and professional crisis experienced by a number of psychologists, engaged in a bitter and increasingly trenchant struggle with the administration and leadership of the American Psychological Association. The continued presence of psychologists at interrogations at the prison in Guantánamo is an outrage so far unmoved by very continuous determined effort and opposition within the APA (see Soldz, Chapter 5, and Reisner, Chapter 6, this volume, for an extensive discussion and timeline for this battle). In the face of an increasing outcry from the membership about the matter, the APA continued to maintain that psychologists' presence there was necessary to protect the detainees. Many of the leaders in that battle with the APA have chapters in this volume (Boulanger, Soldz, Hollander, Thomas, Altman, Reisner, and Summers).

Far from a small internal battle within a profession, the implications of the APA's continuous refusal to sign on to clear statements of opposition to torture have been stunning and deeply troubling. The APA leadership has consistently refused to make clear commitments to the ethical stance of any psychologist with regard to interrogations and a broad range of destructive practices. Perhaps our continued surprise is worth some attention. However anomalous or inevitable these leadership strategies may appear, this crisis within the organization has opened a deep inquiry into the nature of modern warfare and the role of psychologists, but more ominously the role of psychology, and by extension, psychoanalysis, in warmaking.

For anyone following this battle within the APA, there are many striking thunderclap moments. It was eye opening to realize that the American Medical Association, the American Psychiatric Association, and the American Psychoanalytic Association had much clearer commitments to the Geneva Convention than the American Psychological Association, and each of the former groups established an absolute prohibition on participation or even observation in the kind of "interrogations" occurring at Guantánamo. As more historical excavation of the rise of such interrogations occurred, one realized that many within the military and the Federal Bureau of Investigation stepped back from these procedures, even as the APA continued to claim the utility of psychologists' presence at interrogations (Mayer, 2008).

Horror and curiosity were mixed together. What was preventing a group of psychologists, usually the more progressive elements within the mental health world, from finding a clear path to a principled stance on torture? There is the conflict. Many psychologists had thought themselves to be the more progressive forces within the mental health and helping professions. Indeed psychology, in a wide variety of contexts, does make progressive contributions. And yet, our dependence on and embeddedness in the military system has been visible to all of us for half a century. However morally or even pragmatically one approached this matter, something seemed very mystifying here. Torture is unethical and destructive. It does not produce useful information. The harm to the victim and to the torturer radiates out to family, and to the wider society, leaving decades-long, multigenerational effects of shame and rage. As many have argued, these practices, along with many other aspects of American defense policy, have left us radically less safe than ever.

PARADOX

In a sense, this volume emerged as one way of speculating on the stasis within the profession and the professional organizations of psychology. Looking both historically and currently, one sees that psychology is the creature of the military (Zaretsky, Chapter 9, and Gaudillière, Chapter 2, give us a historical picture). Without military research and defense interests, there would be no APA and no professional discipline of psychology of the scale it currently operates at in the United States and perhaps also internationally.

But there is an even more difficult paradox here. The theories and practices of psychology and psychoanalysis are in a sense morally agnostic, equally deployable for peace and growth on one hand, or destruction and malignant practices of domination on the other. This is true whether the theory is behavioral or depth psychological. The depth psychologies, all officially or unofficially indebted to and linked to psychoanalysis, offer the most powerful ideas into how character is formed, how influence is promoted, and how people are induced to act in particular nonrational ways. (Stein, Chapter 13, explicates in psychoanalytic terms the basis for the effectiveness of mind control techniques, as these have been deployed by cult leaders, totalitarian regimes, and others.)

The insight in Jane Mayer's first *New Yorker* piece (2005) may have been startling but once thought about, one can see that the contradiction between psychology and psychoanalysis as handmaidens of war and destructiveness *and* of peace and healing is obvious and ubiquitous. Mayer's most significant finding, for our purposes, was that a program designed to protect soldiers, by arming them with techniques to avoid breaking down under interrogations, could simply be inverted and reversed to break down our enemies. The theory is in this way ethically agnostic, of utility for purposes benevolent and malign.

Looking over the history of mental health, of psychological research and practice and in particular over psychoanalysis, this contradiction repeatedly surfaces. There might be different ways to describe this. Inevitably, one cures with contaminated tools. The more deeply we understand human character, human defenses, and human reactions to trauma, the more we realize that within the same theoretical/clinical enterprise there is the capacity to both help and hurt.

That said, for all the supposed sophistication suggested by the notion of "reverse engineering of SERE techniques," the actual methods of torture used by interrogators that have been most reported in the media—waterboarding, sleep deprivation, stress positions—hardly required knowledge of the theories (e.g., Seligman's learned helplessness) that were said to have inspired them, and have been used by torturers for centuries (see Mayer, 2008). By contrast, the interrogation methods that rely on building rapport with a criminal suspect, eschewed under the recent American intelligence-gathering program and reportedly more effective than torture in gleaning information, actually require a degree of clinical skill and training. In this sense, it seems the Central Intelligence Agency enlisted psychologists less for their supposed psychological expertise than for the gloss of professionalism their participation lent to the grisly proceedings. As one source told Mayer (2008), psychologists were desired "because they [the CIA] wanted some kind of psychological justification for doing what they were doing. They wanted a theoretician to tell them that they could go hard but not seem like brutes" (p. 163).

In a self-reflexive mode, we wonder at the meanings of our invocation of a directive to physicians to introduce this group of essays, most of which have been written by non-MD psychologists. Although a 2008 referendum on the question of psychologists' participation in military interrogations aimed to change this, the code of ethics for psychologists (unlike that for psychiatrists and other physicians) has been porous enough to allow such participation. Might there be elitism (in addition to our justly credited moral concern) in this move to identify ourselves with physicians as we condemn the behavior of some of our psychologist colleagues? Would we, collectively as a group of concerned psychologists, have taken this vocal moral position if our livelihoods were linked to the military, as the livelihoods of so many psychologists currently and historically have been (see Summers, Chapter 8; also Altman, Chapter 7). Might there be other, less apparent motives at work in our wish to identify ourselves, especially perhaps in this particular moment of economic anxiety, with the prestige of medicine, at the same time as we disidentify with our psychologist colleagues? "We're like these people, not like those" (see Botticelli, Chapter 16, on matters of identification and disidentification as they pertain to politics). Rozmarin (Chapter 15) visits these conflicts between the collective and individual level in another context.

FORGETTING: SOCIAL, INSTITUTIONAL, PERSONAL

Historically speaking, psychoanalysis has a mixed record when it comes to questions of war and militarism (Zaretsky, Chapter 9, and Gaudillière, Chapter 2, this volume). Freud's initial response to the outbreak of the First

World War was one of nationalistic enthusiasm. He wrote to Karl Abraham that "for the first time in thirty years I feel myself to be an Austrian and feel like giving this not very hopeful Empire another chance. ... All my libido is given to Austro-Hungary" (quoted in Breger, 2000, p. 234). The war had the effect of cleaving Freud's professional and personal relationships along the lines established by the clashing powers. He remarked to Abraham of their British colleague Ernest Jones, "Jones, is of course our enemy" (p. 237)—even as he maintained collegial contact with him. As the full extent of the destructiveness unleashed by the war became apparent, Freud was horrified by what had been wrought: "[N]o event has ever destroyed so much that is precious in the common possessions of humanity, confused so many of the clearest intelligences, or so thoroughly debased what is highest" (1915/1957, p. 275). Freud, nevertheless, assimilated this horror as merely reflecting one of the unpalatable truths that psychoanalysis had been urging humanity to come to terms with. In a lecture he gave to his B'nai B'rith chapter in 1916, Freud suggested that the war had broken through the general denial of death that prevailed in peacetime. He urged a realistic acceptance of the reality of death, that such an attitude was necessary in order to make "life ... interesting again," for it to "recover its full content" (1915/1957, p. 291), and asserted that war could not be abolished.

Attached as he was to understanding emotional disturbances in adulthood as ineluctably rooted in infantile sexual conflicts, Freud was unable to consider the role of trench warfare and other traumatic combat experiences in producing shell-shocked soldiers. He went so far as to maintain that "When the furious struggle of the present war has been decided, each one of the victorious fighters will return home joyfully to his wife and children, unchecked and undisturbed by thoughts of the enemies he has killed whether at close quarters or at long range" (1915/1957, p. 295). In Freud's case such ideas did not result in the mistreatment of any soldiers, as he treated none. However, given the preeminent influence of psychoanalysis on the profession of psychiatry over the next decades, such ideas contributed to the delayed recognition of the impact of trauma in adulthood. Indeed, it is only in recent years that analysts finally have felt the freedom to do so (e.g., Boulanger, 2007).

There is, actually, an almost lost but important stream of thought and work, inaugurated in the work of Ferenczi and importantly influenced and expanded in the work of various psychiatrist veterans of the First World War. Tausk (1916, 1987), Feigenbaum (1937), Groddeck (1977), and in a different context Bion (1987) add many ideas and insights into the impact of trauma on the mind, ideas that have been somewhat dormant in the treatment and consideration of veterans until the post-Vietnam era. Work on veterans of the Second World War has perhaps been more occluded from contemporary attention. But for an alternative perspective see Kardiner (1947, 1969). Layton (personal communication, August 2009) adds Fairbairn to this list. Tucked into his chapter on the return of bad objects,

Fairbairn speaks of war neuroses as examples of this kind of ferocious reappearance of repressed bad internal objects, triggered by both intrapsychic circumstances and external reality. His work is drawn on in Ruth Stein's chapter on mind control, and we quote his thoughts about war and intrapsychic process at some length. Notice his close attention to the phenomenology of what we have come to know as post-traumatic stress reaction:

> The spontaneous and psychopathological (as against the induced and therapeutic) release of repressed objects may be observed to particular advantage in wartime in the case of military patients, amongst whom the phenomenon may be studied on a massive scale. … The effect of such traumatic situations and traumatic experiences in releasing bad objects from the unconscious is demonstrated nowhere better than in the wartime dreams of military patients. Amongst the commonest of such dreams, as would be expected, are nightmares about being chased or shot at by the enemy, and about being bombed by hostile aeroplanes (often described as "great black planes"). The release of bad objects may, however, be represented in other ways, e.g., in nightmares about being crushed by great weights, about being strangled by someone, about being pursued by prehistoric animals, about being visited by ghosts and about being shouted at by the sergeant-major. The appearance of such dreams is sometimes accompanied by a revival of repressed memories of childhood. One of the most remarkable cases of this kind in my experience was that of a psychopathic soldier, who passed into a schizoid state not long after being conscripted, and who then began to dream about prehistoric monsters and shapeless things and staring eyes that burned right through him. He became very childish in his behaviour; and simultaneously his consciousness became flooded with a host of forgotten memories of childhood, among which he became specially preoccupied by one of sitting in his pram on a station platform and seeing his mother enter a railway carriage with his older brother. In reality his mother was just seeing his brother off; but the impression created in the patient was that his mother was going off in the train too and thus leaving him deserted. The revival of this repressed memory of a deserting mother represented, of course, the release of a bad object from the unconscious. A few days after he told me of this memory a shop belonging to him was damaged by a bomb; and he was granted twenty-four hours' leave of absence to attend to business arising out of the incident. When he saw his damaged shop, he experienced a schizoid state of detachment; but that night, when he went to bed at home, he felt as if he were being choked and experienced a powerful impulse to smash up his house and murder his wife and children. His bad objects had returned with a vengeance. (Fairbairn, 1952, pp. 76–77)

Psychoanalytic ideas were able to play a constructive role in devising therapy for the shattered minds of traumatized soldiers, providing a humane alternative to such routinely practiced "treatments" as electric shock that aimed at getting soldiers back to the front as quickly as possible. Working independently of each other, the British physicians William Brown and W. H. R. Rivers came to see shell-shocked soldiers' symptoms as the products of traumatic experiences in combat that had been sealed off from conscious memory. In this they were drawing inspiration from Freud's *Studies on Hysteria*, which also suggested a treatment method: "catharsis, the open expression of the sealed off memories with a full range of emotions" (Breger, 2000, p. 257). Not being analysts themselves, Brown, Rivers, and several others like them were free to draw on those aspects of psychoanalytic writing that fit with their own observations and could be useful in their treatment of their patients, without the stultifying requirement of loyalty to the master that Freud imposed on those who wanted to remain in his circle. (See Davoine, Chapter 10, for an analysis of the long hand of trauma in several key figures around the period of the First World War; see also Thomas, Chapter 4, for a consideration of the complex, long-term sequelae to trauma, even when remembered and witnessed.)

There is an interesting symposium that appears in 1919, organized by Ferenczi and introduced by Freud. A note is sounded that will reappear throughout the century. Work on the injuries and psychiatric symptoms of soldiers and veterans heats up during wartime and rather shockingly disappears at the armistice or the end of hostilities. As one of our authors, Jean-Max Gaudilliére, points out, the forgetting is itself symptomatic at a social, collective level. Equally powerfully, there is a consistent low-grade resistance to the recognition of the long-term consequences of war, the facts of postwar trauma, and the long shadow of trauma, including collective attempts to recover from trauma (see also Thomas, Chapter 4, this volume and Boulanger, Chapter 3, this volume).

In a certain way, key evolutions in the healing and the malignant use of psychoanalytic and psychological theory arise in the context of particular wars. Each war brings its own language, specialization, and dilemmas. Faradism, shellshock, hysteria in World War I. "Psychoneurosis" as the narrator of *Let There Be Light* (a documentary film made by John Huston) describes 20% of the injuries in World War II. Posttraumatic stress disorder (PTSD), the term that emerges after Vietnam, migrates from military to civilian populations. Boulanger (personal communication, August 2009) notes that in thinking of the evolution of the diagnosis of PTSD we should remember that it was never intended simply to diagnose postcombat conditions. Among the original movers for the diagnosis were Vietnam veterans, women afflicted by postrape reactions, and Holocaust survivors.

The periods pre- and post-World War II are particularly interesting as the sites of psychological research and work that promotes the effectiveness of propaganda. We can see in the seminal work in social psychology (e.g., Rappaport) that ideas of impression management, rhetoric, and suggestion can take psychology (and the related discipline psychoanalysis) into a dark or a light rehabilitative space. The same theoretical ideas (dynamic depth psychology, unconscious fears and drives, the limits and character of defenses) can serve demonic or heroic forces (Grand, Chapter 11; Stein, Chapter 13).

VIETNAM

For many of the generation of the 1960s movements of liberation and social protest (Layton, Harris, Moss, Gaudillière, Zaretsky, Davoine, Boulanger, Altman, in this volume), Vietnam, a war waged on television, a war whose damage to self and other was palpable, was the watershed experience of our young adult lives. PTSD, as a crucial aspect of postwar veteran life, began to emanate to other aspects of psychological and psychiatric care and healing (as so often happened over this century). The late Ted Nadelson, in *Trained to Kill* (2005), wrote a painful memoir of a lifetime healing wounded military minds. One of us (Harris, 2006) reviewed that book:

> Nadelson reports an interchange with a marine Vietnam veteran. With a stark clarity, the soldier describes the thrill, the intense arousal, and stream of feelings that attend on destruction, the felt beauty of annihilating another person. Nadelson asks if the soldier could imagine him, the doctor, feeling such satisfaction. The answer is clear. "No. You? You worry … about people, you would worry about the hole you make" (p. 72). The prose is extremely simple and clear, so the reader feels the moment between the two men as one of terrible recognition. There is an unbridgeable gap between subjectivities alongside a pained connection. Nadelson can feel the chagrin at knowing his own limit. The soldier knows that a crucial human empathy is lost to him. Nadelson frames all his examples within an understanding that wars are rationalized, sanctioned violence, that the mix of masculine socialization and military training and wartime experience has a toxicity, and a potency, that alters everything. Like Kafka's needle, wartime action, killing and harming, amidst a terrifying and dangerous battle, bites deep into body and soul. (p. 1155)

Nadelson's conclusions are sorrowful. He links the vulnerability of soldiers to historical forces and the powerlessness inherent in warfare, and also to gender socialization, to the induction of masculinity through shame and disdain for tenderness and vulnerability, a lethal gap in any soldier's armor. We see this question of identity or interpellation, at the individual level and the collective, in the chapters by Rozmarin, Layton, Hollander, Botticelli, and

Moss. It is also crucial to see, over a century of work on veterans' trauma, both the paucity of help offered to those afflicted and the chronicity of multiple traumas (Altman, 2009). We live amid layers of recurring trauma.

IDENTITY AND WAR TRAUMA

Working on this project, reading our authors, and delving into the literature on war trauma, genocides, and warmaking, and seeing a century of warfare and its sequelae through the lens of psychoanalysis, one thread can be traced throughout. The long hand of trauma, specific to childhood or adult onset (Boulanger, Chapter 3) or intergenerational came to be visible in many contexts, from the aftermath of genocide (Thomas, Chapter 4), the long sequelae to the Holocaust (Botticelli, Chapter 16; Rozmarin, Chapter 15), to the ongoing treatment of veterans (McGoldrick, Chapter 1). Warmaking, injuring others, repairing, all cohere around the role and presence (disavowed or prominent) of shame. Again, the paradox. Identifications can only be constituted through some encounter with shame, limits, and the culture's requirements conveyed through the family and powerful others. Yet identity is threatened whenever shame dominates. We know this from the dissociation and trauma literature. We know this from developmental studies (Lewis, 1995). Whereas guilt may target and organize around particular acts or traits, shame is a full body–mind experience. Shamed, the self implodes; identity itself is shattered.

Can we not see shame everywhere in these stories? As medical psychiatry tools up in the wake of industrialization, the mass technologizing of war at the end of the 19th century, the debates about manliness in a soldier, and the unstable status of psychiatric and psychological symptoms and diagnoses often organize around the adequacy of the soldier's identity, and therefore the threat of shame haunts the enterprise. Nadelson would argue that shame is part of the deep constituting of masculinity, seen in his clinics in the Veterans Affairs hospitals in Boston. And upon injury or trauma, shame accompanies the soldier and his illnesses every step of the way (Grand, Chapter 11; Boulanger, Chapter 3; McGoldrick, Chapter 1).

From Nancy Hollander (Chapter 14) we can see that women are crucial as targets and pawns for warmaking. Through rape and sexual assault, shame with its effects on the collapse of identity, war is waged and trauma is incurred. If the culture colludes, or requires particular formations of self, gendered, racialized, culturally inflected, cultures, through medical and political personnel and through public discourse, have a stake in the maintenance of those identifications. As psychoanalysts, we can both notice the powerful intrapsychic requirements of identification as a deep vulnerability in war, and the social and intrapsychic requirements not to notice this. The cultural amnesias and institutional forgettings are very much the topic of

Layton's essay in this volume (Chapter 18). It is very telling that in looking at resistance, some prosthetic for shame may be in place, some capacity to function in opposition to culture and society and often to family has been fostered and survived (Rozmarin, Chapter 15; Botticelli, Chapter 16).

In "psychologizing" the traumatized soldier or civilian, we are committed to avoiding reduction. The intrapsychic, identificatory aspect of soldiers' experiences are just one element in a complex multileveled story. Shame, we know from the dissociation and developmental literature, comes from powerlessness. The ultimate powerlessness of the soldier is extraordinarily visible in the trench warfare of the First World War, the time of much professionalization of psychoanalysis and psychiatry in the treatment of war neurosis (Zaretsky, Chapter 9). Shame, if you like, exposes the illusions of autonomy, one lynchpin of masculinity.

Shame is an ingredient in the carrying of memory in a somewhat different constellation in Donald Moss' memoir (Chapter 12). In tracing a guilt, shame and pleasure-ridden repeating war story told by his father to him, over almost half a century, Moss explores the complex pleasure in aggression and destruction, and the contagious damage of these pleasures over the generations. These transmissions are integral to masculinity and gender construction, perhaps also to patriotism, in unusual, unexpected forms, and to the repetitive installation of trauma in the psyche.

RESISTANCE

Freud (1927/1957) laid down the shibboleth for psychoanalysis as a basis for political resistance when he asserted in "The Future of an Illusion": "It goes without saying that a civilization which leaves so large a number of its participants unsatisfied and drives them into revolt neither has nor deserves the prospect of a lasting existence" (p. 12). The political power of this position lies in its universalism, its assertion of common human need, the denial of which provokes people to rebel. As Terry Eagleton (1996) put it in a defense of essentialism, "Needs which are essential to our survival and well-being ... [are] politically criterial: any social order which denies such needs can be challenged on the grounds that it is denying our humanity, which is usually a stronger argument against it than the case that it is flouting our contingent cultural conventions" (p. 104). The Frankfurt School theorists found in Freud's drive theory (in its postulation of an opposition between a pleasure-seeking id and the repressive forces of ego and superego) a model for social struggle and transformation. The notion of the id as a repository of animal sexual impulses provided "a psychic and theoretical external point from which to mount a critique of society" (Jacoby, 1983, p. 157). And indeed the work of the Frankfurt School, and of Lacan in France (see Turkle, 1992), provided inspiration for some anti-Vietnam war protesters.

Still, psychoanalysts traditionally have considered themselves to be practitioners of a contemplative discipline, suspicious of action and more likely to pathologize protest than to call for it; consider, for example, Bruno Bettelheim's dismissal of the 1960s antiwar movement as "oedipal acting out" (Zaretsky, 2004, p. 313). On the level of theory, this tendency was reinforced through the 1950s and '60s, especially in the United States, by the dominance of ego psychology, which moved the id into the background at the same time that it (as Zaretsky captures it) "lost the view of the ego as the locus of resistance" (p. 278). Instead the ego became the seat of reason and self-control. Nevertheless, antiwar sentiment survived in other precincts within psychoanalysis. Several essays by Hanna Segal (1987, 1995) are significant not so much for the illumination they provide into the populace's acceptance of the possibility of nuclear war (in familiar Kleinian terms of psychotic fears, paranoid mechanisms, etc.) but more for Segal's assertion of her right to speak as a psychoanalyst of political matters during a period of widespread quiescence and complacency among analysts. As she put it, "I think psychoanalytical neutrality must not be confused with being neutered" (1995, p. 204).

Although psychoanalysis has never been a major force in the world in opposing war, several recent developments in psychoanalytic theory have made it more axiomatic for individual analysts to take antiwar positions and perhaps to influence others to do so, within the limited spheres in which we operate. One such development has been an ever-expanding and increasingly refined appreciation of the impact of trauma on mental functioning. Contrary to Freud's blithe assessment of soldiers returning home from war psychologically unscathed, we now understand much more about the mental damage inflicted on soldiers by their experiences in soldiering, part of a broader appreciation of the impact of traumatic experience in adulthood in general (see Boulanger, 2007 and Chapter 3, this volume).

Second, analysts increasingly have been asserting the moral dimension of psychoanalysis as a theory and practice (e.g., Cushman, 1995; Alford, 1998; Grand, 2000, 2009), an aspect that was explicitly disavowed by Freud in his insistence that psychoanalysis was a science, not a worldview or a moral philosophy (see Abel, 1989). One salient example of contemporary theorizing in this vein has been Jessica Benjamin's work (e.g., 2004) on mutual recognition and intersubjectivity. Hers is implicitly a moral theory, one that defines the good as that which supports separate subjects' recognition of each other as equivalent centers of experience. This position fosters interest in specifying the conditions that facilitate mutual recognition even as it maintains an awareness that such recognition is always at best a temporary achievement, continually subject to breakdown, in which the encounter between two subjectivities degenerates into a struggle for power. With due regard for the caution necessary when moving from the individual to the group level of analysis, this thinking could be taken as a model for the encounter between two parties to a political conflict, perhaps helping to provide a standpoint from which to attempt to intervene (Rozmarin, Chapter 15; Harris, Chapter 17). Such theorizing seems of a piece with the efforts to organize opposition to the war on Iraq (for example, through the creation of Psychotherapists for Social Responsibility in late 2002, in which analysts played a leading role) as well as with analysts' prominence in the effort to force the American Psychological Association to take an ethical stand with regard to psychologists' involvement in CIA and military interrogations (see Summers, Chapter 9; Reisner, Chapter 6; Soldz, Chapter 5).

In the sphere of cultural studies, Judith Butler (2009) recently has drawn on psychoanalytic (especially Kleinian) categories in her effort to understand how it is that some lives come to count as human lives, worthy of mourning if they are lost, while others do not. Contesting the differential valuation placed on Western and non-Western lives as shown, for instance, in the conduct and media representations of current U.S. wars, Butler insists on our interconnectedness as the basis for moral action: "[T]he subject that I am is bound to the subject I am not ... we each have the power to destroy and be destroyed, and ... we are bound to one another in this power and this precariousness. In this sense, we are all precarious lives" (p. 43).

In preparing this volume of essays over the past year and a half, we have been mindful of timing and of the time-sensitive nature of some of our essayists' focus and concern. Perhaps another paradox in this volume is the way we represent the importance of action and political responsiveness. Some of these essays allow us to see the tragedy of inaction, the results of neglect of this topic and of the victims of warfare, civilian and military. Nothing or little changes at the social level and in institutional practices. Yet the essays on the battles with the APA suggest that action and responsiveness are crucial. Even as this book comes into being, the struggle to secure an ethical stance on interrogations is ongoing. And in another context, recent accounts and assessments of sexual violence against women in areas of ethnic or imperialist violence are actually increasing, even as our awareness of the plight of women has also expanded. This paradox—nothing changes over a century of modern warfare, and the situation requires immediate action—is still present as we publish this book. Time stops and time is speeding up. This perhaps constitutes one of the social dimensions of trauma.

ORGANIZATION OF THE BOOK

The essays in this collection, all solicited by the editors, wanting to give a very free hand to authors, have arrived in very different registers. We want to pay attention to the very different tonal and emotional strategies (if that is the right term) our authors have chosen. To recite an oft-noted epithet from feminism: The personal is political. Private stories; narratives in different formats; surrealistic, visionary dreamscapes. Some authors have written powerful, dispassionate essays like legal briefs building the case for the rise of torture and its professional underwriting by psychologists (Soldz, Chapter 5; Reisner, Chapter 6). We find evidence as well of the personal transformations in this professional scandal (Altman, Chapter 7; Reisner, Chapter 6).

There is another aspect of these articles, often subtle, often carried in tone or in footnotes. For the mental health workers treating war trauma, there is the inevitable and bravely borne matter of secondary trauma, adding to the deep costs of care and cure. Writing personally or professionally, our authors track the personal and collective burden, lived in conscious and unconscious processes, of carrying history (Rozmarin, Chapter 15; Moss, Chapter 11; Davoine, Chapter 10; Harris, Chapter 17).

This collection is organized into four sections, though inevitably overlap and interactions occur. We look first at the history and contemporary work on the injuries and repair of soldiers. We look then at the use of psychoanalysis in the service of warmaking and torture, the demonic side. We take up, in a third section, the use of psychoanalysis as a deconstructive tool for understanding warmaking and militarism. Finally, we address, from a sociohistorical, political, as well as a psychoanalytic perspective, the question of resistance.

REFERENCES

Abel, D. C. (1989). *Freud on instinct and morality*. Albany, NY: SUNY Press.

Alfred, C. F. (1998). Melanie Klein and the nature of good and evil. In P. Marcus and A. Rosenberg (Eds.), *Psychoanalytic visions of the human condition: Philosophies of life and their impact on practice* (pp. 118–140). New York: New York University Press.

Altman, N. (2009). Ongoing trauma, ongoing help. Paper presented at the conference of the International Association for Relational Psychoanalysis, Tel Aviv, Israel, June, 2009.

Benjamin, J. (2004). Beyond doer and done to: An intersubjective view of thirdness. *Psychoanalytic Quarterly, 73*, 5–46.

Bion, W. R. (1987). *The long weekend, 1897–1919: Part of a life*. London: Free Association Books.

Boulanger, G. (2007). *Wounded by reality: Understanding and treating adult onset trauma*. Mahwah, NJ: Analytic Press.

Breger, P. (2000). *Freud: Darkness in the midst of vision*. New York: Wiley & Sons.

Butler, J. (2009). *Frames of war: When is life grievable?* London: Verso.

Cushman, P. (1995). *Constructing the self, constructing America: A cultural history of psychotherapy*. Reading, MA: Addison-Wesley.

Eagleton, T. (1996). *The illusions of postmodernism*. Oxford, UK: Blackwell.

Fairbairn, W. R. D. (1952). *Psychoanalytic studies of the personality*. London: Tavistock.
Feigenbaum, D. (1937). Depersonalization as a defense mechanism. *Psychoanalytic Quarterly*, 6, 4–11.
Freud, S. (1957). Thoughts for the times on war and death. In J. Strachey (Ed. & Trans.), *The standard edition of the complete psychological works of Sigmund Freud* (Vol. 14, pp. 273–302). London: Hogarth Press (Original work published 1915).
Freud, S. (1957). The future of an illusion. In J. Strachey (Ed. & Trans.), *The standard edition of the complete psychological works of Sigmund Freud* (Vol. 21, pp. 5–56). London: Hogarth Press. (Original work published 1927)
Grand, S. (2000). *The Reproduction of evil: A clinical and cultural perspective*. Hillsdale, NJ: Analytic Press.
Grand, S. (2009). *The hero in the mirror: From fear to fortitude*. New York: Routledge.
Groddeck, G. (1977). *The meaning of illness: Selected psychoanalytic writings*. New York: International Universities Press.
Harris, A. (1985). Radio Broadcast. Women of Peacemaking Series. *Ideas* (October 1, 8, 5 1985).
Harris, A. (2006). Trained to kill: Soldiers at war by Theodore Nadelson. *International Journal of Psycho-Analysis*, 87, 1154–1157.
Jacoby, R. (1983). *The repression of psychoanalysis: Otto Fenichel and the political Freudians*. New York: Basic Books.
Kardiner, A. (1947). Traumatic neuroses of war. In S. Arieti (Ed.), *The American handbook of psychiatry* (pp. 243–251). New York: Basic Books.
Lewis, M. (1995). *Shame: The exposed self*. New York: Free Press.
Mayer, J. (2005, November 14). A deadly interrogation. *The New Yorker*.
Mayer, J. (2008). *The dark side: The inside story of how the war on terror turned into a war on American ideals*. New York: Doubleday.
Nadelson, T. (2005). *Trained to kill*. Baltimore: John Hopkins Press.
Segal, H. (1987). Silence is the real crime. *International Review of Psychoanalysis*, 14, 3–12.
Segal, H. (1995). From Hiroshima to the Gulf War and after: A psychoanalytic perspective. In A. Elliot & S. Frosh (Eds.), *Psychoanalysis in contexts: Paths between theory and modern culture* (pp. 191–204). New York: Routledge.
Tausk, V. (1969). Diagnostic considerations concerning the symptomatology of the so-called war psychoses. *Psychoanalytic Quarterly*, 38, 382–404.
Turkle, S. (1992). *Psychoanalytic politics: Jacques Lacan and Freud's French Revolution*. New York: Guilford Press.
Zaretsky, E. (2004). *Secrets of the soul: A social and cultural history of psychoanalysis*. New York: Knopf.

Part I

Psychoanalysis and antiwar work

Healing

In this section we draw on the work of clinicians, analyst-activists, and psychoanalysts rooting their work in history and political life. We consider what psychoanalysis offers as a route to healing. These practitioners and theorists are interested in the relief of war suffering for soldiers and for civilians. We see the value and function of a psychoanalytic lens turned toward the immediate and very long-term damage of warmaking.

The diagnosis of posttraumatic stress disorder (PTSD) implies that the trauma is over, providing some comfort and distance for both the survivor and the clinician. Increasingly, many clinicians must deal with people repeatedly exposed to traumatic events or living in traumatized societies. Tom McGoldrick (Chapter 1) uses his unsettling early work with returning American soldiers of the Iraq and Afghanistan conflicts as an illustration of the difficulties and rewards of working with those for whom traumatization is not over. What are the ethical, existential, and personal dilemmas faced by the clinician working with "those who return to the battlefield"?

Jean-Max Gaudillière (Chapter 2) gives us a historical account of the development and the abandonment of psychiatric and psychoanalytic understanding of war trauma. He is writing about the psychic costs of amnesia, and the unconscious communications between patient and analyst, each carrying a forgotten tie to warfare. We think it is important to note that Gaudillière writes from a European perspective, which means he writes in the context of over a century of wars, colonial and domestic. His chapter speaks to his deep conviction that it takes half a century to process a war, and that in repeated and overlapping traumatic battles, the layering of symptoms and half known, unknown losses underlie much pathology in the culture, in patients, and in analysts. Through clinical vignettes, Gaudillière's chapter presents the historical resistance of psychoanalysis to history, and a way to break it, using in the transference, some traumatic spots of the history of the analyst when confronted with trauma and madness brought by a patient as instruments of research regarding erased parts of history.

Ghislaine Boulanger's chapter (Chapter 3) is about the consequences of violence and narrowly missing violent death or witnessing violence but not, in fact, escaping it. Focusing particularly on combat trauma, the author points out that the wounds inflicted by reality in adult life become psychic reality and lead to the collapse of the self. These facts are frequently obscured by politicians, by the media, by psychoanalytic metapsychology, by diagnostic practices, and by our own personal reluctance to face up to the consequences of adult onset trauma. Horror of this magnitude does not happen in every lifetime, but psychoanalytic clinicians must always be alert to the evidence of selves that have collapsed under the weight of terror.

Nina Thomas (Chapter 4) focuses our attention on a different point in the long process of bearing, witnessing, and metabolizing the trauma of warfare. In the particular situation she writes of, the ordeals and horrors of ethnic cleansing, she tracks the long shadow of survivors' searching for accountability and justice. The psychic dilemmas of witnessing and testifying are explored, in a way that requires the suspension of easy idealism. Thomas builds an argument for the need for rigorous work, underwritten by a psychoanalytic understanding, in thinking about forgiveness, its limits, and its possibilities and impossibilities. We feel the very long arm of trauma, and the queasy questions of justice and accountability, even as institutions and individuals work for an ethics of collective responsibility.

Chapter 1

Where is the "post" in posttraumatic stress disorder?
First impressions working with Iraq and Afghanistan soldiers

Tom McGoldrick

CASE ONE

Jim, a 25-year-old Marine, came in two weeks before his second deployment to Iraq. Newly shorn and unnervingly polite, he would have looked more like a 16-year-old high school student were it not for the telltale thousand-yard stare that marks those who have seen entirely too much. He sat stiffly in my office, eyes cast down. He reflected on the ways in which he had changed and wondered if he would ever be the same. He was embarrassed to admit that he was actually anxious to get back to the field where he felt confident and his so-called symptoms were adaptive. On the other hand, he had a tremendous sense of fatalism regarding his redeployment.

THERAPIST: Well, when you come back we can begin to make some sense of what you've been through.
JIM: [*Looking up, making eye contact for the first time in a piercing way*] I'm not coming back.
THERAPIST: [*Speechless*] What do you mean?
JIM: I'm not going to make it this time.
THERAPIST: So why go back?
JIM: My men need me. They will be safer with me.
THERAPIST: So you'll have to stay safe to make sure that they are safe.
JIM: And if not, at least we'll die together.

CASE TWO

Bill arrived for his initial session accompanied by his mother. Bill had been with an Army infantry unit in Afghanistan and served as a sniper with a Ranger team whose regular sniper had been killed. His first combat kills were not only multiple, but up close and personal through the scope of a sniper rifle. He was so thoroughly shaken by his experiences that he sought the help of the mental health team at the battalion aid station despite the threats of his platoon sergeant that only a weak shirker and coward would leave his men. When Bill returned, he was shunned by the other men and given punitive details by the platoon sergeant. Since coming home, he had spent all of his time in the basement of his mother's home with the shades drawn. He barely spoke during the session. It was his mother's opening salvo that I recall. She angrily stormed into my office almost dragging her withdrawn son and said, "What have *you* done to my son?" For a moment I am Uncle Sam, George W., the U.S. Army, and the complacent American people all rolled up into one.

CASE THREE

I had been working with a major in the Army Reserves who was a head nurse in a medical-surgical hospital attached to Abu Ghraib prison, one year after the well-publicized scandal there. A strong, articulate woman in her 50s, she had nearly 20 years in the Army Reserves and was near retirement from the service. She was uncharacteristically late for one of her appointments. I received a call 10 minutes into the session.

LOIS: Sorry I didn't call. I'm in Texas at Ft. Sam Houston. A few of us were ordered to go for advanced training to do emergency work in the field at FOB's (i.e., Forward Observation Bases).
THERAPIST: But you are getting out soon.
LOIS: I guess Uncle Sam has other plans for me.

I have worked with combat veterans for over 20 years. The vast majority have been Vietnam veterans with a smaller group of World War II, Korea, and Persian Gulf vets. In all cases, the war was literally over, although the "war within" still raged on. When people in treatment would say that they felt like they were back in combat, both they, and I, were safe in the notion that the war was removed in time and space. It was both *post-* and *tele-* (i.e., distant in time and space). I began working with the Vietnam vets 10 years after the war had officially ended. Many of the vets had already tried a number of avoidant solutions in drugs, alcohol, and angry confrontations. They had no place to turn other than therapy.

I am also 10 years younger than the Vietnam veterans. When I first started, I became the "kid brother" too young to have faced the draft, and therefore not morally culpable when asked the question, "Were you there?" "Of course not. You're like my kid brother. He would have gone but he was too young." This implies that I would have gone too.

I was also a kid therapist cutting my therapeutic milk teeth with these men. I use this image with all of the parental nurturance that this implies. For these men did become nurturing older brothers who discussed the horrors of war and its aftermath while also trying to protect me from their rage. There was also a tacit understanding, "I went so that you didn't have to." I became Telemachus, the dutiful son willing to hear of his father Odysseus' combat and 10 years of wandering among the monsters of intrusive memories and the lotus eater's trap of drug-induced forgetfulness (Shay, 2000). As the name translates, I too was "far from the battle."

Perhaps every combatant naively hopes that theirs is "the war to end all wars," and that they fought so that their sons and now daughters would not have to. I had a complementary naive fantasy that I was the last of war therapists. Since the Vietnam War was seemingly the last major conflict involving the United States, there would be no need for readjustment counseling or for the Department of Veterans Affairs for that matter. We would peacefully grow old together. In the same way that all trauma survivors long to return to their pretraumatized, innocent selves, I too long for the perhaps false sense of safety and security of wars long ago and far away.

Inherent in the diagnostic term *posttraumatic stress disorder* is the distance of time and space. The therapist and the survivor begin in a safe place to explore explosive material with the reassurance that the trauma is unlikely to reoccur. Indeed the very word *survivor* implies that the person is now no longer at risk; otherwise they would be only potential survivors.

Many of us who treat trauma survivors are now thrust into an unaccustomed role in which the traumatic situation is not post. The survivor is an ongoing participant or is being revictimized in various ways. The traditional safety of time and space evaporates as we, ourselves, go into the war zones to debrief those who are sometimes in the midst of traumatic situations. Or, in the case of 9/11, the trauma is brought to our doorstep threatening our own assumptions of safety.

This state of affairs is familiar to those who work with first responders, as well as those whose clients live in violent inner-city neighborhoods or whose clients are in abusive domestic relationships. Our patients are preparing for, or defending against, the very real probability of future trauma. Indeed, for some, trauma becomes routine, and mundane experiences are regarded as childish fantasies. The traumatized person views us, the "yet to be traumatized," as naive children believing in the fairy tale notions of personal safety, agency, and faith in the possibility of interpersonal healing. The therapist is forced to look at what we all take for granted living day to day. What does this do to our faith in psychotherapeutic theories, practice, professional, and personal ethics? Indeed what does this do to our sanity when we are living in an insanely traumatized world?

Working with those who are in or soon to return to the war zone also opens up the possibility of actively intervening in someone's life perhaps to prevent retraumatization. What are the ethical, therapeutic, and existential dimensions of such interventions or decisions not to intervene?

There is also the issue of the type of work that can be done with those who are on alert to return to the battlefield or the inner city. Is not a certain amount of emotional numbing and guardedness essential for survival? In brief interventions, we may suggest that someone refrain from speaking about past traumas in order to not compromise their defensive alertness. This is certainly a major consideration when seeing first responders at their work site and now especially true for those who face redeployment. Jim, the Marine from the first vignette, was emotionally prepared for battle and for the reality that he might die. It was neither the time nor the place to ask about his first deployment other than in the most cursory way.

Time is a matter in another way. Those who work psychoanalytically are used to working over a period of years, multiple times a week. As the staccato rhythms of my initial vignettes suggest, I often see more people for far fewer sessions, often one or two. For those who stay, their concern is often a circumscribed, practical problem with past traumas or future deployments as stark but often unspoken backdrops. Once this practical concern is dealt with, however, the contact is finished.

In my experiences at the Staten Island Readjustment Counseling Center, run by the Department of Veterans Affairs, the principal presenting problems are marital and family problems especially for older Reservists and National Guard soldiers. In many cases, the predeployment cracks in the foundation of the marriage have widened. In other much more poignant cases, the couples describe a strong, loving relationship now undone much to the surprise of both members of the couple. This is the couple equivalent of the dissolution of the self that can occur in severe trauma (Boulanger, 2007). For younger soldiers the principal presenting problem is trying to begin one's life's work or return to school with the threat of possible future deployments. Other presenting problems are disciplinary problems either in their Reserve unit or on their jobs; substance abuse problems; the noxious problems of hyperarousal (i.e., sleeping problems, feeling tense, and problems controlling anger); driving problems, either driving too slow or too fast, or sudden erratic lane changes as the soldier either consciously or without conscious awareness tries to avoid IEDs (improvised explosive devices) in the road.

Military psychiatry has a different ethic than we are used to as individual practitioners, much more akin to those who work in employee assistance programs. The client is the unit, not the individual. The motto is "Preserve the fighting strength," or what one medic said, "Patch 'em up and send 'em out." The military, fond of acronyms, uses PIES to describe its treatment of those acutely traumatized:

P—Proximity, as close in space to the scene of the trauma as possible (in theory this means at the site of combat).
I—Immediacy, as close in time to the traumatic event as possible.

E—Expectancy of return to the unit, in other words normalizing whatever the extreme reaction is and reassuring people that they will be reunited with their unit.

S—Simplicity, that is, three hots and a cot, short time off the front lines, listening but not inquiring about the soldier's combat experiences.

This method was perfected in World War I by French and British military psychiatrists who found that many men evacuated from the field became worse in the rear hospitals and that the forward infantry units became depleted of essential infantry soldiers. Forward psychiatric centers were established often within 10 miles of the trenches (Jones & Wessely, 2005).

The civilian version of this is the critical incident stress debriefing (CISD) model, in which people who have experienced a circumscribed trauma are "debriefed" often at the site of the traumatic situation. Where it differs is in the S or simplicity part of the acronym. Rather than dealing with the issues of comfort and safety, the CISD model encourages people to speak of their traumatic experience as well as attendant thoughts and feelings (Mitchell, 1983). While initially seen as something that might inoculate the patient from the long-term consequences of PTSD, there is some controversy as to its effectiveness and even some evidence of possible harm that might be done (Cochrane Library, 2006).

For myself in dealing with the returning soldiers, I have found that fantasies of "stress inoculation" or active intervention to "save someone" were all ways to defend myself from the powerlessness and the ineffectiveness that I have felt. Becoming aware of my own avoidance and dissociation has opened up the possibility of some human contact in an impossible situation, in other words the possibility of hope.

Initially, I avoided working with the new returnees by choosing to work with new patients who were Vietnam vets. Every new conflict awakens wounds of all combat veterans particularly those who have never sought treatment. So along with the recently returned veterans our clinic had a flood of Vietnam veterans seeking treatment for the first time. They provided me with a buffer and rationalization against dealing with the fresh wounds of Iraq and Afghanistan. Added to this was the life-affirming joy of becoming a father at a somewhat advanced age with all of the attendant concerns about personal mortality. To see young men and women traumatized in the war was almost too much to bear. Secretly, I was aware of an old fear, that of contagion, now given added weight since not only I, but my children as well, would contract "the illness." I had to protect my children from the horrors of war. What better way then to turn away? I contemplated an early retirement from the Department of Veterans Affairs with the idea that I saw the last war through, let one of the new psychologists deal with this one. To save my son and daughter, I refused to see the sons and daughters coming into the center.

Here group supervision helped. The group members helped to give me permission to take some R and R (rest and recreation) before beginning with the new returnees. They also pointed out that I was sounding just like a soldier not wanting to be redeployed. Realizing I had a choice to "redeploy" or not, my feeling at the mercy of forces beyond my control dissipated. I regained some sense of agency. But how was I to work? I had some more dissociation and grandiosity to reckon with.

In her brilliant historical fiction, *Regeneration*, Barker (1992) describes the work of the British military psychiatrist W. H. R. Rivers during the First World War and his compassionate treatment of shell-shocked soldiers using techniques of psychoanalysis. She is able to describe Rivers's moral quandary. So long as Rivers declared that his patients were unfit for service, they would not be sent back to the front lines. A successful "analysis" would lead to being redeployed, thus underlining a grisly meaning to the word *termination*. In some ways I initially imaged my self as Rivers, helping to keep soldiers from harm's way. But the book makes clear it was not Rivers's decision to make but the soldier's. Barker describes Rivers's work with the poets Siegfried Sassoon and Wilfred Owen. Both men decide to return to the front lines out of love and loyalty for their comrades left behind, and in spite of grave reservations about the morality of the war. Indeed, the dual meaning of "grave reservation" comes into play since Owen did indeed die one day after armistice was declared.

A number of soldiers have come between redeployments with the presenting issue being not wanting to go back. Of all of these I have only written two reports: one in support of a client's wish not to be redeployed and one against a client's wish to be deployed. In all other cases the clients chose to return after discussing as many of the consequences of both redeploying and of getting a psychiatric discharge from the military. We think of the obvious consequence that they might be killed in combat or at the very least retraumatized. But what of the consequences of writing the letter? Unlike Rivers, a military psychiatrist, I do not have the power to hold someone back; I can only give my professional opinion. I also have to warn the soldiers that my recommendation may be disregarded and may lead to punitive results. The soldier would have to go back with the label coward or shirker, thus losing essential group support. If the letter is successful in getting someone out of being redeployed, the personal and professional consequences of this must be discussed. People must live with themselves and their decisions. In *Regeneration*, Sassoon decides that he cannot bear the notion of leaving his men despite his increasing pacifism. Also one's military discharge has a tremendous impact on entrance into certain careers in law enforcement, firefighting, and some government jobs. Seeing the possibility of these hopes dashed has led some to return for another tour.

CASE FOUR

Ed came for therapy for about 10 sessions after his first tour in Iraq. He had been involved in the invasion of Iraq and the rush toward Baghdad. He spent much of the remainder of his tour guarding palaces and oil wells. He returned a year after his therapy ended.

ED: My unit is being redeployed. I only have a year left on my contract. I don't want to go back to Iraq. I was stupid. I joined the Army after 9/11 because I thought we were going after Osama Ben Laden. They said Iraq was involved in this and I believed them. Now I know better. I don't want to go there and be killed for nothing. It's more dangerous there now than during the invasion. And besides I did my time. Can you write me a letter and get me out?

I explain the limits of what I can write and how effective this might be. In this case it worked. Ed was able to finish his contract stateside with the family liaison while the rest of his unit was deployed. Ed filed for and received a Department of Veterans Affairs disability for PTSD.

Another year latter Ed returned quite distraught.

ED: I've been turned down for the NYPD because of my PTSD.
THERAPIST: You didn't tell me that you wanted to be a policeman.
ED: Yes, I've always wanted to be a policeman. Well maybe I can apply to be a corrections officer or join the FDNY.
THERAPIST: I'm sorry to tell you this but all of those applications will ask if you have ever been turned down for any other police department job anywhere. If you lie and they find out you will be dismissed and may even be prosecuted.
ED: That's not fair. I served my country. I went to Iraq. And now I can't be a cop. If I didn't go, I would be able to be a cop. Why didn't you warn me?

Why indeed? Perhaps I was overanxious to "save" someone rather than explore the consequences of all possible options with him.

In the cold and dark of deepest winter all major religions have some ritual celebration of life, light, and incarnation. These celebrations can take on the quality of manic avoidance of the void, but in their essence they signify a historic moment of hope amid absolute despair. Hope is different than infantile wish fulfillment, such as my wish to return to a pre-9/11 world or my wish to make the Iraqi vets disappear by closing my eyes. Hope dawns in very ordinary ways only after we acknowledge the darkness of our current situation without fleeing, trying to fix, cure, or even help. Hope often begins with the acknowledgment of hopelessness. Mitchell (1993) quotes Eliot's (1963, p. 186) beautiful words from "East Coker" in his own profound essay on the dialectics of hope:

> I said to my soul, be still, and wait without hope
> For hope would be hope for the wrong thing. ...
> Wait without thought, for you are not ready for
> thought:
> So the darkness shall be the light, and the stillness the dancing.

Personally, the shift came when I allowed myself the freedom to think about leaving the Readjustment Counseling Center and the new vets. Only by allowing myself to leave could I also acknowledge the choice to stay. As my group supervisor pointed out, unlike the vets, I was not being redeployed. So why stay? Why not? Where can you go to get away from the war in Iraq, the global war on terror, the possibility of future terrorist attacks, your own vulnerability, your own mortality? I realized that I could not look away from the new combatants in the hopes that I could return to a pre-9/11, preinvasion world. A Vietnam vet said to me that after his first firefight he felt like running but stopped when he realized that no matter where he would go he would still be in Vietnam. Similarly, where can we go to flee our own fragile yet precious humanity? The poet Gerard Manley Hopkins (1888) uses the following images to describe the dialectic, "This Jack, joke, poor potshard, patch, matchwood, immortal diamond is immortal diamond." We are both potshard and diamond. The experience of trauma certainly compels one to experience himself or herself as matchwood (i.e., common, ephemeral, disposable, replaceable). Former images of our unique precious subjectivity, spirit, and soul collapse and seem childish fantasies. The glimmer of hope becomes present when we can keep both poles of the dialect alive. We are both potshard and immortal diamond.

I recall my first day working at an inner-city clinic as a new psychology extern. I was handed five charts the size of telephone books. Sitting there with these charts I was overwhelmed with despair. One case was worse than the next with every form of abuse, neglect, and horror beyond my worst nightmares. I felt heavy, unable to move, crushed by this terrible burden. Reading about trauma or treating PTSD is overwhelming. An interesting thing happened. I started working with children, traumatized, yes, but with sparks of life. It is the nature of posttraumatic reactions that the traumatic material is simultaneously dissociated and given absolute weight sucking the life out of all current interactions. Similarly, discussing trauma without its being embodied in the context of someone's whole life can overwhelm and deaden the therapist leading us to turn away as well. The dawning of hope is the recognition that our clients are more than their traumatic experiences and we are more than PTSD therapists.

CASE FIVE

It is my first session with the master sergeant of a local Reserve unit that had experienced heavy combat. As he talked he looked at me and the objects in my office as if to size me up in the same way I was sizing him up. He saw the drawings, the books on trauma. He asked about my experience working with other veterans. In an astute and biting observation he stated:

SERGEANT: Gee, you really have the best of all worlds. Unlike other civilians you get to see what only the military sees without having to experience it yourself.
THERAPIST: [*Choosing not to take the bait*] You are right. I am privileged to be here and at one time it did feel good to hear all of the stories. Now I'd rather not hear the stories but that's what's necessary if you want to work with combat veterans. If it were only for the stories I'd stop now. It is about the men and women who come here.
SERGEANT: Pretty fucking deep, doc. Sounds like love.
THERAPIST: Yes, I love what I do and I come to love the people I work with in the same way I'm sure you cared for your men.
SERGEANT: More than you'd know, doc. More than family. I miss them.

So through some combination of therapy and supervision, I collect my thoughts and have some hope of human contact and faith in the healing nature of this. Where does that leave me?

First, I treat all initial sessions with the Iraqi and Afghanistan vets as if they were the only sessions I may have. This has added poignancy when people are deployed and this may indeed be their only session. Second, I keep in mind that I am only one therapist at one clinic run by the Department of Veterans Affairs. I am present as a fellow vulnerable human trying to do what he can to ameliorate another's suffering. But I am also aware that I may be seen as someone much more powerful: as the all-healing parent; or as the cold, calculating government; or as the uncaring medical community. Third, I try to plant a seed for the future. I suggest that this may not be the time or the place to talk, and I may not be the person to talk to about their experiences. However, that it is important to do so when they are back on terra firma, meaning back from the war zone without the threat of redeployment, and can speak with someone who has the capacity to hear the horror without fleeing in horror. Finally, I reassure them of continued support when they return, be it here with me or in any one of the other Vet Centers with other dedicated professionals. In the age of the Internet and instant access to the war zone there is also the possibility of continued contact and even a few helpful psychic Band-Aids from afar.

CASE SIX

Dave is a slight, 22-year-old Navy corpsman (i.e., medic) who was assigned to a Marine combat unit near Falluja. He explained to me why he decided to enlist.

DAVE: After 9/11 I felt I had to do my part. I knew the whole thing with Iraq was bullshit. I knew that they had nothing to do with 9/11 but I felt it was unfair that some people went to Iraq while most stayed home and watched *American Idol*. On the other hand I had the Catholic thing going on, you know, "Thou shalt not kill." So I became a corpsman. I could help the other Marines and be on the ground. Besides I wanted my last moment in life to be helping someone rather than killing someone.
THERAPIST: I know what you mean.
DAVE: I'm sure you do doc. Say, how long have you been doing this work?
THERAPIST: Since 1983.
DAVE: That's the year I was born.

Dave began treatment because he was unable to drive to his Marine Reserve drills on the open highway for fear of IEDs. He was also fed up with the meaningless busy work that reserve units sometimes do. His chief corpsman was a New York City Fire Department captain who had his own PTSD symptoms from 9/11 and urged him to come to the Vet Center. He described in graphic detail many of his operations including having to care for dear friends.

THERAPIST: How were you able to go on?
DAVE: I just kept my head down and patched up whoever was in front of me.
THERAPIST: That's good advice, Doc, thanks.
DAVE: Sure thing, Doc.

Good advice. Bear witness to the person in front of you at this moment. Stay alive to what you are experiencing including your own numbness and wish to flee. Seek help from colleagues who can remind you of your humanity in all its fragility and strength.

POSTSCRIPT: CASE ONE, REVISITED

Jim, the Marine who emphatically stated that he wasn't coming back, was about to leave the session. Not feeling as though I had made any impact I wanted to leave him with something, for myself if not for him. I pointed to a lithograph on my wall made for the opening of the California Vietnam

Veterans Memorial in Sacramento. It shows two infantrymen running through a rice paddy toward a waiting chopper. In the background is a vintage 1950s-era automobile. I asked Jim to look at this picture.

JIM: Wow that's really interesting. What is that car doing there?
THERAPIST: The artist is a Vietnam vet. Let me read the title, "Extraction from a Hot LZ Leaving behind Classic Ford and Our Innocence."

Jim didn't say anything. He just stared and nodded.

The following week Jim was sitting in the waiting room for the appointment I gave him three days before his redeployment. He was livelier.

THERAPIST: I'm glad to see you. When you said that you weren't coming back I also took you to mean back here with me as well for this session. How are you doing?
JIM: Better. That session last week helped.
THERAPIST: What about it helped?
JIM: Two things. You didn't try to talk me out of going back. And you showed me that picture.
THERAPIST: What about that picture helped?
JIM: I felt like someone else had been through what I'm going through. I don't feel all alone.

At the end of the session he asked if he could get a copy of the lithograph to bring with him in Iraq. I promised I would have a copy sent to him. I contacted the artist in California and received this e-mail:

Hello Tom and thanks,
   Wilco on your request ... That particular image is likely my most well-known and the curious thing about it is that originally I completed the thing without the '56 Ford ... I was suddenly moved to make a very bold and risky move by adding the car to represent my own experience of having at one second been driving it around as an innocent civilian and in the next instant stumbling across a bullet swept field toward my assigned Huey.

Three weeks later I received a handwritten note from Iraq:

Dr. McGoldrick,
   How is everything going with you? Things out here aren't too bad at all. Time has been going by very quickly. The other day I received two copies autographed by the artist of that lithograph, "Hot LZ." Thank you very much. That artwork really gives me a feeling of not being alone. I look at it every day. I've been doing quite well out here. Keeping

a positive frame of mind. ... My family is doing well. They are taking this deployment better than last year's. I think we're all taking it better since I'm getting out after this. Well, I have to get going. I'll talk to you soon and thanks again.

Jim safely returned from his second deployment, is currently a New York City police officer, and is making some sense of his two tours in therapy.

REFERENCES

Barker, P. (1992). *Regeneration*. New York: Dutton.
Boulanger, G. (2007). *Wounded by reality: Understanding and treating adult onset trauma*. Mahwah, NJ: The Analytic Press.
Cochrane Library. (2006, April 20). *Psychological debriefing for preventing post-traumatic stress disorder*. Retrieved from www.cochrane.org
Eliot, T. S. (1963). *Collected poems 1909–1962*. New York: Harcourt, Brace & World.
Hopkins, G. M. (1888). That nature is a Heraclitean fire and of the comfort of the resurrection. In W. H. Gardner (Ed.), *Gerard Manley Hopkins: Poems and prose* (pp. 65–66). Baltimore: Penguin Books.
Jones, E., & Wessely, S. (2005). *Shell shock to PTSD: Military psychiatry from 1900 to the Gulf War*. New York: Psychology Press.
Mitchell, J. (1983). When disaster strikes: The critical incident stress debriefing procedure. *Journal of Emergency Services, 8*, 36–39.
Mitchell, S. (1993). *Hope and dread in psychoanalysis*. New York: Basic Books.
Shay, J. (2000). *Odysseus in America: Combat trauma and the trials of homecoming*. New York: Simon & Schuster.

Chapter 2

Men learn from history that men learn nothing from history

Jean-Max Gaudillière

For patient and psychoanalyst alike, the field of trauma and madness is largely uncharted territory, in need of further research. But clearly one thing we understand as therapists is that in handling the transference of the traumatized or psychotic patient, the ordinary attitude of neutrality is exactly irrelevant because it is always a matter of war, at whatever scale.

As everybody knows, an attitude of neutrality on the therapist's part simply does not work when we are confronted with trauma and madness in a psychotherapeutic setting. One can easily observe the catastrophic effects that follow from the attempt: symptoms persist, social status deteriorates, while the psychoanalyst is reduced to blaming the patient's "resistance." Our late friend Martin Cooperman, who was the medical director at Austen Riggs Center for decades, used to call that situation the "defeating process." When war bursts out into the transference, it is too late to be neutral.

Before presenting the vignette of my encounter with a young patient, I choose first to pay tribute to Francis Braceland, who coordinated psychiatry in the U.S. Navy on an emergency basis, when the United States entered World War II.

Twenty years before, the military health services of the allies of World War I had to face the very new types of psychic casualties and their unexpectedly high numbers, due to new weapons used in that terrible war. Thomas Salmon then coined the four principles of "forward psychiatry": proximity, immediacy, expectancy, simplicity. He added two extra requirements that were very demanding: no diagnosis and no medications, except the ones necessary to stop a violent outburst or to let a patient sleep after too many sleepless nights. For us, these principles remain our guidelines when we are treating disturbed patients who have been discharged from the psychiatric hospital. We remember that the psychoanalysis of psychosis had been developed and put into practice in that difficult and violent setting, where patients and doctors were submitted to the same dangers.

The important work of William Rivers (1918), at Craiglockhart War Hospital near Edinburgh and afterward in London under the falling

German bombs, still has much to teach us even after almost a century of further work with patients labeled schizophrenic or traumatized. Rivers was a crucible of different disciplines: He was a medical doctor, a neurologist, a prominent anthropologist, and held the post of professor of natural sciences at Cambridge University.

> These experiences, along with his work during the war, led him to examine how these different disciplines could learn from each other ... The future psychoanalyst John Rickman also followed this integrationist approach, investigating the implications of psychoanalysis for social and anthropological fields throughout his life. (Harrison, 2000, p. 34)

With Rivers' encouragement, Rickman went to Vienna to work with Freud and became one of the founders of the British Psychoanalytic Society, eventually becoming its president. If we add that he was the analyst of Bion before becoming his colleague during World War II, we understand what kind of interesting genealogy we could trace regarding the relationship between war, trauma, psychoanalysis, and psychosis.

After the end of every conflict, people, including therapists, are eager to forget and to repress, if possible, all the horrors of the preceding time: that is why forward psychiatry, like psychoanalysis, has often been contested by military psychiatrists themselves, once peace has returned. On this model, everything that was learned during the last conflict needs to be discovered again when a new war breaks out. As a result, each conflict in the 20th century has seen a great increase in the number of psychiatrists who treat war trauma, a number reduced to just a few in times of peace. With each new conflict, mental health services always suffer the same lack of resources they did in the last war.

Let us give a geographical and then historical illustration of that point. In Europe we are perhaps more accustomed to such a rhythm, over centuries, when peoples, nations, regions, even villages confronted each other in endless wars. But Americans could think about the concept of "frontier" in order to see if it doesn't characterize the same phenomenon I will try to describe here.

After a while, people are reasonably happy to enjoy life under the protection of their borders. They could call them natural borders, historical borders, linguistic borders, and so on. But history teaches us that every border is the political result of a treaty, and a treaty the juridical result of negotiations between former foes. Peace is only the fragile and temporarily result of former wars.

In fact, we have inherited the same situation in psychoanalysis. During wars, therapists were obliged to invent new tools in order to intervene almost under fire and sometimes actually under fire. But soon after the treaties were signed, these old geniuses are often considered to be monuments: better if they remain in their statues in the museum of history. After the end of combat, their number decreases, their teaching is forgotten; eventually they are treated as embarrassing veterans, often diagnosed politely as bizarre or even psychotic. (Consider for example the case of Bion, who was repeatedly "sacked" as he put it, over the course of his career, notwithstanding whatever honors and presidencies he received. Likewise, Ferenczi was betrayed by his own psychoanalytic family. Firemen don't appear so useful when the fire is over.)

In an article published in 1946 in the *American Journal of Psychiatry*, Francis Braceland distills the psychiatric lessons from World War II. He wrote:

> When the final history of WWII is written, and like all history is entombed in large volumes, to the student who reads it carefully, it will reveal the same lessons which are learned so painfully and at such great cost in all wars. Each successive war necessarily brings with it new problems, but the old ones crop up repeatedly; and, as we encounter them again and again, we wonder why man is so slow to profit by experience and history. Perhaps, the cynic expressed it best of all, when he said "Men learn from history that men learn nothing from history." (p. 587)

* * *

I think of madness as a form of research conducted by patients in the field of catastrophe, in which time and history take on a peculiar status in the transference.

These symptoms express themselves first as a stoppage in the dimension of time. Through the patient's words or actual behavior, even in the extremes of catatonia, time appears out of joint, in a manner that does not allow a reference to the past as an explanation. Perhaps such an explanation is objectively right, but it is useless in the therapeutic relationship, as in such cases as these the dimension of causality itself has failed to flow with the usual course of the arrow of time; if there is no past, no present, no future, there is no more causality. This is true even if from the point of view of an observer, we could quite validly identify the moment when a special event happened one, two, or three generations before. But, as we know very well, in the transference, the point of view of an observer is only the point of the resistance on the side of the analyst. Concerning the event, which apparently caused the crisis, Winnicott (1974) definitively wrote in "Fear of Breakdown": "The underlying agony is unthinkable (p. 102). The unconscious here is not exactly the repressed unconscious of psychoneurosis (p. 102). The patient needs to 'remember' this, but this thing of the past has not happened yet, because the patient was not there for it to happen to" (pp. 103–105).

So the anamnesis with such a patient has no meaning, and the analyst is facing a veritable symptom of time. It is of no use for him to evoke the patient's memories or to echo obvious signifiers. One patient once calmly gave me this recommendation: "Don't wear yourself out. Everything is present. The only thing that matters is the here and now." The suspension of the flow of time here does not refer in a mechanical way to whatever caused the symptom, but renders useless inquiry into the specific temporality of a "death area" that has dropped out of history. I use here the name given by Gaetano Benedetti in 1980 to these holes of time.

Benedetti,* one of the founders of the International Symposium for Psychotherapy of Schizophrenia (ISPS), elaborated this precious notion, clinically and theoretically, during his long life and teaching. He used the expression death area to describe a state of mind in which all the thoughts, all the affects, even all the actual events are polarized by an unrelenting attraction of death. The most original aspect of his presentation concerns the peculiarity that arises when the attraction is extended to the transferential space. Benedetti's technique for extricating oneself from these threatening areas brought by the patient is one of his major contributions to our field.

As a metaphor one might compare these death areas to the astronomical black holes, in which the density of the matter is so high that even light cannot get out of it, preventing one from obtaining any direct information of what happens "inside." The presence of a black hole can only be inferred indirectly, through irregularities measured by astronomers. As psychoanalysts, we are often in the same situation. Benedetti urges us not to be fascinated by the attraction exerted by the "black holes" of such patients, but for the analyst to transform into a positive presence the "irregular" shapes that appear at their borders.

* * *

Now, I want to question my own experience to determine if psychoanalysts are members of that community of men (and women) who learn nothing from history.

In considering this question, I want to take up the matter of transference as it manifests in cases of trauma and psychosis. Whatever the means, the aim of the patient is to drive the therapist to the same kind of area of death, in a movement that Harold Searles once called "the effort to drive the other crazy." The questions asked by the patient, sometimes only by his or her presence, always concern his or her relationship with history not just his or her personal history

---

* For one article translated into English, see Benedetti (1996). For a while, Benedetti worked as a psychoanalyst at Chestnut Lodge in Rockville, Maryland. Unfortunately, his major work, *Alienazione et Personazione nella Psicoterapia della Malatia Mentale* (1980), is not translated in English (but in French, under the title *La Mort dans l'Âme*).

with daddy and mummy (that which, according to Winnicott, constitutes the repressed unconscious of psychoneurosis). But the same question, regarding the relationship to history also implicates the therapist.

To continue our astronomical comparison with black holes, we should recognize that in our work, there is paradoxically no direct utility in speaking *about* such black holes: Everybody knows well enough of the files piling up the misfortunes of a patient, of parents, grandparents, and so on. It becomes more interesting if we can speak *to* those holes. And the program is always the same: If the analyst begins to speak *to* the area of death, it is only as an attempt to answer its question, often spoken vicariously through the patient. Who are you as a therapist, from which catastrophic area? If one answers only with what is printed on one's ID card, it is not equal to the level of the question.

I remember a patient in the hospital, whom the chief psychiatrist asked me to visit him in his room. He was laying on his bed, in his own feces and urine, refusing any food, his eyes burning with fever. Uneasily, I introduced myself with my name and my position on the ward. I will never forget his response: "Myself, I am an encoded of the anti-past."

His strange response reminded me of a lesson learned from our great ancestors: Otto Will, our old friend Martin Cooperman (who died at the beginning of 2006), and Wilfred Bion. Each of them became psychoanalysts out of their confrontation with History. Or, more exactly, they became analysts when the "little" history of their life and family crossed the "big" History of wars, revolutions, and other catastrophes of the social link. Bion (1975, 1982), for instance, wrote bluntly many years after his experience as a soldier during World War I (when he was just 20): "I died at Cambrai in 1917, at Amiens in 1918." Our patients have taught us that this is not a metaphor, an error, or a symptom of Alzheimer's, as some idiots used to diagnose Bion after some of his conferences. Every question a patient asks a therapist can be translated: "Where, and when, did you experience something that transformed you such that you can meet me in these crazy regions?" Our duty, to be at the level of the transference, is to answer and certainly not to stay silent in too-cautious neutrality.

Our patients ask us to join them as subjects of History, under the authority of "historical truth." Freud (1939) himself coined that expression in "Moses and Monotheism,"* which he wrote in exile after the *auto-da-fe* of his books in the Nazified Berlin. In this text we could hear him asking,

---

* "This remarkable feature can only be understood on the pattern of the delusions of psychotics. We have long understood that a portion of forgotten truth lies hidden in delusional ideas, that when this returns it has to put up with distortions and misunderstandings…We must grant an ingredient such as this of what may be called historical truth to the dogmas of religion as well, which, it is true, bear the character of psychotic symptoms but which, as group phenomena, escape the curse of isolation" (p. 85).

"Is anything able to survive, when all the traces are erased, even the traces of the destruction?" This question would reverberate with particular poignancy in light of the fate of the Jews and of psychoanalysis itself in the conflagration of World War II.

Thus the subject of History is not the subject of the repressed unconscious, but the one produced from scratch when confronted by active threats of erasure. My purpose here is to show how this subject of history is transferentially produced as the result of interconnections between the history of an analyst and the history of a patient. Benedetti calls this place of meeting between patient and analyst the transitional subject. When such a meeting takes place, the nonrepressed unconscious is at work; this is the kind of unconscious produced by annihilation.

When Blue Flower,* one of my first patients, suddenly spoke in the big room of the ward of the hospital for the first time after decades of complete silence, I understood that this quasimiracle was somehow linked with my father, who had fought in the same region she spoke about, the very day when she spoke again:

> In the common room of the first psychiatric hospital we worked in, a chronically mute woman patient used to stand near a radiator, always the same one. This was a typical asylum in northern France, placed around a former abbey not far from the front and the cemeteries of the last wars. The analyst got into the habit of joining her each week; he, too, would stand with his back to the wall, describing aloud to her what was going on in the room or sometimes simply whatever came into his mind. This small talk, addressed to the woman as though to himself (since she said nothing) had the effect that, after more than a year, she began to speak and to tell a story as though it had taken place yesterday. (Davoine & Gaudillière, 2004, p. 211)

I presented this vignette in my book (Davoine & Gaudillière, 2004) in a chapter titled "Expectancy: The Trustworthiness of the Other." In my small talks with Blue, I sometimes evoked the presence of my father as a private during World War II. The hospital was situated in the middle of what had been the battlefield, marked by thousands of white crosses along the roads. One can imagine my surprise when my patient's first speech was about that precise issue. I quote it as printed in my book, transcribed after her first speech addressed to me:

> It is wartime. One day, she followed her mother toward the fortifications of the little town where they lived. She caught her mother quickly leaving a German soldier. Back home, her mother had lifted her skirt

---

* About this case, see Davoine and Gaudillière (2004).

over her nakedness and screamed at her astonished daughter, "You want to know what I'm doing? OK, take a look!" From then on, the little girl had cancelled herself as a subject until she reached her forties, ageless. (Davoine & Gaudillière, 2004, p. 211)

It was only later that I came to understand that her father had also been a private during the war, probably stationed a few kilometers from my father. While her inadequately armed father was fighting against an overequipped German army, her mother was prostituting herself to soldiers from the very same army.

When our patients ask about History, they are in fact, only asking us why we became psychoanalysts, why we have chosen to do such bizarre work. Lacan once said that the psychoanalyst is the one who identifies himself with his symptom: a good start, but not the whole story, at least to my mind. You probably know the Latin etymology of the word desire: *de-siderare* (to desire) means to takes one's eyes off the stars (the real stars, not movie stars), but, for instance, the destiny written on the sky like a horoscope we are tempted to contemplate in a quasi-hypnotic state.

Awakened by an interpretation, the patient is supposed to be able to make better use of the eyes and other sensations: He or she is able to desire when the useless repression is removed from his or her itinerary on earth. But when it comes to patients dealing with History, I would prefer to define desire—especially the desire of the therapist—as a momentum coming from these rather dangerous areas in time: wars, revolutions, natural catastrophes—all the circumstances when the symbolic guarantee of the social link is deeply shaken. Through the encounter with these areas, if they are engaged, comes the possibility of sharing the energy and the intelligence that are contained within the madness. The answer to the question anxiously asked by our patients implies a declaration of our choice in favor of the horizons of life.

Let Blue Flower herself comment on how this process transpired between us:

> Some time later, she calmly explained to me what had motivated her to speak: "That day, when you passed in front of me, I heard a big YES, and I knew I could talk to you."
>
> I had said no such thing: I had only shown my eccentric persistence in standing alongside her, attracted by the dignified way she held herself and by her wide-eyed, childlike gaze. (Davoine & Gaudillière, 2004, p. 211)

Hadn't she, like so many others, remained an unyielding child, waiting for our words to allow her the freedom to speak?

* * *

More recently I was consulted by a young man of 25, referred to me by his mother, a psychoanalyst, whom I do not know. This guy appeared in the guise of a dropout, dressed in tatters, shaggy-haired, apparently under the influence of some drug, probably cannabis. It was almost impossible to make any eye contact with him, as his head moved relentlessly left and right in an involuntary movement; even more disconcerting, he was practically unable to connect one word to another to convey any meaning.

I told him briefly the conditions for working with me as a psychoanalyst. After almost 30 years of experience in the psychiatric hospital and private practice, I would not begin any analytical work with him, or with anyone, until he completely ceased using drugs. I consider analytic work impossible in such a case, as treating an addicted patient turns the analyst into an accomplice of the voluntary deterioration of the physical integrity of the patient and also because we will not reach together the liberty of thinking necessary to the process.

Attempting to negotiate with me, he asked, "What if I don't take anything the day of the session or even the day before?" I answered firmly that the issue was not one of neurology but rather a matter of trust between us, and that "we cannot divide and fragment trust." With half a smile, he told me: "If I completely stop the drugs, you might regret it." Good challenge for the beginning of combat, isn't it? Then I explained I was ready to meet him every week, if he was interested by such a challenge, until he could stop using drugs, but that I would not accept any money, since he didn't give me the possibility to work as an analyst.

So he came, week after week, in about the same state of disorganization of that first meeting. Although he had begun studies in philosophy after earning his baccalaureate degree, he had dropped out of the university more than 3 years earlier. He spent his time living with different communities in several squats in the suburbs of Paris looking for drugs, chatting with his cohabitants; sometimes his chats ended in violent physical fights.

In one session, we had the following exchange:

"What are you talking about?" I asked.

"About the Algerian War," he answered.

I instantly lost my concentration. As he moved on to talk of completely different matters, I was calculating dates. Obviously he was too young to have been born during the Algerian War, which started in 1954 and finished in 1962. I asked about his father, who had divorced his mother when he was 8 years old. From my figurings, his father would have been too young to have been drafted for that war. So I asked him about his interest in that troubled period of French history.

"You know," he told me with the same unease, "the tortures, the torturers, the French Army, you know what I mean …"

I knew certainly that this uncomfortable period of decolonization had become a touchstone for some more recent ideologies, reheated by French media around the 2005 uprising of minority youth in the Paris suburbs. At the same moment one could hear almost every day the news of massacres of women and children that took place in Algeria threatened by political troubles. During the uprising, simplistic discourses in the media tried to draw together the two periods in a black-and-white interpretation of violence: "the bad French Army," "the good Algerian rebels." In the aftermath of the Algerian War,* general suspicion had fallen on the shoulders of every private returning from the war. This suspicion had the effect of silencing these soldiers forever, until their children showed up years later in our consultation rooms, with that cruel and distorted equation: Silence equals avowal of torture.

I had stopped listening to my patient. I was remembering a particular moment, when I was myself an adolescent studying after my baccalaureate degree in the early 1960s. For that I had arrived in Paris in the middle of the turmoil of attacks organized by the extreme right, with bombs and murders, and a military rebellion, triggered by opponents to De Gaulle's government, which launched the implementation of the process of decolonization I found myself rather firm on that matter: It was not for me a matter of opinion, because the public opinion was trapped in an impossible dilemma.

During several years, there was a political fight in the country between supporters of the FLN (the Algerian Front of National Liberation) and supporters of the OAS (Organization of the Secret Army), which wanted to maintain a French Algeria administered by a military government. But my attention was captured by an actual area of death, a huge catastrophe having exploded just beside me.

I now recalled my own encounter with the "big" History. My family and I had moved to the north of France when I was 8 years old. Near our new house, in a city where the scars of World War II were still visible, a young neighbor introduced me to the city, to the other children in the neighborhood, even to the strange food they used to eat in that part of France (strange to me, having come from the southern central region of the country). My new friend's name was Claude, and he was 2 or 3 years older than I. His father used to get drunk, and often my own father was called in to intervene physically in their home to stop the violence against Claude's siblings and their mother. Claude dropped out of school before the age of 14, and, unable to find work and to escape that hopeless hell, he decided

---

* The Algerian War (1954–1962) began just after the end of the Indochina War (1946–1954). The wars marked the beginning of the process of decolonization, with dramatic consequences for families settled over there, sometimes since more than a century. The wars introduced a strong and durable division in the political and public life, with social tensions triggered by the return of thousands of dispossessed settlers.

to enlist in the army instead of waiting to be called up. (Military service, starting at age 20, was compulsory in France for every male citizen until 1995.) He came back 3 months later in a coffin. At that moment, my own Algerian war began and ended. Without the deferment I received due to my studies, I could have been drafted myself to fight in Algeria. From that moment on, I was unable to have any intellectual debate about the war with my schoolmates; any conversation about it felt silly, empty, and rhetorical.

I told my story to that drug-addicted and agitated young man in front of me. He gave me a look I hadn't seen before. I continued:

"You know, you told me that 'your' Algerian war has nothing to do with the actual everyday massacres in Algeria. OK. We have concluded that you and your father were both too young to have fought in the 1954–1962 war, and I just explained why I couldn't hear anything coming from partisans of either side of the issue concerning that period. So to whom and about what are you actually speaking?"

At that point he was carefully listening to me. So I continued:

"Now I will tell you about, two other, earlier wars that are part of the history between France and North Africa. The first one took place around 1830, when the French government chased the previous colonizers (namely, the Ottoman Empire, present there as a colonial power since the 16th century) out of Algeria. But that pertains to History; you and I learned that in secondary school, transmitted as a text, as a reference for everybody. So I want to tell you about another event that happened in the same country, about which I didn't know myself until 5 years ago.

"There were ferocious battles in North Africa against the German Army during World War II: Everybody knows about the English victory of Montgomery's troops against Rommel. But after the unilateral armistice with Hitler, the French population in Algeria, as well as in metropolitan France, including the Army, was divided between supporters and opponents of Pétain's government. In 1942, American and English troops landed in North Africa and were immediately confronted with a well-trained and victorious German division. But American troops were unprepared, and many were quickly vanquished. One of the most important centers to treat physically and psychologically wounded soldiers was located in Algiers. As French pupils, and eventually French citizens, we were never taught about that part of our own history, until we read *A War of Nerves** published by the English historian Ben Shephard in 2000, which described the evolution of war psychiatry in the United States and in England. He brought to light the work of military doctors who were in charge of treating the physical and psychological wounds of the American troops in World War II.

---

* Shephard describes without any embellishment the management and population of the American and British Hospital in Algiers in 1943.

"That history had been suppressed from our memory by the laws of amnesty after World War II. No one is aware of that not-so-glorious episode of French history."*

I had just finished my rather long monologue; the young guy was still sitting in front of me, looking at me intensely. He said: "You didn't know that my mother was born in Algeria at the beginning of the '40s."

From that moment, we progressed to the point where he became able to stop using drugs, giving me the possibility of beginning the analytical war, which is the usual outcome in the transference of such traumatic events.

I think my introduction of History into this treatment restored a life to people who had lived and worked in Algeria over several generations, for the most part in harmony with one another: Arabs, Berbers, Jews, Europeans of French and other backgrounds. The social relations that existed among these groups in colonial Algeria cannot be reduced to ones of economic exploitation. When the lived History of real men and women is erased in ideological histories, the real crime is this erasure of their lives and memories.

If I had not taken the opportunity to speak to my patient from my own history, and to challenge the automatic words that emanate from ideology, this clinical vignette, full of detail and specificity, would not be able to support the theoretical/clinical point I'm making here: Speech has the capacity to disguise and erase thought when ideology can provide a kind of odd comfort but requires that one surrender one's ability to think.

If I had responded to him with a feeble yes, saying nothing in an effort to be "neutral," he probably could not have heard my no, which is the first judgment required by the opening of a possible transference, in the case of trauma or madness. No, I don't agree with the common *laissez faire* attitude concerning your addiction and dropping out. Yes, we will begin with a judgment, at least a judgment asserting the existence of a subject, rather than join in silent complicity with an ideology that views the subject as unnecessary, superfluous, to quote Hannah Arendt (1948) in *The Origins of Totalitarianism*.

* * *

In concluding, I come back to literature, to poetry, especially to epics, which are often connected with the topics we are dealing with in our work. "Who are you?" is always the question our patients are asking when they are building a transference. Who are we, to authorize such an inaugural judgment—the judgment that creates the foundation for any possibility of thinking? With the help of a detour through our ancient Greek cultural

---

* Not so glorious, because the huge majority of French civilians were spectators, at best, to the Allies' efforts to treat the English and American casualties in North Africa. Some in the military tried to continue the battle, for instance, through the Résistance, but a lot of the others participated with the collaborationist government of Maréchal Pétain.

background, I would like now to explore the roots of that possibility of thinking when "historical truth" is in confusion.

Near the end of Song XVIII in *The Iliad*, verses 497 to 508, Achilles is entering again the war against Troy, because his friend, his *therapôn* Patroclus (*therapôn* is the exact name given by the poet, meaning the second in combat, the ritual double; see Nagy, 1979), has been killed by Hector, under Achilles' armor. His mother, the goddess Thetys, already knows that he will die in combat against Hector. Nevertheless, and very movingly, she asks the blacksmith god Hephaestus to prepare the most marvelous weapons one can imagine for her son. Included among these is an incredible piece of art, a shield, on which all the features of the Achaean civilization are represented in bronze, gold, and silver. One can trace the movement of human life in cities and during wars through the representation of the primary elements of air, sea, sky, and stars sculpted into the metal.

The scene shifts to a judicial trial. A murder has been committed, probably without any witness, and a judgment has been rendered. One man is ordered to pay a fine for the price of the blood of a man he killed unintentionally. A disagreement ensues over whether the convicted man has paid the fine that was imposed as punishment. The convicted man protests that he has already paid the fine. The people of the city cannot make a judgment about the matter. They are divided into two parties, each shouting in favor of the man sentenced or the party that has been granted reparation. But the heralds are containing the mob. Finally the elders will be the judges, each in turn assuming the stick of authority, and pronouncing his personal judgment.

So the judicial institution needs a higher authority to make its judgment: a judge who will decide who is the better judge. A competition is held to determine who can make the better judgment, with the winner receiving an award of two talents of gold.* Who will decide? The people? No, not in this case. The one who will judge the judges has a name, which appears here in writing, perhaps for the first time in Western literature. His name is Histôr—the same root as in the word *histôria*.

History is what allows us the possibility to think, even after atrocities. Etymologically, *histôria* shares a root with *videre*, Latin meaning "the witness." In the word *witness*, there is the same *wit-* as in *Histôr*. In the Greek verb *oida* ("I know"), we find again the same stem. And if we go from the English *wit* to the German *Witz*, witticism, we can appreciate the continuum going from normality to madness, from neurosis to psychosis, if we are able to find and follow (alongside our patient) the yellow brick road.

---

* In Greek antiquity, the talent was a measure of weight, equivalent to 20 to 27 kg. Anyway, the award seems considerable.

REFERENCES

Arendt, H. (1948). *The origins of totalitarianism.* New York: Harcourt, Brace & Company.
Benedetti, G. (1996). The splitting between separate and symbiotic states of the self in the psychoanalytic dynamic of schizophrenia. *International Forum of Psychoanalysis, 5,* 23–38.
Bion, W. R. (1975). *A memoir of the future: The dream.* London: Karnac.
Bion, W. R. (1982). *The long week-end, 1897–1919.* London: Karnac.
Braceland, F. (1946). Psychiatric lessons from World War II. *American Journal of Psychiatry, 103,* 587–593.
Davoine, F., & Gaudillière, J.-M. (2004). *History beyond trauma.* New York: Other Press.
Freud, S. (1939). Moses and monotheism. In J. Strachey (Ed. & Trans.), *The standard edition of the complete psychological works of Sigmund Freud* (Vol. 19, pp. 1–137). London: Hogarth Press.
Harrison, T. (2000). *Bion, Rickman, Foulkes, and the Northfield Experiments: Advancing on a different front.* London: Jessica Kingley Publishers.
Nagy, G. (1979). *The best of the Achaeans: Concepts of the hero in archaic Greek society.* Baltimore: Johns Hopkins University Press.
Rivers, W. H. R. (1918). The repression of war experience. *The Lancet.*
Shephard, B. (2000). *A war of nerves: Soldiers and psychiatrists in the twentieth century.* Cambridge, MA: Harvard University Press.
Winnicott, D. W. (1974). Fear of breakdown. *International Review of Psycho-Analysis, 1,* 103–107.

Chapter 3

The psychoanalytic politics of catastrophe*

Ghislaine Boulanger

THE TENSION BETWEEN INDIVIDUAL AND SOCIETAL NEEDS

Outside the town hall in Santa Fe, New Mexico, a roughly fashioned adobe monolith commemorates the massacre of a local Native American tribe. It is some 10 feet tall; stuck into the 4 foot plinth on which the monolith rests are fragments of everyday life: children's plastic toys, a sock, a toothbrush, a tattered teddy bear, chipped plates and cracked cups, a family photograph behind broken glass with the frame partly obscured by sandstone. The shocking poignancy of these familiar little objects scattered in and by the enormity of the memorial conveys at a glance a meaning that words alone must struggle to contain. How does one simultaneously describe the personal losses and the vastness of such an event? How does one fit one's understanding of the individual survivor into the larger picture of the catastrophe without losing sight of that individual's own struggles, without objectifying them in some way? How does one keep the individual's plight in mind without submitting to the mind-numbing enormity of a catastrophe?

There is always a tension between an individual survivor's need for a witness, for recognition, understanding, and engagement on the one hand, and society's need to map trends, to categorize, to plan for, and to provide social programs on the other. These needs are at odds with one another. Too often the general obscures the particular. As psychoanalytic clinicians we turn our attention to the particular, to the socks and combs and discarded toys, the evidence of disrupted lives. Our practice concerns the individual. We cannot and should not work from the position of the general, but in the case of adult onset trauma, more than any other psychological condition, there is often pressure to do so.

---

* Portions of this essay are taken from Chapter 10 of *Wounded by Reality: Understanding and Treating Adult Onset Trauma* by Ghislaine Boulanger. © 2007 The Analytic Press. Reprinted with permission.

In this essay, I focus on the tension between a clinical emphasis on individually disrupted lives and the larger societal and political needs that push us to move beyond these reminders of the frailty of the human spirit. Focusing specifically on the military's treatment of combat trauma, I consider some of the social and professional issues that concern and obscure adult onset trauma before returning to the individual treatment of those who have been confronted with the terror of immediate death and survived. At least, they have survived physically but their psychological survival may be in doubt.

Unlike most matters with which psychoanalytic clinicians concern themselves, the private hurts, everyday confusion, disappointments, and dissatisfaction of massive psychic trauma often invites the public gaze and thrusts those clinicians who would treat survivors into an unaccustomed forum. More often than not, the events that give rise to adult onset trauma—on a collective level combat, genocide, natural disasters; on an individual level, torture, rape, violent crime—become a matter of public record. Inevitably such events draw attention and commentary, and attempts to regulate how they will be perceived through the media, through religious institutions, and through politicians who use them for professional capital. Inevitably too these attempts at regulation will impact memories of the event and expectations about how the event should be experienced. These expectations hold true not only for the survivors who have been immediately impacted by the event, but also for those who would treat them.

THE PORNOGRAPHY OF VIOLENCE

Violent encounters destabilize many of those who fall victim to them, as well as many of those who mete out the violence, but they also have a corrosive effect on the psyches of those of us, clinicians included, who stand by and watch, or hear about an event thirdhand or through the media. Knowledge of violence has the potential to destabilize bystanders, distant witnesses who are often many times removed from the immediacy of the disaster.

Toward the beginning of the *Heart of Darkness* (1902/1999), Conrad's protagonist describes what it is like for a man to be confronted suddenly by violence:

> There's no initiation into such mysteries. ... He has to live in the midst of the incomprehensible, which is also detestable. And it has a fascination, too, that goes to work upon him. The fascination of the abomination—you know, imagine the growing regrets, the longing to escape, the powerless disgust, the surrender, the hate. (p. 7)

Violence has a pornography of its own. Like voyeurs, we find ourselves guiltily enthralled by what we see or what is represented in the media or what we hear from immediate survivors. The need for mastery, the need to triumph over the abject, over the object of our fascination who has been stripped of his subjectivity, compels this fascination. But, in the end, as Conrad puts it there is powerless disgust, the surrender that is a longing to escape the ultimate futility of pitting oneself against overwhelming odds. In the end, for many people, and certainly for society as a whole, caught between the fascination of the abomination and the powerless disgust, indifference becomes the most adaptive position.

Is it indifference, a failure of imagination, dissociation, disavowal, or, as Des Pres (1976) slyly puts it, the wish to continue to draw our strength from innocence that allows us to keep the facts of Baghdad and Bagram, of Anwar Province, Guantánamo Bay and Darfur, Gaza and Siderot sufficiently separate from our daily lives that we do not to rise up in continual protest? Recall that the Holocaust was hidden in plain sight during the Second World War and for more than 25 years thereafter until it was sufficiently in the past to become a new object of fascination (see, e.g., Kuriloff, 2008; Prince, 2008).

Clinicians who hope to work effectively with the survivors of adult onset trauma struggle between these two poles; they are caught between the "fascination of the abomination" and "powerless disgust" both of which provide a measure of dissociation. And they are charged with the impossible task of finding a middle ground and engaging individual terror directly and without flinching. Often indifference or dissociation, in the form of irrelevant theory or technique, becomes the preferred psychoanalytic position (Boulanger, 2007).

DENYING THE CONSEQUENCES OF VIOLENCE

We live in a society that fails to acknowledge the long-term consequences of violence. Actively and omnipotently, opinion makers, be they the media, the military, religious leaders, politicians, and sometimes even our own profession, seek to deny or to imply that we can in some way escape these consequences. We do violence to ourselves and to those who have survived when we deny how corrosive life-threatening encounters can be.

The public is led to believe that if violence does not cause death, its consequences will rapidly dissipate. Commentators do not provide a containing function for a traumatized population, but rather a deflecting function. Janoff-Bulman (1985) points to the frequent messages emphasizing that traumatic events serve a particular purpose thereby reestablishing a belief in an orderly and comprehensible world. Linenthal (2001) deconstructs the ways in which local opinion leaders expected survivors of the bombing of the federal building in Oklahoma City to cope with their memories in the aftermath of the attack. He found that the preferred narratives were "progressive" narratives emphasizing religious and political messages. For example, one pastor is quoted as saying that "sometimes evil circumstances can reveal a deep reservoir of goodness that was just waiting to be expressed" (Linenthal, 2001, p. 47). Such socially meaningful but often personally empty narratives emphasized the new opportunities, healing, and rebuilding that emerged in the wake of the bombing. On the other hand, personal narratives that dealt with the ongoing pain of survivors, narratives that sought to portray the full impact and horror of the event were considered "toxic."

Many institutions similarly encourage magical thinking arguing that adversity makes you strong. There is no question that adversity can lead to growth, but there is a breaking point beyond which adversity can and frequently does lead to the collapse of the self (Boulanger, 2007).

A PARADOX

There is a paradox between the public's current fascination of the abomination—the fact that "victims" make good copy and ensure large audiences for the news media—on the one hand, and society's indifference to the long-term consequences of violence on the other. Even as we are invited by the media to indulge our curiosity about those who have fallen victim to contingency, do we not use their fate to lull ourselves into the false belief that if we have not fallen prey to a similar fate, that this is implicit proof that the world is just and orderly? Nguyen (2008) argues that fascination with victims is motivated by the need to become acquainted with cruelty and mortality, to manage death anxiety, and falsely resolve the quest for the meaningful.

Like all paradoxes, the relationship between the short-term fascination with victims and long-term indifference to their difficulties is less contradictory than it first appears. Both fascination and indifference typify dissociative responses to horror, blunting a fuller exploration of adversity. Furthermore, both deny the object of this fascination or indifference a fully realized subjectivity.

THE CONTRADICTORY HISTORY OF POSTTRAUMATIC STRESS DISORDER

The politics of catastrophe and the dangers of the destabilizing forces of violence find expression in the contradictory history and reception of the diagnosis of posttraumatic stress disorder (PTSD). Whether a massive psychic trauma occurs individually or in a group, when individuals face almost certain death and seek refuge in catastrophic dissociation, they are often subject to the diagnosis of PTSD and thus they enter public discourse.

The diagnosis is the legacy of political action taken in the 1970s by Vietnam veterans, by women's groups, and by Holocaust survivors who believed that their particular plight—as, respectively, combat veterans; as survivors of rape, battery, and incest; and as individuals who experienced the Shoah first hand—was not represented in the diagnostic choices that were then available. It is a bitter irony that in its mandate to represent these disenfranchised groups, the diagnosis has become a caricature of what those who originally lobbied for its inclusion in the *Diagnostic and Statistical Manual* (3rd ed.; American Psychiatric Association, 1980) had intended.

The diagnosis has fallen victim to trauma's capacity to enthrall and to repel. It contributes to the tension between the imperative to acknowledge and the danger of indifference into which fascination and disgust inevitably sink. Attempts to capture and categorize reactions to terror seem doomed to failure for too often they trivialize rather than do justice to these reactions. The act of diagnosing or of being diagnosed thrusts any individual into the public arena. When a survivor steps onto or is placed on this contested stage, individual reactions are easily obscured by PTSD's contentious history (see North, Suris, Davis, & Smith, 2009).

Industries have sprung up around the cluster of symptoms labeled PTSD, inevitably shaping, obscuring, and denying an individual survivor's reactions. Arbitrary cutoff points are necessarily established by epidemiologists seeking to determine the incidence or prevalence of PTSD in a certain population. However, the epidemiologists' task is often at odds with survivors' needs.

When the diagnosis is made appropriately it can serve an important role, offering relief to those who had not previously understood that their symptoms form part of a recognizable reaction to an extraordinary event. It becomes a reference point in a world where categories have suddenly collapsed. Too often, though, even when properly made, the diagnosis marks the end of an inquiry rather than the beginning of treatment. With overuse, misuse, and even correct use, the very words posttraumatic stress disorder have become a cliché, a form of avoidance or denial for many mental health professionals. When details of the trauma itself and of its consequences have been categorized and fixed in place, the response to terror is reduced to a formula. Rather than encouraging understanding of the experience, it is forced into recognizable and socially prescribed categories, which discourage further investigation.

In the very act of being labeled, the subject located by this diagnosis has ceased to be a subject, becoming instead an object of curiosity or a statistic. But this is not the only danger of the diagnosis, for while it offers a veneer of legitimacy and opens the gateway to treatment, those who do not reach criterion for the diagnosis itself yet experience the chronic alienation and isolation of posttraumatic symptoms are often denied appropriate treatment, particularly when this treatment is dependent on the military as I describe later. These survivors of violence, whose symptoms are not officially recognized, frequently fail to make a connection between the violence they have survived and their symptoms and, in turn, become indifferent to their own survival.

In her great pacifist essay *The Iliad or Poem of Force*, Simone Weil (1940/2005) writes that violence turns anyone subjected to it into a thing. Regrettably, the diagnosis that was created to acknowledge reactions to violence can compound the objectifying forces of violence.

MILITARY POLICY ON COMBAT TROOPS AND RETURNING VETERANS

A brief review of how the Department of Defense and the Department of Veterans Affairs handle the mental health needs of combat troops offers an opportunity to deconstruct the ways in which the diagnosis can be manipulated to obscure the real and to normalize reactions to horror. This brief review also points to the danger of confounding the needs of an institution, in this case the military, with the individuals that make up that institution.

Recently the Rand Foundation (Tanelin et al., 2008) announced that "more than 300,000 veterans—or 18.5% of those deployed since 2001—now have PTSD or major depression." These are deceptively low statistics and they overlook several facts. Not only is the diagnosis difficult to make with the closed-ended questionnaires used by Rand, many troops may suffer disabling symptoms without reaching criterion for the diagnosis; further, as I describe later, the criterion shifts according to the phase of the war or the availability of resources. In addition, the symptoms of posttraumatic stress disorder often emerge many years after the stressor has ended. There is frequently a period of latency during which symptoms, if they are experienced, are dismissed or numbed with alcohol and drugs. It is to be anticipated, then, that as time passes more and more veterans of the conflicts in Iraq and Afghanistan will become incapacitated by the symptoms of PTSD.

The 18.5% rate of PTSD among returning veterans being reported by the Rand Foundation is only the tip of the iceberg. Abraham Kardiner (1969), whose brief paper describing his work with soldiers during and after the Second World War remains the gold standard for treating combat trauma wrote, "The syndromes described under war neuroses do not appear on the battlefield. All one sees are tired, exhausted and frightened men. It is only later that the full extent may be recognized" (p. 248). Indeed after the Second World War researchers documented cases of gross stress reaction presenting 5, 10, 15, 20, even 30 years after the end of the war (Archibald

& Long, 1968). Today, with news of growing casualties from Afghanistan and Iraq, Vietnam veterans who never sought treatment before are finally realizing they can no longer cope with the intrusive memories of combat about which they have often felt ashamed.

Given politicians' and the public's readiness to deny the long-term effects of violence, these facts are forgotten each time war is declared. The military is complicit in this denial. Young men, and now women as well, continue to be seduced by the rhetoric of war that offers them an opportunity to behave heroically, to serve their country and, if necessary, to die for a worthy cause. Since the First World War military psychiatrists have suggested that it is possible to confront the violence of combat and emerge unscathed psychologically.

During the First World War, military psychiatrist Thomas Salmon (1917) identified ways of avoiding psychiatric casualties in combat. "Proximity, immediacy, expectancy, and simplicity" became the code words for treating soldiers on the battlefield whether in World War II, Korea, Vietnam, and the several Israeli conflicts of the last 40 years. These words mandate treating psychiatric casualties as close to the front line as possible, as quickly as possible, allowing peer pressure to exert its influence. In the short run, this practice reduces psychiatric casualties in combat itself, but it does not reduce the number of troops who claim (and the many who do not dare claim publicly) to suffer psychological symptoms as a consequence.

With these superficial suggestions, military psychiatrists appear to believe that they have found a way to avoid the long-lasting consequences of violence. It is an omnipotent wish; in fact, this practice promotes catastrophic dissociation (Boulanger, 2007), enabling soldiers to endure increasingly terrifying conditions in a numbed and altered state of consciousness, only to find that when the violence is over it is not easy to escape this state and comfortably take up their former selves, for during the period of horror the psyche lost faith with the resilience of its core.

"Battlemind" is the 21st century's version of Salmon's mantra (Adler et al., 2009). Since 2006, the Department of Defense (DOD) has promoted a one-hour class taught by combat veterans to boost resilience before battle by giving troops an idea of what to expect during combat. Parenthetically DOD points out that the reason soldiers did not tell other soldiers what to expect in the past is "because they didn't want to brag" (Huseman, 2008). This is an astounding failure to understand the psychic costs of violence and why veterans prefer not to talk about combat and its aftermath. Only a very few leave combat wanting to brag about it, and those are not the ones who should be educating new troops.

On the return from combat, a second phase of battlemind is intended to help soldiers adjust to life after deployment and thereby reduce the percentage of service members reporting postcombat symptoms by normalizing those symptoms.

The symptoms of posttraumatic stress disorder appear deceptively easy to identify leading to the mistaken belief that naming them is sufficient to enable those who are suffering from them to place the intrusive memories, nightmares, and numbness in context and thus vanquish them. In practice, posttraumatic symptoms are often covert and undramatic; those who experience them become increasingly disabled and isolated, and experience considerable difficulty linking their feelings of alienation, rage, and fear to their survival.

Protracted arguments about which symptoms do or do not form a part of the syndrome labeled posttraumatic stress disorder take place each time the *Diagnostic and Statistical Manual* is revised, as if an individual survivor's experience can be represented or changed by the inclusion of a particular symptom. These deliberations also take place on a local level. In 2005, Vedantam reported that at a Department of Veterans Affairs meeting in Philadelphia the "utility and objectiveness of PTSD criteria and the validity of screening techniques" were being reviewed just as the costs of treating war veterans with PTSD were spiraling. This process had profound implications for returning soldiers in that the outcome of these deliberations led to many who might otherwise have qualified for treatment being turned away.

The political stakes are high when diagnosing posttraumatic stress disorder. Government funding for treatment after a disaster is often contingent on meeting criteria for the diagnosis. Writing in the *New York Times*, Benedict Carey (2005) reported that for those returning from Iraq "whether their difficulties are ultimately diagnosed as mental illness may depend on the mental health services available." One psychologist (Gadberry, 2008) writes that at Camp Pendleton she was told levels of PTSD were almost non-existent. This is not unusual; despite the advent of battlemind, many bases cast doubt on the legitimacy of the diagnosis. At Camp Pendleton, however, chaplains are appointed to counsel the troops because it is claimed that they are more suited to work with the lack of morale and lack of faith that cause PTSD, whereas mental health workers are believed to further damage individual morale. On other bases, troops wait so long for treatment that they become discouraged, or they end up overmedicated but without psychotherapy.

A further deterrent to seeking treatment is the fear of being misdiagnosed, and prematurely discharged without disability, without veterans benefits, and possibly having to pay back a reenlistment bonus.

The question of whether predisposing factors play any part in developing PTSD is constantly debated when the diagnosis is being revised. During a war, when disability claims are growing exponentially, claiming that predisposing factors account for posttraumatic symptoms is routine. A Department of Defense Report to Congress (2007, p. 5) show that 22,656 personality disorder discharges were processed between 2002 and 2007.

It is not known how many of these troops were experiencing posttraumatic stress symptoms possibly exacerbated by preexisting conditions. However, these discharges meant that troops were denied all benefits and disability payments.

Disability payments for mental health claims are always contingent on meeting diagnostic criteria. In May 2008, the head of psychiatry at one Veterans Affairs hospital wrote in a private memo to her staff that was leaked to the media: "Given that we are having more and more compensation-seeking veterans, I would like to suggest that you refrain from giving a diagnosis of PTSD straight out. Consider a diagnosis of adjustment disorder instead."

Clearly decisions about how to characterize and treat the negative reactions experienced by troops in combat, whether these reactions are deemed disabling, dismissed as normal, or attributed to preexisting conditions, depend on the local availability of financial mental health resources. However, in what must be the most cynical manipulation of public opinion intended to send an unambiguous message to troops considering requesting a discharge on psychological grounds, or even seeking treatment for stress symptoms, Gittelman (2003) reported that one Iraqi veteran was brought up on charges of cowardice for making such claims. Charges against this "battle hardened" Green Beret, who showed all the symptoms of PTSD and had to be returned to the United States, after seeing an Iraqi cut in half by machine gun fire, were dropped only when the press got hold of the story.

POSTTRAUMATIC STRESS DISORDER IN PSYCHOANALYTIC PRACTICE

With the relational turn, psychoanalysts have become increasingly political and increasingly aware of a century's failure to provide for many survivors of the harsh realities of life within our theory and our practices. Our discourse is becoming more inclusive of these underserved populations, not only combat veterans but also minorities (Altman, 1995), the poor (Cushman 1995; Layton, 2005), immigrants (Boulanger, 2004; Perez Foster, Moskowitz, & Javier, 1996), homosexuals (Butler, 1990; Corbett, 1993), or women and those of uncertain gender (Harris, 2005). It is important to recognize that many members of these groups are also more vulnerable to violence and more likely to have faced destructive situations in which they feared for their lives than those who have traditionally sought out psychoanalysis or psychoanalytic psychotherapy (Satcher, Friel, & Bell, 2007). On a concrete level, the urban poor are exposed to more violence and more of them are minorities because members of this group disproportionately volunteer for military service. Some have fled persecution in order to find sanctuary in America, only to live with constant reminders of the violence they have fled. Women are frequently the targets of life-threatening violence because they have in some way transgressed personal, family, or group norms. Erikson (1976, 1994) demonstrates that the poor are more likely to live in the path of natural disasters or in neighborhoods where they are exposed to government and corporate environmental neglect. Whether they are African Americans in the 9th Ward of New Orleans whose homes and lives were disproportionately lost during the hurricanes of 2005, or mining families in Buffalo Creek living in the path of an industrial accident, or Native Americans whose community is threatened by a toxic spill, or troops returning from Iraq and Afghanistan who have been denied necessary treatment, these populations are often silenced by shame and confusion at the failure of the world to acknowledge their experience.

Now that psychodynamic clinicians have turned their attention to these underserved people, it is important to bear in mind that if these confrontations with catastrophe and violence are not directly addressed in treatment, they can take a psychic toll that lasts a lifetime. As relational psychoanalysts increasingly appoint themselves spokespersons for those who have not been well served by the mental health profession as a whole and by psychoanalysis in particular, it is incumbent on us as clinicians to develop an awareness of the insidious and corrosive impact that violence and terror have on adults just as they do on children.

Independently of other psychodynamics, and recognizing how frequently the survivor is indifferent to his own survival, if clinicians do not undertake the work of restoring the collapsed self, of enabling those who finally find treatment to pick up the threads of their lives and to weave themselves back into the fabric of society, they are ensuring that the legacy of adult onset trauma will be visited on future generations. Time and again those who are alert to trauma's psychic toll find that their gaze is directed to its relentless spiral through sociopolitical movements, across temporal barriers, transmitted from one generation to the next, and always already embodied and enacted on a personal level (Davoine & Gaudilliere, 2004; Faimberg, 2005; Grand, 2000).

Our particular psychological discipline has taught us the danger of evading what lies beneath the surface, knowing that finding a cognitive or a metapsychological rationale for the reaction to horror does nothing to heal that reaction, but it does increase dissociation both on an individual and on a societal level. In a painfully familiar feedback loop, this dissociation further blunts our capacity to sustain an understanding of horror. However, psychoanalysis is less practiced at looking beneath the surface when it comes to adult onset trauma (Boulanger, 2007). Psychoanalytic theory has provided a very narrow range of interpretations for the sense of destruction found there. Briefly put, our discipline has relied on arguments such as psychic conflict, developmental arrest, and, most obviously, a childhood trauma, and situated the source of the problem, not in the recent terrifying event, but in childhood.

For the most part psychoanalytically trained clinicians are called upon to work with survivors of violence and horror after the symptoms have become chronic. Chronic posttraumatic symptoms impoverish the internal object world and consequently impact relationships, constrict thought and affect, compromise the ability to symbolize, take the meaning out of work and leisure, collapse the self and obliterate the sense of belonging.

There are many reasons why survivors do not seek immediate treatment or do not get the kind of treatment they need. I described earlier how society at large has mistaken expectations about a rapid recovery from catastrophic violence. Politicians' and the media's mandate to spin the news and to shape public perception emphasizes uplifting conclusions and simple dichotomies: good guy, bad guy, evil villain, noble victim, survivor hero. A survivor who does not feel heroic, a victim who does not feel ennobled by his experience, has a difficult time finding a place for his self-doubts in this discourse (see also Nguyen, 2008; Thomas, 2005).

Survivors themselves are often afraid of seeking the kind of treatment that will require reliving horrifying events. Ashamed not to have been able to withstand the assault, not to have "bounced back," these survivors remain silent about their catastrophically altered sense of reality. They believe that their experience has cast them beyond the reach of society's walls. Sometimes they have sought treatment but the treatment modality has been too superficial.

Immediate treatment is frequently short term and based on exposure therapy protocols. Research (e.g., Keane, 1995) has shown this to be an effective treatment in dealing with acute reactions to violence when undigested intrusive memories are often the most overwhelming symptoms. In this case, 30% to 70% respond successfully to exposure therapy alone. Short-term treatments reduce symptoms and, if they take place sufficiently close to the time of the trauma and address the symptoms of catastrophic dissociation, they reduce the likelihood that the traumatic reaction will become chronic. There is no question that if terrifying memories are handled sensitively, with a practiced ear for those who are close to being flooded or who are being avoidant, an initial debriefing of the event can begin the process of integrating the traumatic experience. Putting words to recently dissociated experience can determine whether a particular survivor will require further professional help.

The longer survivors wait to seek treatment, the more convinced they become that they have suffered irreparable damage, the more their psychic disenfranchisement becomes a reality.

With or without immediate intervention, epidemiological studies have found that about one quarter to one third of those exposed to massive trauma develop long-term psychological reactions. However, epidemiological studies are constrained by diagnostic practice and neither captures individual experience. Here I return to the dilemma described at the beginning of this essay, the necessity not to be blinded by the larger picture of a catastrophic trauma, not to lose sight of the painful evidence of individually disrupted lives. This is the psychoanalytic clinician's role.

Long-term treatment with survivors of adult onset trauma demands that rigorous attention to the patient's experience must take precedence over any preexisting theory of mind. Invoking early conflict, developmental arrests, or childhood trauma as an explanation for the alienation of adult onset trauma is tantamount to blaming the victim, which is a political position. It should not be a psychoanalytic one.

In his address at the 44th International Psychoanalytic Congress in 2005, Vinar rejected the "lineal development which mechanically subordinates current experiences to infantile neurosis. Extreme experiences, as in the case of war and torture, are able to shape and reconfigure the existing organization of the psyche" (p. 323). But this opinion is not shared by many in the psychoanalytic community (e.g., Stimmel, 2003).

Most of the time our work as psychoanalytic clinicians is with more prosaic passions and pains; it is with the small endurances of everyday life that can overwhelm and defeat a sense of liveliness. Adult onset trauma takes us into a different arena, out of the quotidian into the custody of horror, where an adult has faced the random indifference of nature or been confronted with the terror of immediate annihilation at the hands of a familiar or unfamiliar aggressor. There must be room in our psychoanalytic work for both the quotidian and, when and if it is present, for the immensity of horror, and within the immensity of horror for the individual context. If the painful magnitude of adult onset trauma is not treated in and of itself, there is a danger of kicking over the traces of the trauma, taking that swift terror of death and burying it in the mundane, confusing it with the little indignities of everyday life.

There is always a relationship between the survivor's dynamics, the psychological impact of the traumatic event itself, the psychological consequences and meaning that event assumes, and current symptoms. To overlook any of these variables and their interaction with one another is to fail the patient. The trauma must be contextualized. If it is given short shrift, the patient feels misunderstood and blamed, her ordeal minimized. On the other hand, if the trauma is emphasized but its psychic consequences are not considered and understood in and of themselves, the patient continues to be overwhelmed by aspects of internal experience that have not been articulated and that therefore remain inchoate and incomprehensible.

Close consideration of the trauma narrative is necessary, but it is not sufficient. In his deconstruction of the narratives given by Holocaust survivors, Langer (1991) writes, "Oral Holocaust testimonies are doomed on one level to remain disrupted narratives. ... Instead of leading to further chapters in the autobiography of the witnesses, they exhaust themselves in the telling" (p. xi). This is where psychodynamic treatment is of vital importance. We must ensure that the survivors we treat, who share their narratives with us, do not exhaust themselves in the telling. We must take our work with the survivor beyond the trauma, filling in the subsequent and previous chapters in the patients' autobiographies to which Langer refers. Psychodynamic treatment with survivors of adult onset trauma is not about memorializing the trauma; it is about learning to live with the knowledge of the trauma, not being lived by it.

CONCLUSION

This chapter is about the consequences of violence and narrowly missing violent death or witnessing it but not, in fact, escaping it. In this examination of the phenomenology of horror, I have insisted that the wounds inflicted by reality in adult life become psychic reality and lead to the collapse of the self. These facts are frequently obscured by politicians, by the media, by diagnostic practices, and by our own reluctance to face up to the consequences of adult onset trauma. Horror of this magnitude does not happen in every lifetime, but psychoanalytic clinicians must always be alert to the evidence of selves that have collapsed under the weight of terror.

REFERENCES

Adler, A., Bliese, P., McGurk, D., Hoge, C., Castro, C. (2009). Battlemind debriefing and battlemind training. *Journal of Consulting and Clinical Psychology*, 77 (5): 928–940.
Altman, N. (1995). *The analyst in the inner city: Race, class, and culture through a psychoanalytic lens*. Hillsdale, NJ: The Analytic Press.
American Psychiatric Association. (1980). *Diagnostic and statistical manual of mental disorders* (3rd ed.). Washington, DC: APA Press.
Archibald, H. C., & Long, D. M. (1968). Persistent stress after combat: A twenty year follow-up. *Archives of General Psychiatry*, 119, 317–322.
Boulanger, G. (2004). Lot's wife, Carey Grant, and the American dream: Psychoanalysis with emigrants. *Contemporary Psychoanalysis*, 40, 353–372.
Boulanger, G. (2007). *Wounded by reality: Understanding and treating adult onset trauma*. Mahwah, NJ: The Analytic Press.
Butler, J. (1990). *Gender Trouble*. NY: Routledge
Carey, B. (2005, November 26). The struggle to gauge a war's psychological cost. *New York Times*. www.nytimes.com/2005/11/26/health/26psych.htme
Conrad, J. (1999). *Heart of darkness*. New York: The Modern Library. (Original work published 1902).

Corbett, K. (1993). The mystery of homosexuality. *Psychoanal. Psychol.*, 10: 345–357.
Cushman, P. (1995). *Constructing the self, constructing America*. Reading, MA: Addison Wesley.
Davoine, F., & Gaudilliere, J.-M. (2004). *History beyond trauma*. New York: Other Press.
Department of Defense. (2007). Report to Congress. www.veteransforamerica.org
Des Pres, T. (1976). *The survivor: An anatomy of life in the death camps*. New York: Oxford University Press.
Erikson, K. (1976). *Everything in its path*. New York: Simon & Schuster.
Erikson, K. (1994). *A new species of trouble: The human experience of modern disasters*. New York: Norton.
Faimberg, H. (2005). *The telescoping of generations: Listening to the narcissistic links between generations*. London: Routledge.
Gadberry, S. (2008). Personal communication.
Grand, S. (2000). *The reproduction of evil: A clinical and cultural perspective*. Hillsdale, NJ: The Analytic Press.
Gittleman, J. (2003). Green Beret on trial for cowardice. *New York Times*. www.nytimes.com.2003/11/06/us/soldier-accused-as-coward
Harris, A. (2005). Gender as soft assembly. Hillsdale, NJ: The Analytic Press.
Huseman, S. (2008) Battlemind prepares solidiers for combat. www.2009military.com/military-news-story (*editor, dates are discrepant online*)
Janoff-Bulman, R. (1985). The aftermath of victimization: Rebuilding shattered assumptions. In C. R. Figley (Ed.), *Trauma and its wake*: Vol. 1. *The study and treatment of posttraumatic stress disorder* (pp. 15–35). New York: Brunner/Mazel.
Kardiner, A. (1969). Traumatic neuroses of war. In S. Arieti (Ed.), *The American handbook of psychiatry* (pp. 245–257). New York: Basic Books.
Keane, T. (1995). The role of exposure therapy for the psychological treatment of PTSD. *NCP Clinical Quarterly*, 5(4): 225–243.
Kuriloff, E. (2008, April). *Theory as trauma: A foray into applied psychoanalysis*. Paper presented at the meeting of Division 39 (Psychoanalysis) of the American Psychological Association, New York.
Langer, L. (1991). *Holocaust testimonies: The ruins of memory*. New Haven, CT: Yale University Press.
Layton, L. (2005). Notes toward a nonconformist clinical practice. *Contemporary Psychoanalysis*, 41, 419–429.
Linenthal, E. T. (2001). *The unfinished bombing: Oklahoma City in American memory*. New York: Oxford University Press.
North, C., Suris, A., Davis, M., & Smith, R. (2009). Toward validation of the diagnosis of Posttraumatic Stress Disorder. *American Journal of Psychiatry*, 166(1), 1–8.
Nguyen, L. (2008). *Psychoanalysis with torture victims*. Paper presented at the meeting of Division 39 (Psychoanalysis) of the American Psychological Association, New York.
Perez Foster, R., Moskowitz, M., & Javier, R. A. (Eds.). (1996). *Reaching across boundaries of culture and class: Widening the scope of psychotherapy*. Northvale, NJ: Jason Aronson.

Prince, R. (2008). *Theory as thick description*. Paper presented at the meeting of Division 39 (Psychoanalysis) of the American Psychological Association, New York.

Salmon, T. (1917). *The care and treatment of mental diseases and war neuroses in the British army*. New York: War Work Committee of the National Committee for Mental Hygiene.

Satcher, D., Friel, S., & Bell, R. (2007). Natural and manmade disasters and mental health. *Journal of the American Medical Association, 298*(21), 2540–2542.

Stimmel, B. (2003). Tragedy and technique: Stability when the earth shakes. *Psychoanalysis and Psychotherapy, 20*, 97–118.

Tanelin, T., Jaycox, L. H., Schell, T. L., Marshall, G. N., Burnam, M. A., Eibner, C. Invisible Wounds Study Team. (2008). *Invisible wounds of war: Summary and recommendations for addressing psychological and cognitive injuries*. Santa Monica, CA: Rand Center for Military and Health Policy Research.

Thomas, N. (2005). The use of the hero. In Y. Danieli & R. Dingman (Eds.), *On the ground after September 11: Mental health responses and practical knowledge gained* (pp. 394–400). New York: Haworth Press.

Vedantam, S. (2005, December 27). Spiraling costs of PTSD treatment for war veterans. *Washington Post*. www.google.com/#h1=en&source=p&q=shankar

Vinar, M. (2005). The specificity of torture as trauma: The human wilderness when words fail. *International Journal of Psycho-Analysis, 86*, 311–333.

Weil, S. (2005). *The Iliad or the Poem of Force*. New York: New York Review of Books Classics. (Original work published 1940).

Chapter 4

Whose truth?
Inevitable tensions in testimony and the search for repair

Nina Thomas

> You can live comfortably with the details once they have acquired academic relevance.
>
> Antjie Krog
> *Country of My Skull* (1999)

Minja* is 57 years old though she looks far older. She is a stout woman now, but in 1993 when she was released from the concentration camp she weighed barely 70 pounds. As the Bosnian Serb military attacked their town, Minja, her husband, and their three children took refuge in the surrounding snow-covered woods. Soon after, she became separated from her husband and teen-aged son. Minja, her 12-year-old daughter, and 3-year-old remained in the woods along with other women, children, and old men from their town who also tried to hide from the attacking Serb forces. The snow provided little cover, however. Many were killed during the firefight that night and the next, including the mother of the two young boys whom Minja grabbed to shelter along with her own children. It was not long before the Serbs captured Minja as well as many of the others. They were taken to a women's prison. When she names the place she shivers and momentarily gasps for breath. It was there that she was nightly gang-raped by soldiers, beaten with rifle butts and the steel-toed boots of her captors, and starved.

One night after being raped by four, five, maybe more men (she doesn't remember how many), and while she was being led back to her cell, the commander of the prison pulled her by the hair and sneering darkly into her face said: "Tell anyone and I will find you in five countries." In fierce defiance of his threats made several years before, Minja told me her story of surviving, being found by a cousin in Western Europe who united her with her husband and son after 4 years. She testified before the International Criminal Tribunal for the Former Yugoslavia in The Hague, insisting to me

---

* To preserve the confidentiality of people who have shared their stories with me I have changed their names and disguised the identifiers of their histories unless they are a part of a public record. Where perpetrators are concerned, I use their names as taken from court transcripts or other public documents.

that she had not been afraid to be in the courtroom with her attacker, nor to answer the questions of the lawyers and justices.

The assertion of her fearlessness is part of her defiance of the soldier's threats, as much, too, as having survived the brutal conditions of her captivity. She is proud to have testified. She even notes that on the plane from Sarajevo to Amsterdam the defense counsel sat near her and her companion (another inmate of the infamous prison), though this time she acknowledges being afraid. "When they took us in the airplane, the defense attorney was sitting behind us. When we came out of the airport, he took a cab and that cab followed our cab. And I said: 'Oh my God, he's going to kill us.' Then I said, 'It doesn't matter if he kills us.'"

I understand Minja's "Oh my God, he's going to kills us," juxtaposed against "It doesn't matter if he kills us," as her struggle against succumbing to the soldier's threats which had previously silenced her, as they were intended to do.* The waxing and waning of Minja's efforts to reclaim her agency in resistance to domination is what Laub (1992) described as characteristic of the testimony giving of concentration camp survivors. Her silence had been so complete that she had never shared with the other women in her cell, nor they with her, what happened to them in those nights. Each knew but would not dishonor her neighbor by asking. In testifying to her experience Minja is, in Laub's framework (1992), "breaking out" of silence and of the helplessness and powerlessness of the camp. Such stories and struggles are repeated in the experiences of tens of thousands of survivors of the Balkan war, those who have chosen to testify as well as those who have not.

Minja is one among the number of survivors of the Balkan war whom I interviewed in an effort to understand the processes underlying "healing" from the trauma of war, though I find the idea of healing personally repugnant. The word suggests that one recovers from the horrors of such evil as being subjected to inhuman treatment by people who, as was often the case in Bosnia, had once been your neighbors and with whom you shared birthdays, holidays, and other celebrations. I don't believe one recovers from such violent transgressions. Rather, one is more or less able to dispel their perpetual "tyranny" (Laub, 1992), by being able to narrate the experience in the presence of another, about which I will say more later in this chapter.

My interest in the question of how countries, and more to the point, how individuals recover from political atrocity has developed over the course of more than 10 years of working in international settings. In early 1997, about 15 months after the signing of the Dayton Accords that brought

---

* Although it would be valuable to consider the threat speech holds for both perpetrator and victim, particularly as it represents the creation of a "third" (Benjamin, 2004), to do so would take us far afield of the task at the moment. I expect to pursue this area in further writings on the significance of postwar testimony for perpetrator, victim, and witness.

an end to the nearly four years of open warfare in the Balkans, I made my first trip to Bosnia to consult and offer training to local mental health workers. That initial contact has spawned ongoing relationships with several agencies in Bosnia that provide services to the forgotten casualties of that war—survivors of concentration camps and torture—whose psychic wounds have not been dressed in the postwar effort to recover. What forms the text of this chapter are portions of interviews I held with survivors in Bosnia and in South Africa, portions of testimony they or others have given either before the International Criminal Tribunal for the former Yugoslavia (ICTY) or the South African Truth and Reconciliation Commission (TRC), and excerpts from my field and interview notes. I bring my psychoanalytic perspective to the subjects of "truth" as lived and reclaimed in these settings, as well as to the differing and at times contradictory aims of an individual's as compared with a nation's "healing."

PSYCHOLOGICAL FALLOUT

In this chapter I address the psychological fallout for survivors when different processes of societal repair are undertaken in the aftermath of a country's ethnic conflict, war, or state-sponsored repression, and what a psychoanalyst can add to our understanding of these. I use the term *fallout* with purpose, intending to convey that it is not only in the immediate impact that the effects are felt but also in the later and possibly in the long term that the unanticipated and potentially iatrogenic consequences become evident. As psychoanalysts we can be attuned to such potentialities. If we are engaged in making meaning and disentangling coconstructed enactments, identifications, and shifting self-states, and even more in creating the psychoanalytic space for "the third" in Benjamin's (2002, 2004) formulation of that idea, then what we have to offer holds the promise of being valuable to the processes involved in testifying to history. I will focus largely on the experiences of Bosnian survivors of the war in that country who gave testimony before the ICTY using parallels from the South African apartheid experience and witness testimony before the TRC for what lessons can be learned from each.

How countries fare in moving out of social violence involves politico-legal and psychosocial dimensions. The question has confronted dozens of regimes around the world for decades. Some would say more than half the countries of the world have faced or are facing the issue (Rosenberg, 1999). Most often one or both of two methods are pursued in the interest of establishing (a) a democratic, constitutionally stable government within which (b) respect for human life and protection of human rights are taken as fundamental tenets, and (c) the truth about the past may be established, as a consequence of which recovery and reconciliation may occur.

The methods I refer to are first, truth commissions, most often tasked with producing both an official truth about past events and reconciliation between former enemies, and second, war crimes tribunals, which prosecute the perpetrators of human rights violations.* Such tribunals typically are established or facilitated by international bodies like the United Nations has done with the International Criminal Tribunal for the Former Yugoslavia (ICTY), for Rwanda (ICTR), and for Cambodia along with several other countries. The rationale that underlies tribunals is that justice will take the place of vengeance.

Truth commissions and trials or tribunals are established by very different methods, operate with different systems, and usually have vastly different mandates. They are each, however, fundamentally judicial undertakings. Each is often the outgrowth of complicated negotiated settlements between entrenched political systems (typically repressive regimes) in opposition to other embattled political forces eager for the installation of rights and privileges. Thus truth commissions and tribunals are effectively political compromises intended to advance political agendas. To adequately critique the two endeavors is beyond the scope of this chapter. Those who are interested in pursuing the subject further would be well rewarded by immersing themselves in the rigorous scholarship of writers from such different disciplinary perspectives as the clinical psychologist and former researcher for the TRC, Brandon Hamber (2002, 2003, 2009); the anthropologist and long-time researcher about the TRC Richard Wilson (2002; as well as the work of Hamber & Wilson, 2002); the psychologist Kaminer and his colleagues (Kaminer, Stein, Mbanga, & Zungu-Dirwayi, 2001); human rights activists and the directors of the Human Rights Center at the University of California, Berkeley, Stover and Weinstein (2004) among others. I will, however briefly, describe some of the principal elements and points of controversy about truth commissions and war crimes trials in the interest of understanding how these processes affect the survivor-witnesses—both perpetrators and victims—who participate in them. In addition, in the history of human rights struggles, the South African TRC occupies an iconic place, which itself warrants careful analysis and possible reconsideration. I offer some hypotheses for why it may be that the TRC has come to be a signifier of such importance, largely to Western sensibilities,† so much so

---

* Each of these methodologies has uncanny resonance in the United States, in 2009 particularly, as calls for both—truth commission and war crimes tribunal—have been made within the U.S. Congress and the media by which to address the events associated with the war in Iraq and the related constriction of human rights within the United States and in its pursuit of foreign policy abroad.
† The South African TRC both implicitly and explicitly privileged the Christian ethic of confession as redemptive and frequently utilized the language of Christian sacrifice in its proceedings, for example, characterizing those who died in the "liberation struggle" as "heroes" who had sacrificed themselves for future generations (cf. Wilson, 2001).

that its history has virtually eclipsed any awareness of the 20 similar commissions that preceded it and the nearly 30 that have followed.

ORIGINS OF THE SOUTH AFRICAN TRC

The mission of the TRC was laid down in the country's interim Constitution of 1993, achieved after lengthy and difficult negotiations between apartheid government principals including the then ruling National Party and the leading opposition group, the African National Congress, as well as several other smaller parties aligned with one or the other side (the Inkatha Freedom Party being the most notable of these, headed by Prince Mangosuthu Buthelezi). The act of Parliament establishing the TRC (The National Unity and Reconciliation Act, No. 34 of 1995) set out a very broad mandate of establishing the truth of past victimization in order to achieve reconciliation among all the peoples of South Africa.* Indeed, blazoned across a banner that hung behind the commissioners at public hearings was the slogan: "Revealing is healing," underscoring the commission's aim to gather the stories by which a history of the apartheid era could be written that would resist revision. Truth telling thus was a fundamental trope in the discourse of South Africa's recovery.

For many people outside the country, the South African TRC is the paradigm for such commissions. My hypotheses about how it comes to its status in the world's consciousness of truth commissions† are several; not least that it reflects the leadership of the TRC by Nobel Peace Laureate Archbishop Desmond Tutu, a renowned peace and antiapartheid activist. Equally, its functioning was brokered in the course of negotiating Nelson

---

* The act expressly calls for, among other things, "establishing as complete a picture as possible of the causes, nature and extent of the gross violations of human rights which were committed during the period ... including the antecedents, circumstances, factors and context of such violations ... establishing and making known the fate or whereabouts of victims and ... the granting of amnesty to persons who make full disclosure of all the relevant facts relating to acts associated with a political objective committed in the course of the conflicts of the past during the said period; affording victims an opportunity to relate the violations they suffered; restoring the human and civil dignity of such victims by granting them an opportunity to relate their own accounts of the violations of which they are the victims, and by recommending reparation measures in respect of them" (National Unity and Reconciliation Act, No. 34, 1995).

† The South African TRC it is often mistakenly thought of as the first such undertaking. Although its mandate and the conditions under which it was established are significantly different—an act of parliament as compared with presidential mandate—the first truth commission in 1983 was held in Argentina in response to the years of military dictatorship in that country known as the period of the "Dirty War." Even less well known is that a truth commission was established in the United States in 2005 in Greensboro, North Carolina, to investigate the murder in 1979 of five union activists by the Ku Klux Klan and members of the American Nazi Party, and the role played by the police both in the attacks and in the coverup of those responsible.

Mandela's release from Robben Island after nearly 26 years in prison, much of it in solitary confinement. The establishment of the TRC marked just how great a political shift had taken place within South Africa. The TRC began its functioning at the opening of Nelson Mandela's tenure as the first democratically elected and first black president of South Africa, a time in which there was enormous hopefulness about the "rainbow nation" and broad-based enfranchisement in the social and political life of the country.

The TRC was the most ambitious of the truth commissions that preceded it or have come since; and the only one to include subcommittees on amnesty, reparation and rehabilitation, and human rights violations (South African Truth and Reconciliation Report, 1998a). Initiated by an act of Parliament, the life of the commission extended from 1995 through 2002, having been extended beyond the original cutoff date of 1998 to complete all the amnesty petitions it had received, more than 20,000 victim statements, and 8,000 petitions for amnesty. Further, I suggest that the standing of the TRC is due at least in part to the fact that many (about 10%) of the hearings it held were public and regularly televised nationally, occasionally internationally. As Madeleine Fullard, a former researcher for the TRC noted: "Those public hearings are what really has remained in the mind of the public more than our truth commission report" (personal communication, April 2003). She goes on: "The way the truth commission communicated to South Africa about what its goals were, who could come to it, what kinds of cases, the public hearings were the most important vehicle for conveying what the truth commission was about and reaching as wide an audience as possible in a way that a report could never have done." As a result, the South African TRC, and its commissioners and witnesses became the public face of a national enterprise of contrition and "repair," though that public face is by far not representative.

> April 2, 2003: At the Johannesburg airport I am waiting for my connection to Cape Town. The man seated next to me is reading an Afrikaans newspaper. I immediately feel myself filling with suspicion and distrust. A black African man and a white woman check us into the plane. I am aware of their races and wonder if the assignment is in fact intentional by race and gender. Can you *not* be conscious of race in this society?

In the interest of full disclosure let me acknowledge the bias I bring to this discussion. I believe that individual survivors' needs are often subordinated to an inappropriate degree to the aim of national reconciliation and recovery, within the structure of these tribunals and truth commissions. Postconflict countries recovering from their contested pasts depend on the willingness and hopefulness of survivors to share their stories. In so doing they risk not only psychic but very real physical costs. More than 300 witnesses at the International Criminal Tribunal for Rwanda were killed (Neuffer, 2001). Witnesses before the ICTY whom I interviewed repeatedly reported having been threatened either directly or indirectly. In reporting on her work about the South African TRC that parallels my own work in Bosnia (Thomas, 2005), Byrne (2004) also noted that many of those who had testified had been threatened (see also Feitlowitz, 1998, on parallel experiences in Argentina following the Dirty War in that country). Although witnesses before the ICTY are offered the option of testifying under "protection," meaning they may be assigned pseudonyms and have the choice that, during the 30-minute delay in televised broadcast of the proceedings, their voices and faces are electronically "covered," the protection is not always effective. During my attendance at The Hague there were several times when I witnessed attorneys using the names of a witness family members, for example. In another instance, one "protected witness" reported that a member of the defendant's family had threatened her mother-in-law. And a third example derives from an interview I conducted with an elderly, dignified, white-haired Bosnian Muslim who had spent 2 years in a hospital in another country recovering from the wounds he had sustained during the war in Bosnia. Kresimir arrived carrying a well-preserved packet of papers tied up with string from which he extracted one sheet that he slowly read to me. It was the report of the local police to his claim of having been threatened by repeated phone calls warning him that he would be killed if he testified. My interpreter translated: "This is confirmation that on that day the police came about the anonymous telephone calls ... 'If you show up in F___, 15 guns will be waiting for you. You won't get out alive.' After that, he hung up." These examples make dramatically clear how much a debt is owed by the entire nation to those who would engage such risks, one that is only repaid when careful consideration and provision are extended, not only for the basic safety of witnesses, but also for individual and social healing.* While other researchers have not, to my knowledge, directly addressed the issue of the potential physical danger to prospective witnesses, many have written about the psychological costs involved (Byrne, 2004; Hamber, 2002, 2009; Kaminer et al., 2001). I suggest that financial and professional support to underwrite victim-focused efforts would begin

---

* Clearly these are not necessarily separate domains. For example, the Rwandan government has, along with the international community, instituted "gacacca courts" to address the enormous number of those charged with crimes of genocide in the 1994 murders of more than 800,000 people in that country. The gacacca courts operate by traditional codes of adjudicating conflict in an attempt to restore individuals to their communities. The distinction is perhaps more appropriately made between individual and political healing.

to address the complicated path of derailing the "doer–done to" dynamic (Benjamin, 2004). I detail this further later in this chapter.

The TRC conflated truth with reconciliation and in so doing oversimplified the complex vectors involved in surviving years of human rights abuses and social violence.* Further, as Hamber (2009) notes, it collapsed the individual process of recovering from trauma with the national one. What must be remembered is that the truth commission process of recovery and reconciliation was a politically brokered arrangement that included amnesty for perpetrators as an inducement for their testimony and for yielding power in the interest of political stability. The conflation of the interpersonal concept of forgiveness with the legal concept of reconciliation within the logic of the proceedings laid the basis for many people's ultimate disappointment with the commission. The experience of one widow of apartheid-era violence serves as a counterweight to the forgiveness of others. One of the assassins of what came to be known as the Craddock Four† sought her forgiveness, asking for a 15-minute meeting with her. "You have teased our grief for nearly 12 years ... and you think you can reconcile in 15 minutes," she said upon meeting him (Meredith, 1999).

Hers is only one of a significant number of such expressions of dissatisfaction with the explicit demand that the perpetrator and victim reconcile at the culmination of the Human Rights Violations hearings. Another even more dramatic example occurred during the amnesty petition hearings of Craig Williamson, a former double agent of the South African Security Police.‡ One of Williamson's victims, Marius Schoon, testified at Williamson's amnesty hearing about the brutal murder of Schoon's wife and daughter by a letter bomb sent by Williamson. During that hearing Williamson's attorney said to Schoon: "I tender Mr. Williamson to you at the next adjournment of these proceedings, in the spirit of reconciliation, do you accept this tender?" Schoon's response was a straightforward "no," adding that he thought it was "unfair and embarrassing" to be called on to reconcile with Williamson (*South African Truth and Reconciliation*

---

* See in particular Hamber (2002, 2009); Hamber and Wilson (2002); Posel (2002); Posel and Simpson (2002); Wilson (2001).
† This was a notorious and, as the amnesty hearings made clear, an especially gruesome assassination involving South African Security Police and the notorious figure "Prime Evil," Eugene de Kock. At the hearings the details of what happened to the victims at last became known, freeing their families of what Hamber and Wilson (2002) have called the "liminal" state of not knowing and always hoping for information about, if not the return of their loved ones' bodies. Former TRC researcher Fullard has said, in referring to this case: "Even if we never solved another case in the whole truth commission that case would stand in quality" (personal communication, April 2003).
‡ Like the case of the Craddock Four referred to earlier, the amnesty petition by Williamson in the case of the murders of Schoon's family aroused considerable rancor toward the TRC proceedings (Brittain, 2000). Many considered Williamson's testimony less than complete but even more, the murder of two academics and a 6-year-old girl could not, they argued, be justified as politically motivated as was required to establish the basis for amnesty.

Commission Hearings of the Amnesty Committee, 1998b). Examples of similar expectations that victims would reconcile with their perpetrators abound throughout the records of the South African TRC proceedings. Despite the extensive amnesty hearings with their explicit requirement that full and complete confessions be given of politically motivated crimes, victims often expressed the feeling that perpetrators had not fully disclosed the truth of what happened (Byrne, 2004) from which survivors could take some relief of knowing the fate of their loved ones if not entirely why or by whom they had been singled out. Such persistent absence maintains the "liminal" state that Hamber and Wilson (2002) describe as the place of the "living dead," that of uncertainty and of the "uncanny." Gampel (2000), expanding on Freud's (1893/1955) original formulation of the concept of the *unheimlich*, refers to this liminality as the dread and horror of social violence (p. 49) that leave the survivor in an ongoing state of uncertainty and distrust, of unspoken and unmediated trauma. In the "speaking" of their experiences to a compassionate other, the survivor comes, again using Laub's (1992) and also Boulanger's (2007) formulations, to know, to inhabit, and become a witness to her own experience. For some, the TRC commissioners and public audience and the judges of the ICTY were those compassionate listeners, so much so that as Boraine, vice chairperson of the Truth and Reconciliation Commission, reports of one victim-witness that, after having testified and telling the story of his ordeal, said he "had the first good night's sleep in many years" (personal communication, April 24, 2001).

Such a comment, however, warrants unpacking to extract its many potential meanings. As psychoanalysts we may question how to account for the remark Boraine reports. Is it a consequence of the victim's having given testimony, the effect of the catharsis of his suffering and survival? Is it what the victim *anticipates* commissioners want to hear has occurred? That is, experiencing the empathy and concern of Truth Commission members, is this the witness's way of repaying that concern? Is the witness responding to a feeling of "indebtedness" to the commission? It is in the layered psychological domain that psychoanalysts can have a significant impact on the sociopolitical and legal processes that are increasingly being undertaken around the world. By offering our informed understanding of the underlying complexities involved in the provision of testimony by witnesses, victims and perpetrators, and the multiple psychological layers within such undertakings (e.g., both conscious and unconscious motivations) not least where amnesty hangs in the balance, we are able to establish the basis for the importance of psychological provision as an accompaniment to the judicial process. Little attention was given to the expectations victim-witnesses brought to the proceedings, whether those were truth commission or tribunals, which is, at least, partially responsible for the frequent instances of disappointment victim-witnesses reported (cf. Byrne, 2004). Many in South

Africa expected reparations, assistance with medical attention, education, or housing. From my work in Bosnia it is evident that specific expectations are less clear beyond the abstract idea of "justice" although witnesses there too expected some form of compensation. As a consequence, unarticulated and unmet expectations became one source of frustration for victim-witnesses. Further, financial resources in such countries as Bosnia and South Africa or other places in which tribunals and truth commissions take place, do not permit the wide-scale provision of individual psychotherapy, nor is such intervention the most useful in cases of such wholesale social violence. Kaminer et al. (2001) and others have made similar arguments for the importance of psychological interventions and particularly for the invaluable work of victim support groups in facilitating survivors' reclaiming their lives. Incorporating support of such survivor–perpetrator interactions in postconflict recovery processes is one example of what I refer to as acknowledgment of the debt owed to victim-witnesses for incurring the psychic and physical dangers of providing testimony.

THE USES OF TRUTH

Whether the choice be a truth commission or war crimes tribunal like that in The Hague or in the Rwandan tribunal held in Arusha, Tanzania, the politico-legal aim in the period of transition out of conflict to whatever lies ahead is to retrieve a country's history from its previously disavowed space by including once excluded voices. The proceedings of the Bosnian war crimes tribunal in The Hague have made it impossible, for example, for the continued denial of events like the massacre of Srebrenica—the largest massacre in Europe since the end of the Second World War, or that it was Serbian not Bosnian attacks that were made on the Sarajevo marketplace. But even these events are subject to disavowal. Until the summer of 2008, Radovan Karadzic, the Bosnian Serb leader and one of the architects of the war in that country, had eluded efforts to capture him and bring him to trial in The Hague on charges of genocide, crimes against humanity, extermination and forced deportation during the siege of Sarajevo, and the massacre of more than 7,000 men and boys in Srebrenica. In a 2009 BBC interview, his daughter Sonja Karadzic bitterly contested her father's responsibility for the crimes he is charged with, alleging that "[U.S. President] Clinton determined then [1993] that Srebrenica would be the location of the crime and that there would have to be a minimum of 5,000 victims in order to justify NATO air strikes" (Little, 2009).

Similarly, in South Africa the TRC helped establish the factuality of atrocities that had been denied by the regime, for instance, the circumstances of the brutal murder of the Craddock Four and those of the symbol of the antiapartheid struggle, Steven Biko. Despite its many shortcomings, for example, what has been called its "perpetrator-friendliness" ("They got amnesty. We got nothing."*), I have often heard from South African friends and colleagues that if there were no other outcome of the TRC than that white South Africans could no longer claim "we didn't know," that alone would have made the truth commission worthwhile. Others too point to positive outcomes of the hearings processes, for example, the witness noted earlier who spoke of at last being able to have a good night's sleep once he had testified. Hamber (2009) cogently argues that in attempting to arrive at a hegemonic history of events, truth commissions do not consider the writing of history as an ongoing process informed by multiple subjectivities and intersubjectivities, as well as politics.

At the same time that truth recovery may be a necessary component of a nation's emergence from conflict, particularly with respect to writing a history that contains the experiences of those previously excluded, it raises multiple problematics at the individual level. Notable, first, is the challenge to victims' truths that occurs in the process of adjudicating history, and second, the consequences for victims when their personal tragedies are used for public ends. The processes of truth commission and tribunals become rooted in a one-person psychology, what Benjamin (2004) has examined through a framework she has called appropriately "doer–done to." (I will expand on this shortly.) Further, the equation of national with individual healing or the claim that individual truth telling is itself healing represents the misapplication of therapeutic constructs to a larger public context.†

> August 22, 2003, Sarajevo, Bosnia: In the midst of Kresimir's testimony I think, please don't talk about sexual torture. Please don't talk about sexual torture. I don't think I can listen further if you do.

The tension between the private and the public is frequently described by victim-witnesses as one basis for their feeling betrayed by the South African TRC and to a lesser but still significant degree by those who have appeared before the ICTY and the ICTR. Survivors are understandably sensitive to

---

* Father Michael Lapsley, antiapartheid activist and survivor, personal communication, April 2003. Lapsley's comments were, further, that the TRC was at risk of going down in history as "perpetrator friendly" because of the barter of amnesty for truth. Others too, including Wilson (2001) and Hamber (2009), note the charge of the commission's "perpetrator-friendliness."
† This is a dimension of the TRC process that has been vigorously critiqued by Hamber (2009), Wilson (2001), Byrne (2004), and others. It also overlooks the intersubjective nature of the process of giving testimony and its impact on the survivor-witness, perpetrator-witness, public attendee, as well as the television audience.

their stories being appropriated for others' benefit. As one participant in the South African TRC expressed:

> All these foreigners who came along and have written their books, done their Ph.D.s, etc. on the backs of many victims. But they have not paid anything or given anything to the victims, not even taken them for a cup of tea, or lunch, or a dinner or something like that. So, *ja*, I think there was a lot of exploitation during the whole process. (Picker, 2005, p. 12)

Or, from the story of Thandi Shezi (Dube, 2002):

> I am not going back there. Pray to God that I am not asked to appear before the TRC again. Yes, going to the TRC was a victory. It was a victory in that I found the courage to confront my rape. It gave me a platform to share my grief. It made me talk. Hopefully, I will heal in time.
>
> But going before the TRC also feels like I exposed myself to more abuse. It feels like I was abused all over again. With the TRC, it felt like all they wanted was my story, I felt used. There was no support system to help me heal. From the very day of my presentation, I cursed ever going before the TRC. (p. 128)

Shezi's feeling exploited is, in my experience, shared by many others, particularly when survivors are left with often unarticulated if not unconscious expectations of what their providing statements would yield.* Such frustration likely lies embedded within the sense that the commission was perpetrator friendly. Byrne (2004), for example, makes the powerful point that there is no word for *amnesty* in three of the languages in which TRC statements were conducted (Zulu, Sotho, and Tswana). The closest word to *amnesty*, according to her findings, is *forgiveness* and is therefore likely to have caused, at a minimum, misunderstandings and frustrated expectations about the aims of the commission. Equally, she notes one of her subjects who remarked: "It just made me realize that the perpetrators seemed to have more rights than the victims because all that they were saying was lies, they were not telling the truth, but in the end they got amnesty" (Byrne,

---

* Although victim-witnesses were asked during the human rights violation hearings and amnesty hearings what they wanted from the TRC, as Lapsley points out (personal communication, April 2003), what they want in 2003 is very different from what they might have wanted in 1998. In 1998 they may have wanted money for housing, education, or medical care, to know the truth of what happened to their loved ones. It is unclear what they may want now years after testifying. As I suggest later in this chapter, the writing of history is an ongoing process. So too, I believe, is the working and reworking of the meaning of testifying in survivors' lives.

2004, p. 246). In Bosnia too, expressions of deep resentment are common among those who testified before the ICTY when they compare their own impoverished and forgotten circumstances with what they portray as the comparative comfort of defendants, even those who have been convicted. Survivors envision their perpetrators living lives of comfort in foreign jails with good food, medical care, and heat. And they are not wrong.

At the same time as Shezi was resentful of the process of the South African TRC, she also expressed considerable ambivalence about her involvement. As though she has no volition in the matter when indeed she does, Shezi says: "I pray to God I am not asked to appear before the TRC again." Can there be a more profound statement of ambivalence? Compelled to testify by internal forces that are unclear, yet she also experiences a sense of "victory" over the abuse she suffered. I want to underscore that this is only one deponent's commentary on the process of giving testimony, albeit one that is repeated numbers of times. It contains the painful reality that creating a trauma narrative within a public space involves the exposure and appropriation of an intensely personal experience. At the same time Shezi expresses a sense of victory. We might conjecture that part of that victory, albeit ambivalently acknowledged, is her having broken out of the isolation wrought by her trauma.

Similar expressions of ambivalence are found in the narratives of survivors of the Bosnian war who testified before the ICTY. The phrase so often repeated in answer to my question of why they had testified was "because of the truth." "So the world will know what happened here." "The truth, to tell what happened to us in prison in Foca." In saying "because of the truth," they express the profound belief that the truth will produce justice, a vindication of the injustice that had been perpetrated against them. But their ambivalence about the process is revealed at the same time in their remarks: "He didn't get enough," in response to one perpetrator's receiving a sentence of 28 years. "He should be shot in a public place." None of the witnesses I spoke with felt the sentence their perpetrators received was sufficient "for what he did to us." From our psychoanalytic perspective, such ambivalence is expectable, even normal, particularly in the context of the much vaunted positive inflection of the truth commission and tribunal processes.

Too often, however, the politico-legal processes of truth commissions and tribunals fail at containing the personal meaning of such profound events survivors have lived through. Shezi's powerful indictment detailed earlier reflects the change in methodology within the TRC over time. Originally victims' statements were taken in unstructured interviews, which could last for several hours. To speed up the process, statement takers were directed to spend 30 to 45 minutes with victims after which coders transformed the narrative into a 48-item checklist that served as the basis for defining gross human rights violations. This alteration in methodology (Hamber, 2009; Wilson, 2001) and knowledge production, as Posel (2002) refers to it, ultimately denuded events of their subjective meaning and context. Forensic truth thereby came to supersede subjective or narrative truth as the hegemonic one by which events would come to be characterized. I will expand on the different categories of truth below.

RES PUBLICA/RES PRIVATA

From the moment of the first act of violence toward a victim of war (and I am using *war* here to include all the conditions I enumerated earlier), a person is no longer a subject but a political object. To paraphrase Humphrey (2002), in the context of war, violence toward "victims is never about them just as individuals, but as *politically selected* [italics added] victims, as political signs" (see also Hamber, 2003). In using her testimony as the basis for national recovery, the objectification of the victim continues. She is a political object whose subjective experience revealed through testimony constitutes the "theatre" that Posel (2004) so ably describes in her critique of the South African TRC.

The mandate of the TRC enumerates four types of truth—forensic or factual, subjective, social, and healing—that are presumed to be unique and identifiable. Posel (2002) compellingly argues against the possibility of separating one from the other typology of truth when she says: "The grounds for differentiating the four types of truth are poorly specified and remain rather opaque" (p. 155) She goes on: "the marker of 'healing truth' (putting truth in the context of human relationships) seems largely to reiterate the criterion for social truth (reflecting the social norms of human relations)" (p. 155). The truth commission's effort to assemble a presumably impartial, "objectified," ultimately statistical record of gross human rights abuses from which to write the country's apartheid history is what Posel (2004) has called "science" (p. 12). But the TRC set out as well to provide a space in which perpetrators and survivors could tell their stories. The Human Rights Violations hearings constituted what Posel (2004) refers to as "theatre" (see also remarks of Fullard noted earlier). It is in this conflict of "science versus theatre," "history versus subjective experience" that some of the difficulties with the TRC have arisen, not least as described regarding Shezi's painfully wrought testimony, the sense of abandonment by the TRC, or the refusal to acquiesce to the demand to forgive or to reconcile that are made both implicitly and explicitly.

In recounting her story, Shezi noted an essential dilemma for many survivors who appeared before the TRC and the ICTY; that is, confronting the perpetrator who denies her experience. She says:

> Van Heerden sat there and looked me in the eye and said he did not know me. He said he did not recall beating and torturing me. ... [F]or him to

deny knowing me, that is impossible to swallow. I am the person he beat up, and tortured with electric shocks and acid water. Yet he stood there and denied my existence. How am I supposed to forgive and forget when I am not recognized by my perpetrator? (Dube, 2002, p. 127)

"How am I supposed to forgive and forget when I am not recognized by my perpetrator?" I will return to Shezi's question in a moment.

Like the stories of their South African counterparts, the narratives of the victims of the Balkan war are also interrogated, their stories used as "evidence." Here lies another instance of the clash between historical and subjective truth. Amnesty in South Africa was predicated on full and complete confession of all details relevant to a crime committed as a political act. But, we may ask, is it truly a confession when the unique suffering of the victim is not recognized and responsibility for the harm to that individual acknowledged. I would contend no.* In the absence of recognition of the other's subjectivity there may be political "confession" but not intersubjective relatedness. In adjudicating truth, as, for example, in the ICTY, the perpetrator's effective challenge of the victim's experience potentially destabilizes reality for her and thereby contributes, at a minimum, to her dissatisfaction with the "justice" rendered.

An example from the ICTY echoes the story of Shezi. It is drawn from my interview with Minja, the woman whose experience I described at the beginning of the chapter.

> One of the lawyers, his nickname was Zaga. He asked me questions and then I answered about how many times they raped me during that night. That was in December. His lawyer said he wasn't there in that period and he didn't do anything. You can't say that he wasn't there from the first hour of the camp. ... There were 30–35 women who were testifying against him. I know the one, she was 16 [she says naming other girls whom he raped and giving their ages]. It's like he was saying I'm a liar.

"It's like he was saying I'm a liar." The frustration and rage at the contest of Minja's truth is almost palpable in her statement. That her rapist denies her undermines her attempt at justice in its most basic aspect—through recognition. The judicial process then rests on the adjudication of truth,

---

* One might also wonder if, in the context of committing the crimes for which the perpetrator is seeking amnesty, he or she was not in a dissociated state that precludes the registration of the individual humanity of the victim. Thus the failure to acknowledge the other may conceivably reflect the inability to psychically bear such registration. In any event, failure to do so has significant consequences for the victim-survivor whose willingness to testify before the TRC may be said to be motivated by seeking acknowledgment. I am grateful to my colleague Jill Salberg for suggesting this idea.

the finding of one to blame. There is no "third" here (Benjamin, 2002, 2004), no space for reconciliation in the absence of acknowledgment.

What Buur has called the "bureaucratic production of truth" (Fullard, personal communication, April 2003) as has been detailed in the privileging of the TRC's compilation of statistical data by which to establish a pattern of political abuse (what, as was previously noted, Posel, 2004, has referred to as science), compounds the victimization of the survivor and robs him of his subjectivity. The question of what restores that subjectivity is to my mind what needs to concern us most as we attend to the multiply layered recovery from political violence. I suggest that what is prerequisite for subjectivity is to restore a sense of agency that has been destroyed by the trauma of political violence. Agency involves having will for and initiating action as well as feeling recognized and acknowledged. In this I am characterizing *acknowledgment* as the confirmation that "this happened to you" and *recognition* as seeing the human in the other, "seeing myself in you." Benjamin (2004) describes related ideas in her discussion of the doer–done to dialectic constituted by a persistent cycle of blame and self-justification. The processes of repair for a society and the individuals contained within it must include the possibility of moving beyond self-righteous claims and disidentifications. In my opinion, projects that bring together victims and victimizers (often indistinguishable from one another except to themselves) have been more successful than truth commissions and tribunals* in achieving the outcome of moving beyond blame and self-righteousness.

Beyond doer and done to also lies the witnessing "Other." In the realm of intractable violence, that Other is the international community. Steve Rukongi, general manager of Urwintore, a Rwandan theatre group that has performed the Peter Weiss post-Holocaust play *The Investigation* in Rwanda and around the world, has said about the tribunals for his own country that their importance is "as a symbol of international recognition of the crime of genocide. We need such a symbol in order for reconciliation to occur" (personal communication, February 8, 2009). The impact of social violence is in the realm of the "unthinkable." Truth commissions and tribunals break the silence to the crimes that have occurred and begin, albeit with some glaring missteps and failures, to make them thinkable. They constitute a space where, though contested and problematic in all the

---

* See Institute for Healing of Memories" (http://www.healingofmemories.co.za/), a South African group begun to bring survivors of antiapartheid violence from all sides together; To Reflect and Trust as well as the Peace Research Center in the Middle East initiated by the late Daniel Bar-On to bring together the children of Holocaust survivors and those of Nazi perpetrators; and the Sustainable Peace Network of the Glencree Survivors and Former Combatants Programme (http://www.glencree.ie/site/sustainable.htm) that brings together survivors from both sides of "The Troubles" in Northern Ireland.

ways I have detailed, survivors can begin to reclaim their history. It is this that is an essential dimension of the restoration of subjectivity.

> March 19, 2001, The Hague Tribunal for the Former Yugoslavia: I see that like me, the judges too look tired and glazed over. I know that I am having difficulty staying with the testimony of witnesses and focus on the translators who change every half hour or so. If I do continue to be attuned I will start to cry. How am I to process what I am listening to?

I return now to the question Shezi posed: "How am I supposed to forgive and forget when I am not recognized by my perpetrator?" The failure to recognize works intersubjectively. If she is unrecognized by him, is he unrecognized to her and each of them to themselves? Without the capacity and opportunity for mutual recognition, the victim remains rooted in a one-dimensional space of victimhood, which has its complement in one-dimensional "perpetratorhood." (Please forgive my neologism. I have not yet found a more elegant way of expressing it.) In such circumstances, traffic remains one way, to use Benjamin's framework (2004).

The danger embedded within the various truth recovery processes undertaken to date is that truth and healing, forgiveness, and reconciliation have been conflated. In my work I have repeatedly heard the expression of intense ambivalence about the ICTY and the TRC. There are different agendas involved in the politico-legal as distinct from the social and psychological. The debt owed to survivors must include provision for both sets of agendas with neither overpowering the other lest the privileging of one power structure over another be repeated. Finally, by making it possible for both participants to become human to one another a higher order process is initiated. That is, the space is created within which forgiveness and reconciliation can truly begin. To my understanding that is the meaning of *ubuntu*.*

ACKNOWLEDGMENTS

I would like to thank the Psychoanalytic Society of the New York University Postdoctoral Program in Psychotherapy and Psychoanalysis for its generous award of a Scholars' Grant that greatly facilitated my work on this chapter. In addition, I am grateful for the careful reading and suggestions of several of my colleagues, Elizabeth Goren and Jill Salberg, and editors, Adrienne Harris and Steven Botticelli.

---

* *Ubuntu* is a Xhosa language word meaning essentially "a person is a person in the context of other people."

REFERENCES

Benjamin, J. (2002). The rhythm of recognition: Comments on the work of Louis Sander. *Psychoanalytic Dialogues, 12*, 43–53.

Benjamin, J. (2004). Beyond doer and done-to: An intersubjective view of thirdness. *Psychoanalytic Quarterly, 73*(1), 5–46.

Brittain, V. (2000, June 13). Outrage over amnesty for apartheid killer. *The Guardian*. Retrieved March 10, 2009, from http://www.guardian.co.uk/world/2000/jun/13/victoriabrittain

Boulanger, G. (2007). *Wounded by reality: Understanding and treating adult onset trauma*. Mahwah, NJ: The Analytic Press.

Byrne, C. C. (2004). Benefit or burden: Victims' reflections on TRC participation. *Journal of Peace and Conflict, 10*(3), 237–256.

Dube, P. S. (2002). The story of Thandi Shezi. In D. Posel & G. Simpson (Eds.), *Commissioning the past: Understanding South Africa's Truth and Reconciliation Commission* (pp. 117–130). Johannesburg, South Africa: Witwatersrand University Press.

Feitlowitz, M. (1998). Night and fog in Argentina. In M. Feitlowitz (Ed.), *A lexicon of terror* (pp. 40–74). New York: Oxford University Press.

Freud, S. (1955). Studies on hysteria. In J. Strachey (Ed. & Trans.), *The standard edition of the complete psychological works of Sigmund Freud* (Vol. 2). London: Hogarth Press. (Original work published 1893).

Gampel, Y. (2000). Reflections on the prevalence of the uncanny in social violence. In A. C. G. M. Robben & M. M. Suarez-Orozco (Eds.), *Cultures under siege: Collective violence and trauma* (pp. 48–69). Cambridge, UK: Cambridge University Press.

Hamber, B. (2002). Symbolic closure in post-conflict societies. *Journal of Human Rights, 1*, 35–53.

Hamber, B. (2003, October 11–13). *Flying flags of fear: The role of fear in the process of political transition*. Paper presented at the Risk, Complex Crises & Social Futures Conference, Amman, Jordan.

Hamber, B. (2009). *Transforming societies after political violence: Truth, reconciliation, and mental health*. New York: Springer.

Hamber, B. & Wilson, R.A. (2002). Symbolic closure through memory, reparation, and revenge in post-conflict countries *Journal of Human Rights, 1*(1), 35–53.

Humphrey, M. (2002). *The politics of atrocity and reconciliation: From terror to trauma*. New York: Routledge.

Kaminer, D., Stein, D. J., Mbanga, I., & Zungu-Dirwayi, N. (2001) The Truth and Reconciliation Commission in South Africa: Relation to psychiatric status and forgiveness among survivors of human rights abuses. *British Journal of Psychiatry, 178*, 373–377.

Krog, A. (1999). *Country of my skull: Guilt, sorrow, and the limits of forgiveness in the New South Africa*, New York: Three Rivers Press.

Laub, D. (1992). Bearing witness. In S. Felman & D. Laub (Eds.), *Testimony: Crisis of witnessing in literature psychoanalysis, and history*. New York: Routledge.

Little, A. (2009, January 26). Sonja Karadzic defends her father. *BBC World News*. Retrieved Feb. 13, 2009, from http://news.bbc.co.uk/2/hi/europe/7851417.stm

Meredith, M. (1999). *Coming to terms: South Africa's search for truth*. New York: Public Affairs.

Neuffer, E. (2001). *The key to my neighbor's house: Seeking justice in Bosnia and Rwanda*. New York: Picador.

Picker, R. (2005). *Victims' perspectives about the human rights violations hearings* (Research report). Johannesburg, South Africa: Centre for the Study of Violence and Reconciliation.

Posel, D. (2002). The TRC Report: What kind of history? What kind of truth? In D. Posel & G. Simpson (Eds.), *Commissioning the past: Understanding South Africa's Truth and Reconciliation Commission* (pp. 147–172). Johannesburg, SA: Witwatersrand University Press.

Posel, D. (2004). Truth? The view from South Africa's Truth and Reconciliation Commission. In N. Tazi (Ed.), *Keywords: Truth*. New York: Other Press.

Posel, D., & Simpson, G. (2002). The power of truth: South Africa's Truth and Reconciliation Commission in context. In D. Posel & G. Simpson (Eds.), *Commissioning the past: Understanding South Africa's Truth and Reconciliation Commission* (pp. 1–13). Johannesburg, South Africa: Witwatersrand University Press.

Rosenberg, T. (1999). Afterword: Confronting the painful past. In M. Meredith (Ed.), *Coming to terms: South Africa's search for truth* (pp. 327–370). New York: Public Affairs.

Stover, E., & Weinstein, H. M. (2004). *My neighbor, my enemy: Justice and community in the aftermath of mass atrocity*. New York: Cambridge University Press.

South African Truth and Reconciliation Commission Report. (1998a). *Truth and Reconciliation Commission South Africa, Report* (Vols. 1–5). Capetown, ZA: Juta and Co.

South African Truth and Reconciliation Commission Report. (1998b). *Hearings of the Amnesty Committee. Amnesty hearings, decisions and transcripts: On resumption. 5 November 1998–Day 4*. Retrieved June 29, 2004, from http://www.doj.gov.za/trc/amntrans/1998.htm

Thomas, N. K. (2005, March 18). *Discussion of Hamber's "Psychologizing the Nation."* Paper presented at the NYU Relational Colloquium, New York.

Wilson, R. A. (2001). *The politics of truth and reconciliation in South Africa: Legitimizing the post-apartheid state*. Cambridge, UK: Cambridge University Press.

Part 2

The paradox
Psychology's militarism

In this section, we present essays from psychologists/psychoanalysts engaged in the critique and struggle in opposition to the American Psychological Association's policy on psychologist's presence in "interrogations," primarily though not exclusively at Guantánamo. We look through the eyes of psychoanalysts active in opposing the institutional policies and the entrenched resistance to change in relation to an enforced ethical stance against torture. We look at the other side of the paradox: Psychology as the handmaid of militarism and warmaking.

Stephen Soldz (Chapter 5) takes his readers through the agonizing and deeply frustrating process of trying to hold the APA responsible for its policies regarding torture and interrogation. His chapter illustrates the difficulty in forcing this accountability, the amnesias and obfuscations that have characterized the organizational response to its members' demands. In a certain way, from the crisis within institutional psychology and in the groups resisting and protesting APA policy, the genesis of this book arose. Soldz's chapter is notable for its extensive documentation of the politicizing process of opposing the APA on torture, for the use of new media (blogs and the Internet), and for the unsettling conclusions that extensive presentation of facts and argument have not impacted APA policy and the administration's stance. This chapter and this section of the book, we feel, illuminate how deadly our powers of disavowal and denial are, how costly and endless the task of remembering and finding accountability turn out to be.

Steven Reisner's chapter (Chapter 6) describes how his psychoanalytic self played an important part in how he listened, thought, and was finally moved to action to expose the role of the APA in implementing the Bush administration's interrogation program for detainees accused of participating in terrorism. The chapter reads as a detective story, as Reisner searches for clues to put together a picture of the APA's complicity, all the while becoming increasingly outspoken in his opposition to what was going on and demanding accountability from the APA. The chapter provides a personal and comprehensive history of the unfolding of the controversy, from one of the major players in bringing it to light.

Neil Altman (Chapter 7) represents the multiplicity of his experience of the controversy within APA over torture with a "one-person play" in which a dialogue takes place between various aspects of the author. While part of the author is outraged and enraged over APA's collusion with the government, another part is keeping an eye on the diplomacy necessary to win over members of the APA governing body who start out in disagreement with him. As a psychoanalyst, Altman also tries to keep in mind that disavowal of a terrified and potentially sadistic part of himself may fuel his outrage with those who openly support, and rationalize, the existence of detention centers where human rights are violated. Believing that effective political action requires the ability to identify with one's adversary, this chapter represents Altman's effort to keep all aspects of himself, even those that ultimately he rejects, in play.

Frank Summers (Chapter 8) asks why the American public seems to have accepted developments like the invasion of Iraq and the revelation of the torture of detainees with so little outcry. He finds a partial explanation in the theory of American exceptionalism—the widespread belief that the United States is uniquely virtuous and powerful among nations. Tracing the history of this attitude over the course of American history, Summers shows how Americans' embracement of this idea makes the population vulnerable to manipulation by leaders with expansionist and aggressive agendas. Maintaining that psychoanalytic understanding can help make sense of the historical and political realm, Summers suggests how in its grandiosity the United States is like a narcissistic patient in its insistence on its specialness, denial of its vulnerability, lack of consideration of the perspective of others, and tendency to lash out aggressively at perceived (or real) injuries to its sense of itself. He believes that moving past this grandiosity will require leadership that helps move Americans to a more tempered view of their country, one in which pride can be taken in realistic accomplishments at the same time that the destructive and self-defeating consequences that follow on the pursuit of grandiose goals, can be appreciated.

Chapter 5

Psychologists defying torture
The challenge and the path ahead

Stephen Soldz

It is by now generally accepted that the George W. Bush administration engaged in a systematic policy of detainee abuse that sometimes amounted to torture and often consisted of cruel, inhuman, or degrading treatment, both of which are banned by the United Nations Convention against Torture and Other Cruel, Inhuman or Degrading Treatment or Punishment (Office of the United Nations High Commissioner for Human Rights, 1984, p. 2), signed by the United States. If there had been any doubt that U.S. detainee abuses were the result of a systematic policy, that doubt has surely been erased by publication of detailed reports from the Defense and Justice Departments Inspectors General (Office of the Inspector General of the Department of Defense, 2006; Office of the Inspector General of the Department of Justice, 2008), and the Senate Armed Services Committee (2008a, 2008b, 2008c; Levin, 2008a, 2008b, 2008c). Journalist Jane Mayer (2008) and others have placed the ultimate authorization of the most brutal of these abuses directly in the White House (Greenburg, Rosenberg, & Vogue, 2008a, 2008b).

As information about this policy and its implementation has gradually emerged, it became clear that military and Central Intelligence Agency (CIA) psychologists played central roles in the development and implementation of these abusive tactics. Simultaneously, the American Psychological Association (APA), while publicly opposing torture, took extraordinary measures to encourage and protect psychologist involvement in U.S. national security interrogations at the sites where abuses were rampant. As a result, a movement emerged among APA members and other psychologists to oppose this pro-interrogations policy. By the end of 2008, this movement had succeeded in winning a major change in APA policy, while failing to change those organizational factors that likely contributed to the rejected APA policy. In this chapter, I discuss the path ahead for the anti-torture movement among psychologists. But first I summarize the evidence regarding psychologists' important role in detainee abuse and the APA history of inaction regarding that abuse.

TYPOLOGY OF INTERROGATION SETTINGS

U.S. interrogation operations can be somewhat arbitrarily conceptualized as having six, somewhat distinct, tracks. The first track was the CIA's program of secret prisons—the so-called black sites (Council of Europe Committee on Legal Affairs and Human Rights, 2007)—and the "enhanced interrogation" program that occurred in those prisons (Eban, 2007; Eban, Olson, & Goodman, 2007; Mayer, 2007, 2008; Mayer & Goodman, 2007). The second track was the abuses at the prison at Guantánamo Bay, where hundreds of detainees, many bought for sizeable bounties, were imprisoned, deprived for years of any rights of appeal, and abused. The third track was the interrogations conducted in Iraq and Afghanistan by field operatives of the military's Special Operations Command, along with that of other government agencies (including the CIA and the Defense Intelligence Agency), and various contractors from profit-making firms. The fourth track was the tactical field interrogations undertaken, again in Iraq and Afghanistan, by relatively low-level and often minimally trained and inexperienced military personnel seeking rapid intelligence from potential "insurgents." Tony Lagouranis (Lagouranis & Mikaelian, 2007) gives a good sense of the experience of many of these interrogators. The fifth track is the rendition program, whereby people were kidnapped and taken to other countries where many were apparently subjected to torture by agents of those countries. Finally, in a sixth track, U.S. intelligence personnel sometimes collaborate with agents of other powers in interrogations of individuals in the control of those powers (Brewer & Arrigo, 2009).

In each of these settings, major abuses of detainees occurred.* From information currently available, it appears that much of this abuse, including most abuses conducted in tracks one and two, was officially sanctioned in the form of interrogation techniques authorized through official memos (Greenberg & Dratel, 2005; Jaffer & Singh, 2007) and/or by high officials in the Pentagon (Sands, 2008), CIA (Eban, 2007; Mayer, 2008), and White House (Greenburg et al., 2008a, 2008b; Mayer, 2008). Other abuses, especially those in tracks three and four, were created in the field and were apparently tolerated by a range of military and civilian officials (Human Rights Watch, 2006; Lagouranis & Mikaelian, 2007; Schmitt & Marshall, 2006).

The most systematized torture regime was that in the CIA's secret black site prisons where those CIA detainees suspected of being "high value" al Qaeda members were imprisoned. Thanks to the leak of the International

---

* I am in no way suggesting that all interrogations in all of these settings were abusive. In some of the settings, such as tactical interrogations, major abuses may have been the minority. Many military personnel had internalized the Geneva Conventions and followed them, even absent the active support of the chain of command.

Committee of the Red Cross (2004) report on the treatment of 14 CIA prisoners (Danner, 2009a, 2009b) and the release by the Obama administration of crucial Office of Legal Counsel memos describing the CIA's interrogation techniques in the course of rationalizing their legality (Bradbury, 2005a, 2005b, 2005c; Bybee, 2002), we now have a vivid picture of the techniques used by the CIA.

As Mark Danner (2009b) pointed out, one can get a feel for the CIA enhanced interrogation program by examining the section titles for the Red Cross report (International Committee of the Red Cross, 2007, p. 2):

1. Main Elements of the CIA Detention Program
    1.1 Arrest and Transfer
    1.2 Continuous Solitary Confinement and Incommunicado Detention
    1.3 Other Methods of Ill-Treatment
        1.3.1 Suffocation by water
        1.3.2 Prolonged stress standing
        1.3.3 Beatings by use of a collar
        1.3.4 Beating and kicking
        1.3.5 Confinement in a box
        1.3.6 Prolonged nudity
        1.3.7 Sleep deprivation and use of loud music
        1.3.8 Exposure to cold temperature/cold water
        1.3.9 Prolonged use of handcuffs and shackles
        1.3.10 Threats
        1.3.11 Forced shaving
        1.3.12 Deprivation/restricted provision of solid food
    1.4 Further elements of the detention regime

Waterboarding, the deliberate infliction of temporary drowning, was utilized on at least three prisoners; this controlled drowning apparently was induced 83 times in one prisoner and 183 times in another (Bradbury, 2005c; emptywheel, 2009).

Some interrogations at Guantánamo, at least during the years 2002 to 2004, resembled those at the CIA black sites, but were somewhat less systematically brutal. Thus, sleep deprivation usually took the form of moving prisoners (along with the accompanying time-consuming shackling) from cell to cell ever few hours to disrupt sleep, referred to as the "frequent-flyer program," rather than chaining them to the ceiling (Frakt, 2009; Freeze & Akkad, 2008; White, 2008). Techniques reported in use there included painful stress positions, extremes of heat and cold, forced nudity, prolonged isolation, sleep deprivation, having females smearing (fake) menstrual blood on prisoners, abusing Korans, and sensory overload in the form of strobe lights and/or loud music (Physicians for Human Rights, 2008;

Senate Armed Services Committee, 2009). One of the most bizarre aspects of life at Guantánamo was the manipulation of the number of sheets of toilet paper prisoners could have, including its total removal, along with other "comfort items," as punishment (Joint Task Force Guantánamo, 2003; Mickum, 2007).

Many of the techniques used at Guantánamo were then adopted for use by Special Forces in Iraq and Afghanistan (Office of the Inspector General of the Department of Defense, 2006; Senate Armed Services Committee, 2009; Soldz, 2008a), combined at times with more traditional beatings and other forms of brutality (Physicians for Human Rights, 2008). One Iraq unit had the slogan "No blood, no foul," suggesting that any abuse that did not leave marks was acceptable (Human Rights Watch, 2006; Schmitt & Marshall, 2006; Soldz, 2008a). Abusive techniques were also widely used by CIA and private contractor interrogators in Iraq, leading many young traditional military interrogators to believe that these techniques were the way "real professionals" dealt with prisoners; former Iraq interrogator Tony Lagouranis (Lagouranis & Mikaelian, 2007) describes his personal experience of this process in detail.

PSYCHOLOGISTS

It has become clear that psychologists were critical agents, enlisted in the design and implementation of abusive interrogation techniques in tracks one and two of the U.S. detention system. It also seems likely that psychologists also played important roles in track three, though less information is publicly available on these roles. Further, the release of the Office of Legal Counsel torture memos (Bradbury, 2005a, 2005b, 2005c; Bybee, 2002) has clarified that psychologists and other health professionals were critical to the legitimation of abusive techniques, guaranteeing the techniques putative "legality" by the presence of the professionals (Fink, 2009; Warrick & Finn, 2009). I have summarized the evidence regarding psychologists' involvement in detainee abuse in a series of publications (Olson, Soldz, & Davis, 2008; Soldz, 2008a, 2008e, 2009a, 2009c; Soldz & Olson, 2008b; Soldz, Reisner, & Olson, 2007a), as have numerous other authors (Eban, 2007; Ephron, 2008; Mayer, 2005, 2008), which I won't duplicate here. I will, however, briefly summarize the information that has emerged.

When the first high-level al Qaeda detainee, Abu Zubaydah, was captured and taken to a CIA black site, a team of interrogators, reportedly led by psychologist James Mitchell, was flown in to direct his detention and interrogation (Eban, 2007). Mitchell and colleague Bruce Jessen were former psychologists from the military's Survival, Evasion, Resistance, and Escape (SERE) program (Otterman, 2007; Soldz et al., 2007a), part of the Joint Personnel Recovery Administration (JPRA), which trains U.S. military members at high risk of capture how to resist breaking if captured by "forces that do not abide by the Geneva Conventions" (Office of the Inspector General of the Department of Defense, 2006, p. 23). JPRA provided consultation to the CIA on the interrogation of Zubaydah and provided a more formal 2-day training for the CIA in early July 2002 (Senate Armed Services Committee, 2009).

According to Eban (2007) and Mayer (2007, 2008), Mitchell and Jessen and the firm they created (Morlin, 2007; Steele & Morlin, 2007) supervised many of the detentions and interrogations in the CIA's enhanced interrogation program in the black sites and trained many, if not most, of the CIA's interrogators. This program was based upon the "learned helplessness" theory of psychologist and former APA President Martin Seligman, who, under CIA invitation, presented the theory to the Navy SERE School in spring 2002, with Mitchell and Jessen in attendance (Mayer, 2008; Seligman, 2008). While Seligman denies any awareness of the uses to which his work would be put, his appearance at the SERE school at this time raises questions as to why he thought he was invited and what he did there (Mayer & Sullivan, 2008; Soldz, Olson, Reisner, Arrigo, & Welch, 2008).

Life in the black sites consisted of a total manipulation of the environment, along with the utilization of brutal physical and psychological techniques to induce a sense of complete dependence and break down any ability to resist interrogators. The Council of Europe investigated the black site prisons in Europe and found total isolation was used as an initial step in breaking detainees:

> A common feature for many detainees was the four-month isolation regime. During this period of over 120 days, absolutely no human contact was granted with anyone but masked, silent guards. There's not meant to be anything to hold onto. No familiarity, no comfort, nobody to talk to, no way out. It's a long time to be all alone with your thoughts. (Council of Europe Committee on Legal Affairs and Human Rights, 2007, p. 52)

In other sites, prisoners were forced to stand for many days at a time by being shackled to the ceiling clad only in a diaper (International Committee of the Red Cross, 2007). Interestingly, in the Office of Legal Council torture memos, this technique is described simply as a form of "sleep deprivation," ignoring the intense pain from being forced to maintain such a position for many days (Bradbury, 2005a; Soldz, 2009b). In one case, a prisoner's artificial leg was removed to force him to stand only upon the other leg while shackled (International Committee of the Red Cross, 2007; Mayer, 2008). Additional techniques applied under Mitchell and Jessen's direction to induce a sense of helplessness included forced nudity; semistarvation; repeatedly throwing prisoners against the wall; enclosing them in tiny, totally dark boxes designed to make standing or sitting impossible, and breathing difficult; being immersed in freezing water; and being subjected to constant white noise interrupted only by loud music.

BEHAVIORAL SCIENCE CONSULTATION TEAMS (BSCTS)

Similar to the CIA program, when top Pentagon officials initiated the program of abusive interrogations at Guantánamo, they turned to the JPRA, the SERE parent agency, and to SERE psychologists for assistance in selecting and training for these interrogations (Baumgartner, 2008; Flaherty, 2008; Senate Armed Services Committee, 2008a, 2008c, 2009; Soldz, 2008e). And the military set up a consultancy service for interrogations involving psychologists and psychiatrists, initially at Guantánamo and later Iraq and Afghanistan, titled Behavioral Science Consultation Teams (BSCTs; Ephron, 2008; Marks, 2005; Marks & Bloche, 2008; Mayer, 2005).

The BSCTs helped design the program of largely psychological abuse that characterized Guantánamo. As the BSCT explained at a crucial October 2002 planning meeting: "What's more effective than fear-based strategies are camp-wide, environmental strategies designed to disrupt cohesion and communication among detainees. [That is the] environment should foster dependence and compliance" (Senate Armed Services Committee, 2008a, Tab 7).

Acting on this insight, likely derived at a September 2002 Fort Bragg training (Office of the Inspector General of the Department of Defense, 2006; Senate Armed Services Committee, 2008c), the BSCT instructions over the years have emphasized the role of the detention environment as a crucial aspect of detention procedures. As Colonel Morgan Banks (2005b) expressed in his 2005 instructions to BSCT psychologists: "The psychologist's goal is to assist in helping make sure that the environment maximizes effective detainee operations. The psychologist can assist in making sure that everything that a detainee sees, hears, and experiences is a part of the overall interrogation plan" (pp. 2–3). Similarly, as then Surgeon General Kiley expressed in his 2006 memorandum on the BSCTs: "The goal [BSCT environmental consultation] is to ensure that the environment maximizes effective detention and interrogation/debriefing operations. BSCs can assist in ensuring that everything that a detainee sees, hears, and experiences is a part of the overall interrogation plan" (p. 11).

Thus, in the only (partially) publicly available BSCT consultation, in September 2003, a BSCT psychologist recommended for 16- or 17-year-old

Mohhamed Jawad, when his interrogator found him addressing pictures on the wall and crying for his mother: "He appears to be rather frightened, and it looks as if he could break easily if he were isolated from his support network and made to rely solely on the interrogator," the BSCT said. As described in *Newsweek*:

> The psychologist recommended that Jawad be moved to a section of the prison where he would be the only Pashto speaker, and be moved again if he somehow began to socialize in his new block. The psychologist also suggested that interrogators emphasize to Jawad that his family appeared to have forgotten him: "Make him as uncomfortable as possible. Work him as hard as possible." (Ephron, 2008)

Further evidence of disturbing environmental manipulation on the recommendation of BSCTs emerged from a former Guantánamo interrogator: "According to a former interrogator at Guantánamo who was interviewed at length by a lawyer, behavioral scientists control the most minute details of interrogations, to the point of decreeing, in the case of one detainee, that he would be given seven squares of toilet paper per day" (Mayer, 2005).

Adding to a picture of BSCTs as collaborators in a regime of abuse, several sources have reported that BSCTs used information in detainees' medical records to identify weaknesses such as phobias that could be exploited through increasing fear (Lewis, 2004, 2005; Marks, 2005; Slevin & Stephens, 2004). There have also been repeated accounts by detainees that they were forcibly drugged at Guantánamo (Stein, 2008; Warrick, 2008). Interestingly, at least two BSCT psychologists were among the tiny number of psychologists trained in administering psychoactive drugs.

Psychologists participated in tracks three and four of the U.S. abusive detention and interrogation regime, though far less material is available as to their activities. There were BSCTs in Iraq and Afghanistan, subject in later years to the same operating instructions as were Guantánamo BSCTs. In 2008, newly released sections of the Church (Church, 2005) report of detainees were interpreted by the American Civil Liberties Union (2008) as providing evidence that BSCT psychologists participated in abuses, whereas Stephen Behnke (2008), ethics director of the APA, interpreted these same materials as support for the claim that the psychologists prevented abuses, a claim that I (Soldz, 2008a) found implausible.

Guantánamo commander General Geoffrey Miller recommended the creation of a BSCT as part of his "GTMO-ization" program for Abu Ghraib when he visited in August and September 2003 (Miller, 2003; Senate Armed Services Committee, 2009). This recommendation was acted upon. There was a BSCT at Abu Ghraib prior to the scandal becoming public in April 2004 (Zagorin, 2005); this BSCT was reportedly headed by a psychiatrist, Major Scott Uithol (Bloche & Marks, 2005; Taguba, 2004a, p. 19). The organization chart of the Abu Ghraib Joint Interrogation and Debriefing Center (JIDC; Taguba, 2004a, p. 19) shows the BSCT high up in the hierarchy of the interrogation center. Colonel Thomas Pappas, the military intelligence chief at Abu Ghraib, described to investigator General Antonio Taguba the medical and psychiatric monitoring of interrogations there (Taguba, 2004b). It is thus likely that the BSCT, though not psychologists in this case, was complicit in the numerous abuses committed at Abu Ghraib (Hersh, 2004a, 2004b) with command approval (Hersh, 2004a, 2007). These interrogations were authorized to use such techniques as sleep deprivation (euphemistically called "sleep management," perhaps to skirt concerns that sleep deprivation was illegal), environmental manipulation (such as loud music or strobe lights for hours on end or freezing temperatures on naked detainees), stress positions, and the use of military dogs (Taguba, 2004a).

PROFESSIONAL ASSOCIATION RESPONSE

When the role of psychologists and psychiatrists in Guantánamo abusive interrogations emerged in 2004, pressure started to build upon the professional associations of psychologists and psychiatrists to take a stand against participation in these abuses. The American Psychiatric Association (2005) expressed concern about the reports of psychiatrist involvement in Guantánamo abuses:

> The American Psychiatric Association ... is troubled by recent reports regarding alleged violations of professional medical ethics by psychiatrists at Guantánamo Bay. APA is reviewing issues related to psychiatry and interrogation procedures and plans to develop a specific policy statement in the near future.

Later the American Psychiatric Association (2006) instituted a clear policy of nonparticipation in interrogations, abusive or not: "No psychiatrist should participate directly in the interrogation of person[s] held in custody by military or civilian investigative or law enforcement authorities, whether in the United States or elsewhere."

Not long thereafter, the American Medical Association (2006) issued a policy banning any direct physician participation in interrogations of individual detainees, stating:

> Physicians must not conduct, directly participate in, or monitor an interrogation with an intent to intervene, because this undermines the

physician's role as healer. Because it is justifiable for physicians to serve in roles that serve the public interest, the AMA policy permits physicians to develop general interrogation strategies that are not coercive, but are humane and respect the rights of individuals.

The American Psychological Association (APA), in contrast, initiated a process to protect the involvement of psychologists in interrogations at Guantánamo and elsewhere. APA officials appointed a Presidential Task Force on Psychological Ethics and National Security (PENS Taskforce), allegedly to examine whether participation in national security interrogations was ethical. To this taskforce the APA appointed a majority, 6 of 10 members, from the military-intelligence community. Five of these members had direct involvement in interrogations at Guantánamo, in Iraq and Afghanistan, or at the CIA black sites (Coalition for an Ethical Psychology, 2008; Society for the Study of Peace Conflict and Violence: Peace Psychology Division 48, 2005). Thus, at a minimum, these six members had a conflict of interest in that their careers could be severely damaged should the task force decide that participation in national security interrogations was unethical for psychologists. Further, the task force had a number of observers who had high-level ties to the intelligence community or the Bush administration.

One of the observers had been a psychologist at the National Security Agency. Another was a former Bush White House official. Several more were APA lobbyists with the military-intelligence establishment. Additionally another observer was a senior APA official whose wife, a military psychologist, had served on a BSCT at Guantánamo. As two retired counterintelligence operatives explained to dissident task force member Jean Maria Arrigo, it was no accident that these lobbyists and former officials had high-level connections that outranked those of the military task force members, putting those members on notice that any deviation from official policy might not remain confidential (Arrigo, 2007). One of these former counterintelligence operatives, David DeBatto,

> interpreted the PENS task force process as a typical legitimization process for a decision made at a higher level in the Department of Defense (DOD) [editors' insert]. Because of the hierarchical structure of the DOD, he said, it was absolutely impossible that the six DOD members of the task force participated as individuals bringing their expertise and judgment to the policy issues at hand for [inaudible]. He said that they were certainly there as representatives of the decision maker. And because the decision maker's decision had to be sustained, had to prevail, a quorum of DOD members was necessary, rather than just one or two to express DOD concerns.

> The presence of the APA Science Policy observers, DeBatto said, was a standard intimidation tactic to insure the DOD task force members stayed in line. As funding lobbyists and recipients, they were strictly beholden to DOD interests. In effect, they outranked the DOD task force members because of their high-level connections.
>
> The reason for the several task force observers, instead of just one intern in the corner with a notepad, DeBatto said, would be to represent the perspectives of various agencies to the decision maker, so as to broadly legitimize the prior decision—again, a very standard scenario that counterintelligence operatives know about. (Arrigo, 2007)

Not surprising, given its composition, the task force never examined most of the ethical issues involved (Coalition for an Ethical Psychology, 2008; Soldz, 2009a), such as whether psychologists, bound by an ethics code to a "Do no harm" standard (American Psychological Association, 2002), could ethically participate in interrogations that may cause harm. Rather, participation was simply assumed. And the "Do no harm" standard was simply dismissed. For example, then APA President-Elect Gerald Koocher, the editor of a journal and author of a major text on psychological ethics, told the task force members on his e-mail list:

> In many of the circumstances we will discuss when we meet the psychologist's role may bear on people who are not "clients" in the traditional sense. Example, the psychologist employed by the CIA, Secret Service, FBI, etc., who helps formulate profiles for risk prevention, negotiation strategy, destabilization, etc., or the psychologist asked to assist interrogators in eliciting data or detecting dissimulation with the intent of preventing harm to many other people. In this case the client is the agency, government, and ultimately the people of the nation (at risk). The goal of such psychologists' work will ultimately be the protection of others (i.e., innocents) by contributing to the incarceration, debilitation, or even death of the potential perpetrator, who will often remain unaware of the psychologists' involvement. (Coalition for an Ethical Psychology, 2008; Psychological Ethics and National Security Task Force, 2009, p. 13)

Thus, in Koocher's view, there was little problem with psychologists "contributing to the incarceration, debilitation, or even death of the potential perpetrator, who will often remain unaware of the psychologists' involvement." No task force member objected to this formulation. As this material was kept from the public and other APA members; the "Do not harm" standard here died a secret death at Koocher's hands with no discussion.

The PENS report reasserted the APA leadership position that psychologist participation in national security interrogations at Guantánamo and elsewhere was not only ethical but critical:

> Psychologists have a valuable and ethical role to assist in protecting our nation, other nations, and innocent civilians from harm, which will at times entail gathering information that can be used in our nation's and other nations' defense. The Task Force believes that a central role for psychologists working in the area of national security-related investigations is to assist in ensuring that processes are safe, legal, and ethical for all participants. (American Psychological Association, 2005b, p. 2)

Notice that, through this statement, the APA aligned the psychology profession with the nation's military and national security apparatus. It also unwittingly makes social and political decision-making central to psychology. After all, if psychologists are to protect "our nation ... [and] innocent civilians from harm," they logically should determine which activities are likely to play such a protective role. In fact, one could argue that they are ethically obligated to perform such an analysis. Thus, many critics, including many retired military leaders, have argued that the unprovoked attack of and occupation of Iraq, or the existence of the Guantánamo detention facility intentionally kept outside of U.S. and international law, place our country and its citizens at greater risk. And it is incontrovertible that the U.S. invasion of Iraq placed its "innocent civilians" at greater risk, given the extensive epidemiological evidence demonstrating that hundreds of thousands, and perhaps considerably more (Burnham, Lafta, Doocy, & Roberts, 2006; Iraq Family Health Survey Study Group, 2008; Keiger, 2007; Opinion Research Business (ORB), 2008; Roberts, Lafta, Garfield, Khudhairi, & Burnham, 2004; Soldz, 2004), have died as a consequence of that invasion. Are the task force and the APA board that approved the report suggesting that psychologists have an ethical obligation to oppose the Iraq war in order to protect those hundreds of thousands of "innocent civilians"? Are they suggesting that psychologists oppose the continuation of the Guantánamo detention facility, in order to reduce the danger to U.S. citizens resulting from the anger generated by its existence throughout much of the world? Or is the APA, rather, creating a specious "ethical" rationale for the psychology profession becoming a servant of whatever the administration in Washington claims is necessary for "national defense"? Unfortunately, none of these issues were even discussed by this "ethics" task force or in any other APA-created forum, despite the fact that the report was issued during the reign of the Bush administration, regularly conceded, even by APA leaders and task force members in private, to be one of the most militaristic and law-defying administrations in U.S. history.

Also included in this statement from the PENS report is the rather odd statement: "a central role for psychologists working in the area of national security-related investigations is to assist in ensuring that processes are safe, legal, and ethical for all participants" (American Psychological Association, 2005b, p. 2). This formulation poses that psychologists are "safety officers," preventing interrogations from drifting over the legal limit into abuse or torture. The safety officer conceptualization appears to have been borrowed from Colonel Banks's (2005a, 2005b) instructions for the BSCT, which in turn borrowed from the quite different safety officer role played by psychologists in the SERE program, where psychologists are responsible solely for assuring safety without the conflicting demands of extracting information (Doran, Hoyt, & Morgan, 2006; Otterman, 2007). It seems likely that this safety officer concept for health professional interrogation consultants in fact derives from the role that these professionals play in legitimizing abuse in the Office of Legal Counsel torture memos (Fink, 2009). The PENS report provides no reason to believe that psychologists have any special expertise in serving as safety officers, and fails to explain why legal personnel, such as the military JAGs (judge advocate generals), are not better assigned the task of keeping interrogations "safe, legal, and ethical."

This formulation also ignores that such a safety officer role may very well place everyone involved at greater risk. Imagine the interrogation situation. An interrogator decides to ratchet up the physical or psychological pressure on a detainee. The interrogator might ordinarily worry about where the line was demarcating his (or her) actions from abuse. But the interrogator knows that someone else, a psychologist, is designated as the safety officer. So the interrogator can relax and do whatever until the psychologists says stop. The psychologist, however, is an outsider determined to prove that she (or he) is part of the team, not one of those "fuzzy-minded shrinks" who doesn't understand the military mission and that war is hell. So the psychologist holds back as the interrogator gets rougher. In this scenario, which is psychologically more plausible than that implicitly proposed by the PENS report, "all participants" are placed at higher risk by the purported presence of the psychologist.

I must immediately clarify that I don't actually believe that BSCT or CIA psychologists aiding interrogations served primarily as safety officers. Rather, all extant evidence is that their primary role was to identify detainee weaknesses for exploitation. In any case, to assure that interrogators did not go over the line of authorized ("legal") techniques served no protective function at all. Over the last 4 years it has become clear that the techniques authorized under standard operating procedures were themselves abusive. Military investigation after investigation has concluded that the vast majority of abusive interrogation tactics—hypothermia, sleep deprivation (called sleep manipulation), excruciatingly painful stress positions—among others, were completely legal and authorized. The mythical safety officer psychologist would (and the BSCTs did) allow them to continue.

The PENS report established the essential claim, repeated ad nauseam, that emphasized the safety officer role for psychologists. Thus, in fall 2007, the APA public relations office issued a frequently asked questions (FAQ) document that asserted: "Based on years of careful and thorough analysis, APA has affirmed that psychology has a vital role to play in promoting the use of ethical interrogations to safeguard the welfare of detainees and facilitate communications with them" (American Psychological Association, 2007a; cf. Coalition for an Ethical Psychology, 2008).

Despite repeated repetitions of essentially the same assertion, the APA has yet to produce any evidence of this "careful and thorough analysis" it allegedly spent years conducting. Thus, it did not respond to a request by ethics experts Pope and Gutheil to clarify the nature of this analysis (Behnke, Gutheil, & Pope, 2008; Pope & Gutheil, 2008).

OPPOSITION TO APA POLICY

The APA position supporting psychologist participation in interrogations at Guantánamo and elsewhere did not sit well with many members. Opposition arose, both within APA's governing Council of Representatives and from outside activists. Increasingly, activists utilized professional publications, the press, and Internet media to publicize their dissatisfaction with APA's position.*

In response to criticism, the APA passed two antitorture resolutions (American Psychological Association, 2006, 2007c) while simultaneously insisting that psychologists continue participating in Guantánamo and CIA interrogations (American Psychological Association, 2007a, 2007b). Both of these resolutions drew heated criticism from critics as inadequate and loophole ridden (Coalition for an Ethical Psychology, 2008; Soldz, 2006b, 2006d, 2007a). The second of these resolutions was so flawed that the APA modified it after withering criticism (American Psychological Association, 2008b; Benjamin, 2007; Soldz & Olson, 2008c). In summer 2007, the APA

---

* Altman (2008); Arrigo (2007); Arrigo, Thomas, Rubenstein, Anders, & Goodman (2007); Benjamin (2007); Benjamin & Goodman (2007); Bond (2006, 2008a); Coalition for an Ethical Psychology (2008); Costanzo, Gerrity, & Lykes (2006); Ephron (2008); A. Goodman (2008); D. Goodman (2008); Horton (2007); "Human wrongs" (2007); Jacobs (2007b); Järnefors (2008); Jaschik, (2007a, 2007b); Levine (2007); Lott (2006); Morlin (2007); Olson (2006); Olson & Soldz (2007); Olson et al. (2008); Pipher (2007); "Psychologists and torture" (2008); Reisner (2007); Shinn (2006, 2007); Soldz (2006a, 2006b, 2006c, 2007b, 2008b, 2008c, 2008d, 2008e, 2009b); Soldz & Assange (2007); Soldz & Barahnona (2007); Soldz & Olson (2008a, 2008b, 2008c); Soldz, Reisner, & Olson (2007b); Soldz et al. (2007a, 2008); Summers (2008); Valtin (2006); Welch (2008); Woolf (2007); Zeller (2007).

Council, on recommendation of the board, decisively defeated a proposed moratorium on psychologist participation in interrogations at U.S. detention facilities operating in violation of international law (Altman, 2006, 2007; Benjamin, 2007; Woolf, 2007).

The movement to change APA policy continued to grow. In 2006, a movement started to withhold dues from the APA until the association's polices changed (withholdapadues.com, 2008), eventually gathering over 300 members plus an unknown number who withheld dues independently. A number of prominent psychologists resigned in protest (Jacobs, 2007a; Pope, 2008; Shinn, 2007). After the 2007 moratorium defeat, noted psychologist-author Mary Pipher returned an award she had been given by the APA.

In 2008, activists utilized a never-before-used provision in the APA rules to propose a referendum to ban psychologist participation in detention sites operating outside of or in violation of international law or the Constitution (American Psychological Association, 2008a). (Unlike the previous moratorium, this measure banned not just interrogation support but all psychologist activity at the detention centers, other than treatment of U.S. military personnel.) The measure easily acquired the necessary number of signatures and was put to a vote in summer 2008. In mid-September it was announced that the referendum had passed decisively, with 59% of the vote (Soldz & Olson, 2008a).

WHERE TO FROM HERE?

Psychologist-advocates for social justice won an enormous victory with the passage of the referendum. This success culminated years of struggle by hundreds of activists within and outside the APA. It constituted a major rebuke of the Bush administration's policy of legalized torture in which psychologists played major roles designing, implementing, standardizing, and disseminating abusive techniques. Referendum passage also provided a stunning rebuke of the APA leadership's covert, and sometimes overt, accommodation of that policy.

It is not by accident that the referendum victory came as the world was counting the last days and hours of the Bush administration. As predicted earlier by Bryant Welch (2008), a former head of the APA's Practice Directorate, the APA ended the Bush era with the policy that it should have had at the beginning of the torture regime. Opponents of the 2008 referendum largely claimed to support its "laudable goals" and only to object to its alleged poor execution of those goals. What a far cry this was from the opposition to previous efforts to reform the APA interrogations policy, such as the 2007 moratorium effort. Those efforts were met with perverse claims that psychologists were needed to keep detainees safe. "If psychologists withdraw, people will die," the person chosen by the APA board to present its 2007 antitorture resolution told the APA Council. Such rhetoric has largely gone.

Nonetheless, the struggle against U.S. torture and detainee abuse is far from over. Psychologists opposed to U.S. torture have many tasks ahead of us both inside and outside the APA, as I briefly outline next.

REFERENDUM IMPLEMENTATION

The first task is to see that the referendum is implemented as was intended by those voting in favor: Psychologists do not belong at Guantánamo or the CIA black sites unless they are working directly for detainees or independent nongovernmental organizations (NGOs). An implementation committee, which included referendum authors as well as APA members opposed to the referendum, met in fall 2008 to discuss a number of details. The APA Council of Representatives accepted that the referendum was in effect at its February 2009 meeting and APA public statements sometimes acknowledge that the referendum is in effect. As of this writing, however, the APA has not made a statement suggesting that psychologists should not serve at Guantánamo, the detention facility at Bagram Air Base in Afghanistan, or any actual detention setting.

The committee criterion to determine when a detention facility is operating outside or in violation of international law or the Constitution are vague and are subject to rival interpretations. Similarly, criteria will be needed to determine when psychologists are "working directly for" the detainee. Finally, we need to pressure the APA to make the referendum policy enforceable as part of the ethics code. Making it enforceable also would make it policy for state licensing boards that use the APA ethics code as their standard.

But more important, it is crucial that we work with the DoD, the new Obama administration, and Congress to incorporate the referendum into government policy. We need to get the DoD to remove the Behavioral Science Consultant (BSC) psychologists from whatever U.S. detention facilities remain after Guantánamo is closed.* And we also need to see

---

* The APA referendum would ban members from participating in "illegal" detention centers. As the Obama administration revises U.S. detention policies, it may be harder to make the case that remaining detention centers, such as Bagram Air Base in Afghanistan, are illegal. In any case, the referendum will not necessarily change DoD policy. The DoD could get around it through various means, including encouraging their psychologists to leave the APA, or by simply ignoring or reinterpreting its meaning. Thus, Marks and Bloche (2008) report that several psychiatrists have been trained as BSCT members despite American Psychiatric Association and American Medical Association policies against such participation. Interestingly, the 2006 memorandum on BSCT operations by Army Surgeon General Kiley (2006) quotes the medical association policy but manages to transform its meaning into almost its exact opposite. No such transformation was needed for the then APA policy.

the creation of a new mental health system for detainees at Guantánamo and other detention centers, one that is completely independent of all command pressure. If DoD is nonresponsive, we should work with sympathetic voices in Congress to accomplish these goals through legislation; hopefully, this will be part of a forthcoming revision of the entire military detention and interrogation system. In this effort, we can unite with the numerous military and intelligence professionals who have vigorously opposed Bush administration torture policies. Indeed, military attorneys and interrogators have been among those leading the fight against these abuses, while military psychologists have, unfortunately, remained silent or prevaricated.

It is less clear how to impose the referendum policy on the CIA, which requires the agency to pull psychologists from any involvement in the black site detention centers. Perhaps the referendum will only be implemented if the CIA's entire black site detention system and enhanced interrogation program remains shut, as may result from President Obama's actions. Here again, we can seek collaboration with Congress, as well as with CIA veterans who have publicly opposed CIA torture. As a first step, under a new administration, the CIA and all other government agencies should be required to follow the interrogations protocols of the Army Field Manual (United States Department of the Army, 2006). The Field Manual is far from perfect, allowing, for example, the use of abusive techniques—isolation/segregation, sleep deprivation, and so-called "fear up" in which detainee fears may be radically exacerbated (Kaye, 2009). Therefore, the Field Manual must be modified at the same time it is made a uniform standard for all detainees. Simultaneously, we should work to restrict psychologists' roles to those consistent with the referendum and our ethical obligations.

ROLES

The APA referendum, although a major success, does not resolve all major flaws in APA policy toward participation in interrogations. The referendum limits the types of detention *settings* in which psychologists can work; those at which fundamental human rights are violated are off limits. It fails to deal with the question of what, if any, involvement psychologists should have in interrogations—that is, the question of what *roles* are appropriate or not appropriate for psychologists to take on (Soldz & Olson, 2008a, 2008b).

Many of us would grant that psychologists may have a valuable role to play in training of interrogators, for example, at the Army training center at Fort Huachuca (Arizona), and in psychological screening of potential interrogators. However, some of us believe—along with many senior interrogators—that the APA should follow the American Medical Association and American Psychiatric Association in banning its members from any direct role in the interrogations of individuals. We believe that intelligence interrogation inherently involves "exploitation" of detainees for intelligence potential, as the military describes its activities. This exploitation inherently conflicts with the helping role of practicing psychologists, as represented by Principle A of the APA's ethics code: "Psychologists strive to benefit those with whom they work and take care to do no harm" (American Psychological Association, 2002). Section 3.06 of the code states "Psychologists refrain from taking on a professional role when personal, scientific, professional, legal, financial, or other interests or relationships could reasonably be expected to … expose the person or organization with whom the professional relationship exists to harm or exploitation." And section 3.08 states: "Psychologists do not exploit persons over whom they have supervisory, evaluative, or other authority such as clients/patients, students, supervisees, research participants, and employees." The essence of intelligence is what intelligence professionals correctly call exploitation of assets, as seen, for instance, in the CIA's infamous *Human Resource Exploitation Training Manual* (1983). This type of exploitation simply does not benefit those being exploited, nor can it be said to be harmless. It also violates the explicit prohibitions on exploitation in the ethics code. Psychologists should take the ethics code seriously and bow out of all direct involvement in intelligence interrogations, whether or not these interrogations are abusive.

We also need to discuss whether the APA should go further and join our medical colleagues in banning all direct involvement in interrogations, including domestic and nonmilitary interrogations not involving national security. Law enforcement interrogations pose a number of issues distinct from intelligence interrogations. But there still are potential conflicts with the ethical aspirations of our profession that need to be resolved. The disastrous involvement in intelligence interrogations should serve as a warning. Following the strategic aims of those in authority at the expense of independent, ethical obligations places psychologists in a perilous position. Yet, placing the responsibility for ethical behavior primarily upon individuals is unrealistic. Any system requiring individuals to become heroes in order to behave ethically will fail.

At present, the APA ethics code and ethics office have not been up to the task of offering the clear guidance necessary to help operational psychologists navigate the myriad ethical minefields that are inherent in such dual roles. Therefore, at the very least, we need serious discussion regarding the ethics of psychologists' participation in individual interrogations. We need to develop new ethical principles and standards to decide first if such work can be ethically performed, and, if so, to then provide the guidance that would be required.

ORGANIZATIONAL REFORM

When the 9/11 attacks occurred, the APA leapt to offer the services of psychologists to the evolving national security state (American Psychological Association, 2003, 2004, 2005a, 2005c; American Psychological Association & FBI Academy, 2002; Coalition for an Ethical Psychology, 2008; Davis, 2008; Summers, 2008). Support for participation in interrogations was simply one manifestation of this effort.*

Leadership in change efforts fell largely on groups that had no standing within APA, such as the WithholdAPAdues movement (2008), Psychologists for an Ethical APA (2008), the Coalition for an Ethical Psychology (2008), and Psychologists for Social Responsibility (2009a). These groups were comprised of individuals who, for the most part, had never before participated in APA governance. Yet, when a referendum representing the position advocated by these groups was eventually submitted to the entire APA membership for a vote, an overwhelming majority, 59% of those voting, repudiated the years-long interrogations policy of the leadership. The official representatives had apparently not represented the majority of those concerned about this issue.

This situation is a sign of serious organizational problems within APA. An organization that is so out of touch with its members needs radical reform. The association's deep and long-standing relationship to the military-intelligence establishment (Soldz, 2006c; Summers, 2008) needs serious examination. After the PENS listserv was leaked to the press in May 2009, three organizations of psychologists called for an independent investigation of links between the APA and the military-intelligence establishment (Coalition for an Ethical Psychology, 2009; Psychologists for an Ethical APA, 2009; Psychologists for Social Responsibility, 2009b). Such an investigation of the APA could be free standing or, ideally, connected with the Commission of Inquiry discussed later. At the same time, the Nobel Prize-winning NGO, Physicians for Human Rights, called for an investigation of Pentagon–APA ties by the Defense Department Inspector General (Physicians for Human Rights, 2009). These investigations are essential to clarify the extent and nature of these ties as a starting point for serious reform of the APA.

As discussed later, the APA's ethics policies need serious rethinking. But reform must go further. The organization should closely examine

---

* Although this effort was led by a few key staff members and elected APA officials with national security connections, most segments of the organization went along. The Council of Representatives never questioned or dissented from the initiative. And the 2007 moratorium effort was opposed by every official committee that was asked for an opinion. Only a small minority of divisions dissented from parts of the policy. For a brief period, the Divisions for Social Justice (DSJ) organized opposition, but ultimately this force succumbed to organizational intimidation. After that, only a few individual council members opposed the reigning policy from within the official structures.

the relationship between psychology and human rights. And institutional reforms are needed to make the organization more responsive to the opinions and wishes of its members.*

In 2008, APA dissidents ran one of their own, Steven Reisner, for president. Reisner was the only one of the five candidates who opposed the APA's disastrous interrogations policy (Ephron, 2008). He was the only one who took a leadership role in reform efforts. And he was the only one who is not from the APA inner circle. He had not become socialized to believe that there is only one way to conduct business. Reisner topped the ballot in the nomination phase but eventually lost to an APA insider. This loss stymied efforts to reevaluate the relationship between the APA and the military-intelligence establishment.

The future of the APA is uncertain. If the membership does not remain active in pushing for change, the APA will likely return to business as usual. As in the wider society, only an active and informed membership can provide a counterweight to the forces of inertia and the status quo. However, a number of dissident members have resigned their membership, giving up, for the moment, efforts to reform the APA from within. It is up to those members who remain to insist that their professional organization represent the best of our society, rather than compromise with the worst.

PUBLIC RECKONING

Those who cannot remember the past are condemned to repeat it.

—George Santayana

Over the last 8 years, horrifying acts have been committed in the name of both U.S. citizens and psychologists. The last several years have featured one revelation after another about the U.S. program of torture and detainee abuse and of psychologists' roles in that program. Yet we are far from having a complete picture of the extent or nature of the program, or of the institutional failures that enabled its development. To fail to investigate and understand these harsh realities is to acquiesce to their reemergence in the next political crisis.

---

* One problem is that the council tends to include many of the same individuals over time, as its members simply circulate among divisions they can represent. Thus, as Bryant Welch (2008) pointed out last summer, council members tend to identify more with one another and with APA leadership than with the membership they ostensibly represent. And the lesson gets absorbed that only those who do not make major waves will be accepted into this elite club. Perhaps some form of term limits or other efforts to open APA governance are needed.

The United States needs a commission of inquiry, with subpoena power, to examine our recent history of torture and detainee abuse. The commission must expose the legal rationales that justified detainee abuse, the official actions that authorized abuse, the institutional arrangements that implemented abuse, and the unofficial mechanisms that facilitated the spread of abusive behavior to large sectors of the forces on the ground in all three major theaters of the so-called global war on terror.

Psychology, along with the other health professions and social sciences, need its own investigatory process. This process ideally would be a subcommittee of a nationally authorized commission of inquiry. If national action is not forthcoming, the professions themselves should create such a process.

One possibility is that prominent health professionals, along with organizations such as Physicians for Human Rights, Physicians for Social Responsibility, Psychologists for Social Responsibility (PsySR), the Torture Abolition and Survivors Support Coalition International (TASSC), and the International Rehabilitation Council for Torture Victims (IRCT) could collectively form such a health professionals commission of inquiry. This effort should also embrace members of the social sciences—including anthropology—which has its own experiences with secret intelligence activities by U.S. national security institutions in the human terrain systems fieldwork program and other initiatives (Glenn, 2009; Price, 2008).

One major task of this truth commission would be to reveal the role of psychologists in the design, implementation, standardization, and dissemination of U.S. torture and detainee abuse. But the commission would have to look further. Members of all the health professions abetted our nation's program of torture and detainee abuse. At every torture and detention site, health professionals witnessed—and ignored—multiple signs of abuse. Most health professionals looked the other way when their patients presented with symptoms of abuse. The Physicians for Human Rights study, *Broken Laws, Broken Lives* (2008), describes one doctor who treated a detainee during torture and told the torturers, "If you go on torturing him in this way, he will die" (p. 85), and then left the patient to his fate. Another detainee describes being treated by a doctor and then how he "heard the doctor say 'continue' [to the interrogators]" (p. 21). Guantánamo, psychologists and physicians failed to diagnose posttraumatic stress disorder among detainees subjected to torture and abusive conditions of detention, documenting instead that their abused patients suffered from preexisting personality disorders (Kennedy, Malone, & Franks, 2009; Physicians for Human Rights, 2008). Physicians participated in the often unnecessary, unethical, and deliberately abusive forced feedings of hunger strikers, who were protesting abusive treatment (Miles, 2009).

Further, despite fine-sounding position statements, *none* of the health professions made significant efforts to stop the abuse. As far as I am aware, until the Bush administration was headed for the door, none of the health professions or the social sciences actively condemned the legalized torture openly being committed by our government. Therefore, the commission of inquiry must examine the professions themselves and the institutional, organizational, policy, and ethical failures that allowed such widespread abetting of government torture.

PERSONAL ACCOUNTABILITY

As well as finding and publicizing the facts, we need accountability for those who created and aided the torture system, both for national leaders and participating psychologists. At the level of national leadership, psychologists should support the various efforts by human rights advocates to bring accountability to those in the administration, military, and intelligence agencies who authorized, legitimized, and operated the torture system. On the level of accountability for participating psychologists, the only real avenue, at present, is to bring ethics complaints against specific psychologists. So far, such efforts have been unsuccessful.

Ethics complaints have been filed with the APA Ethics Committee against at least two psychologists. The APA has failed to decide the case of one of these psychologists, Major John Leso, for at least 34 months since the first complaint was filed (Bond, 2008b). In this case an official interrogation log posted on the Web by *Time* magazine (ORCON [Authoring agency classified by Originator Control], 2003) documents his presence during an obviously abusive interrogation, an interrogation that has been proclaimed to meet the legal standard for torture by the Pentagon official in charge of the military commissions at Guantánamo (Woodward, 2009). Official meeting minutes released by a Senate Committee document active participation in a crucial meeting planning abusive interrogation strategies (Senate Armed Services Committee, 2008a, Tab 7; 2009), strategies which, when used in combination, recently were deemed to meet the legal definition of torture by the top Pentagon official responsible for the military commissions (Woodward, 2009). Instead, apparently, of conducting an investigation of its own, the APA Ethics Committee has sent repeated calls for clarification and further information to the psychologist filing the complaint, including such petty harassing measures as refusing to accept a Web URL for documents from the Senate Armed Services Committee, insisting that the complaining psychologist print them and submit a hard copy. The impression given by the Ethics Committee treatment of the complainant is that it is putting the entire burden of investigation onto the complainant, rather than utilizing the considerably greater resources of the APA.

In the other case, the Ethics Committee refused to even open an investigation, citing lack of evidence, despite considerable evidence warranting investigation (Soldz & Assange, 2007; Soldz et al., 2007b). That psychologist is closely connected to several members of the top APA leadership and has himself served in various leadership roles in APA; he was given a major award by an APA division just last summer, after the complaint was rejected, and was elected president-elect by another division. The ability of the APA to fairly conduct investigations in such cases can reasonably be doubted.

For years, APA leaders have stated that they will take strong action against any psychologist found to have participated in abuse. In a 2006 debate on Democracy Now!, then APA President Gerald Koocher stated:

> I wish I had the assurance that Jane Mayer and that Dr. Reisner apparently have that there are A.P.A. members doing bad things at Guantánamo or elsewhere, because any time I have asked these journalists or other people who are making these assertions for names so that A.P.A. could investigate its members who might be allegedly involved in them, no names have ever been forthcoming. (A. Goodman, Koocher, Reisner, & Xenakis, 2006)

And, in a letter to the American Civil Liberties Union (ACLU), the APA's ethics director stated "APA will adjudicate any allegation that an APA member has engaged in unethical conduct" (Behnke, 2008). The impotence of such claims is demonstrated by the organization's failure to aggressively pursue the complaints now that names have been given and complaints filed. As the evidence became overwhelming that psychologists were crucial actors in designing and implementing the U.S. torture regime, the APA has, so far, failed to reach any findings at all in any ethics case involving detainee abuse.

The efforts to pursue accountability through state licensing boards have also been unsuccessful. In every case where complaints were filed against psychologists or doctors, the boards have refused to even open a case, citing various arguments for lack of jurisdiction.

Traditional ethics mechanisms thus failed to respond to a systematic program of torture in which psychologists played key roles. It is reasonable to conclude, then, that these ethics mechanisms are inadequate to the challenge of adjudicating instances of horrifying government-sponsored abuse.

While the entire system needs reexamination in light of its dismal record in response to state-sponsored torture, one aspect of ethics enforcement urgently needs immediate reform. The statue of limitations for participating in torture or cruel, inhuman, or degrading treatment needs to be extended. For most offenses the APA has a 4-year statute of limitations on complaints by members and a 5-year limitation on nonmember complaints. State ethics committees also have similar limitations. As this is being written, the APA limitation has been exceeded for any torture participation before 2004. Given the great secrecy with which interrogations are undertaken, in many cases torture participation will not become evident for many years postabuse. The statue of limitations for torture, and cruel, inhuman, and degrading treatment should be considerably raised or abolished. The APA conveniently failed to act on this despite calls for change before many of the most egregious abuses would be immunized.

Beyond minor revisions in the current system, the profession must either repair the current ethics accountability system or develop alternative, more effective, mechanisms for accountability for such abuses as participation in state-sanctioned torture, where accountability requires the profession to act counter to government policy. As part of this process, psychology should consider adopting ethical guidelines that preclude psychologists from practicing in situations—for example, in classified or other secretive environments like national security detention centers—where it is extremely difficult to hold them accountable for their actions. Ethics standards that are impossible to enforce due to government secrecy are, as we have learned through their failure to constrain psychologist participation in detainee abuse, no standards at all.

ETHICAL DISCUSSION

The failure of traditional mechanisms for ethics enforcement to respond to this national and professional crisis also points toward a larger crisis in psychological ethics. Many of the APA leaders in psychological ethics were among the strongest promoters and defenders of the APA's "policy of engagement," encouraging psychologists to participate in interrogations even after repeated reports of systematic abuse in U.S. interrogations. These leaders included Gerald Koocher, former APA President, editor of the journal *Ethics & Behavior*, and author of a leading ethics textbook; Stephen Behnke, the APA's ethics director and steadfast spokesman for the APA interrogation policy; and Olivia Moorehead-Slaughter, chair of the APA's Ethics Committee and the PENS "ethics" task force. Only one person traditionally associated with psychological ethics, former APA Ethics Chair Ken Pope, publicly criticized the APA for failing to enact adequate ethical requirements to prevent psychologists from being involved in detainee abuse and for failing to provide justification for the APA's oft-repeated claim that psychologists help keep interrogations safe and ethical (Behnke et al., 2008; Pope, 2008; Pope & Gutheil, 2008, 2009).

Clearly, the organizational mechanisms and intellectual traditions that deal with the types of informed consent therapists should obtain from patients, or researchers from college students, are inadequate to deal with psychologists' aiding a program of state-sponsored torture. If we do not confront these failures we will be unprepared for the next major crisis.

As part of this reevaluation of psychological ethics, the APA needs to quickly address the obvious weaknesses in the ethics code (American

Psychological Association, 2002). Section 1.02, which allows psychologists to follow government laws, orders, and regulations in conflict with ethics—the "Nuremberg defense"—should be dropped (Pope & Gutheil, 2009). Rather, if exceptions are to be allowed to certain ethical standards, these exceptions should be individually specified. Current proposals to add "except when fundamental human rights are violated" to 1.02 are themselves problematic due to interpretational ambiguity of "human rights," allowing the creation of additional loopholes protecting abusers. They should, rather, be scrapped in favor of the strategy of dropping the section entirely and explicitly specifying situations where sections of the ethics code need not be obeyed.

Not currently on the agenda, but also requiring serious attention are revisions to the 2002 ethics code on research ethics, which look as if they permit government-sponsored abuse. For instance, section 8.05, added in 2002, which allows researchers to dispense with informed consent "where otherwise permitted by law or federal or institutional regulations," is deeply disturbing and needs immediate revision. This provision would allow psychologists to participate in research on imprisoned detainees, among other abuses.

Similarly, section 8.07 permits too high a threshold for allowing the use of deception in research: "Psychologists do not deceive prospective participants about research that is reasonably expected to cause physical pain or severe emotional distress." The phrase "severe emotional pain"—changed from "unpleasant emotional experiences" in the prior draft of the ethics code (American Psychological Association, 1992)—was added in 2002. It eerily echoes the definition of psychological torture in the United Nations Convention Against Torture: "'torture' means any act by which severe pain or suffering, whether physical or mental, is intentionally inflicted on a person" (Office of the United Nations High Commissioner for Human Rights, 1984). Surely a research procedure should not need to meet the legal definition of torture for disqualification as an ethics violation. In addition to revising this abhorrent ethics standard, we need to analyze the processes whereby informed consent was abolished for government-sponsored research and the definition of torture became incorporated into the ethics code.

More broadly, we need to undertake a public dialogue seeking a social consensus on psychological ethics in national security settings. While consensus may be difficult to achieve, the process of dialogue will at a minimum clarify important factors at play in this work. Psychologists working in such settings need a coherent moral foundation at the base of their professional identity rather than a hodgepodge of constraints and permissions.

In furtherance of these efforts, a group of psychologists affiliated with Psychologists for Social Responsibility and intelligence professionals are preparing a *Psychology and Military Intelligence Casebook on Interrogation Ethics* (Arrigo, Soldz, Bennet, Long, & Davis, 2009). This project is bringing together psychologists, other social scientists, military interrogators, other intelligence professionals, and ethicists in an effort to elucidate the complex nexus of ethical, practical, and institutional issues affecting psychologists working in national security settings. The casebook aims not at definitive answers but at clarifying the issues that must be dealt with in any serious ethical discussion of such work.

THE DARK SIDE REMAINS

Activist psychologists have achieved an amazing feat in transforming APA policy. In the process, we created a broad, decentralized movement. We brought together many individuals and organizations that collectively were able to successfully challenge the largest mental health organization in the world. This movement shines as a beacon to other activists, showing what a democratic participatory polity can accomplish. It has been noticed by many around the world who are trying to shake off the despair generated by the "global war on terror." It encourages those struggling to transform violent, authoritarian institutions and cultures in the United States and elsewhere. It is not by accident that blogger Andrew Sullivan, an early participant in the fight against torture, headlined his notice of the referendum victory "Know Hope" (Sullivan, 2008).

However, our task is far from over. I will end with a cautionary note. In the wider society, the fight against torture and human-rights abuses is neverending. With luck, we will soon put an end to the Bush administration's experiment with legalized torture in national security interrogations. But U.S. support for torture likely will not totally end. After all, U.S. government torture has a long pedigree (McCoy, 2004, 2006, 2009). Intelligence work, by its nature, occurs in the shadows, away from public oversight. Further, as the scholar Darius Rejali (2007) revealed in his magisterial work *Torture and Democracy*, modern forms of torture, including psychological torture, through their lack of clear, telltale signs, were designed precisely to avoid democratic oversight.

Only continual vigilance, combined with cultural change, can remove our nation from the list of those conducting or condoning torture. We must remember that we live in a society where torture is a prominent feature, not only at national security sites, but in domestic settings as well. Many of our prisons and our urban police departments are institutions where abuse is routine, and such abuses are widely accepted by those in the broader society. Some locked mental health facilities have practices that inmates and their representatives report are tantamount to torture. Abuses, sometimes amounting to torture, are all too commonly tolerated in other institutions ranging from boot camps for adolescents to nursing homes for the elderly. Confronting torture in these settings will likely require different strategies than those used to deal with abuse in national security detention centers. In the end, only a cultural change will transform the willingness to accept torture as a fact of life in the contemporary United States. Those of us who fought so hard against the Bush torture regime must now turn to the task of dismantling the many facets of abuse in our society. Psychologists can and should help transform a culture tolerant of abuse to one where abuse is unacceptable.

REFERENCES

Altman, N. (2006). *Resolution for a moratorium on psychologist participation in interrogations at US detention centers holding foreign detainees, so-called "enemy combatants": Summary and overview.* Retrieved August 24, 2007, from http://www.apa.org/ethics/pdfs/2006moratoriumresolutionsummaryandoverview.pdf

Altman, N. (2007, February). *A moratorium on psychologist involvement in interrogations at US detention centers for foreign detainees.* Retrieved August 24, 2007, from http://www.apa.org/ethics/pdfs/resolution22307.pdf

Altman, N. (2008). The psychodynamics of torture. *Psychoanalytic Dialogues, 18*(5), 658–670.

American Civil Liberties Union. (2008, April 30). *Newly unredacted report confirms psychologists supported illegal interrogations in Iraq and Afghanistan.* Retrieved April 30, 2008, from http://www.aclu.org/safefree/torture/35111prs20080430.html

American Medical Association. (2006, June 12). *New AMA ethical policy opposes direct physician participation in interrogation.* Retrieved August 21, 2006, from http://www.ama-assn.org/ama/pub/category/16446.html

American Psychiatric Association. (2005, June 27). *APA statement on psychiatric practices at Guantanámo Bay.* Retrieved March 5, 2007, from http://www.psych.org/news_room/press_releases/05-40psychpracticeguantánamo.pdf

American Psychiatric Association. (2006, May 22). *APA passes position statement barring psychiatric participation in interrogation of detainees.* Retrieved July 26, 2008, from http://www.psych.org/MainMenu/Newsroom/NewsReleases/2006NewsReleases/06-36positionstatementoninterrogation.aspx.

American Psychological Association. (1992). *Ethical principles of psychologists and code of conduct.* Retrieved October 10, 2007, from http://www.apa.org/ethics/code1992.html

American Psychological Association. (2002). *Ethical principles of psychologists and code of conduct.* Retrieved February 27, 2007, from http://www.apa.org/ethics/code2002.html

American Psychological Association. (2003, July). *Science policy insider news: APA works with CIA and RAND to hold science of deception workshop.* Retrieved September 19, 2007, from http://www.apa.org/ppo/spin/703.html

American Psychological Association. (2004, October). *Science policy insider news: Science policy staff meet with psychologists in counterintelligence.* Retrieved October 27, 2007, from http://www.apa.org/ppo/spin/1004.html

American Psychological Association. (2005a, October 23). *APA President Ronald F. Levant visits naval station at Guantánamo Bay.* Retrieved June 12, 2006, from http://www.apa.org/releases/gitmo1023.html

American Psychological Association. (2005b, June). *Report of the American Psychological Association Presidential Task Force on Psychological Ethics and National Security.* Retrieved June 12, 2006, from http://www.apa.org/releases/PENSTaskForceReportFinal.pdf

American Psychological Association. (2005c). *Science policy insider news: Psychology and human intelligence.* Retrieved October 28, 2007, from http://www.apa.org/ppo/spin/1005.html

American Psychological Association. (2006, August 9). *2006 resolution against torture and other cruel, inhuman, or degrading treatment or punishment.* Retrieved September 13, 2006, from http://www.apa.org/governance/resolutions/notortureres.html

American Psychological Association. (2007a, September). *Frequently asked questions regarding APA's policies and positions on the use of torture or cruel, inhuman or degrading treatment during interrogations.* Retrieved October 9, 2007, from http://www.apa.org/releases/faqinterrogation.html

American Psychological Association. (2007b, September 25). *Psychology and interrogations: Statement of the American Psychological Association submitted to the United States Senate Select Committee on Intelligence.* Retrieved October 24, 2007, from http://intelligence.senate.gov/070925/apa.pdf

American Psychological Association. (2007c, August 19). *Reaffirmation of the American Psychological Association position against torture and other cruel, inhuman, or degrading treatment or punishment and its application to individuals defined in the United States code as "enemy combatants."* Retrieved August 24, 2007, from http://www.apa.org/governance/resolutions/councilres0807.html

American Psychological Association. (2008a, August 1). *2008 APA petition resolution ballot.* Retrieved September 19, 2008, from http://www.apa.org/governance/resolutions/work-settings.html

American Psychological Association. (2008b, February 22). *Amendment to the reaffirmation of the American Psychological Association position against torture and other cruel, inhuman, or degrading treatment or punishment and its application to individuals defined in the United States code as "enemy combatants."* Retrieved March 27, 2008, from http://www.apa.org/governance/resolutions/amend022208.html

American Psychological Association & FBI Academy. (2002). *Countering terrorism: Integration of practice and theory.* Retrieved October 27, 2007, from http://www.apa.org/releases/countering_terrorism.pdf

Arrigo, J. M. (2007, August 20). APA Interrogation Task Force Member Dr. Jean Maria Arrigo exposes group's ties to military. *Democracy Now!* Retrieved September 9, 2007, from http://www.democracynow.org/article.pl?sid=07/08/20/1628234

Arrigo, J. M., Soldz, S., Bennet, R., Long, J., & Davis, M. (2009). *Psychology and military intelligence casebook on interrogation ethics.* Retrieved January 19, 2009, from http://www.pmicasebook.com/PMI_Casebook/Home.html

Arrigo, J. M., Thomas, N., Rubenstein, L., Anders, E., & Goodman, A. (2007,

June 1). *"The task force report should be annulled."* Member of 2005 APA Task Force on Psychologist Participation in Military Interrogations speaks out. Retrieved June 1, 2007, from http://www.democracynow.org/2007/6/1/the_task_force_report_should_be

Banks, M. (2005a). Purpose of psychological support to interrogation and detainee operations. In *Providing psychological support for interrogations: Unofficial records of the American Psychological Association Task Force on Psychological Ethics and National Security, June 25–28, 2005*. Stanford, CA: Archives of the Hoover Institution on War, Revolution, and Peace, at Stanford University.

Banks, M. (2005b). The ethics of psychological support to interrogation. In *Providing psychological support for interrogations: Unofficial records of the American Psychological Association Task Force on Psychological Ethics and National Security, June 25–28, 2005*. Stanford, CA: Archives of the Hoover Institution on War, Revolution, and Peace, at Stanford University.

Baumgartner, D. J., Jr. (2008, June 17). *Written testimony to the Senate Armed Services Committee*. Retrieved June 21, 2008, from http://www.senate.gov/~armed_services/statemnt/2008/June/Baumgartner%2006-17-08.pdf

Behnke, S. (2008, May 15). *Letter to Amrit Singh of the American Civil Liberties Union*. Retrieved December 10, 2008, from www.aclu.org/pdfs/safefree/2008_0515_apa_lettertoaclu.pdf

Behnke, S., Gutheil, T. G., & Pope, K. S. (2008). Detainee interrogations: American Psychological Association counters, but questions remain [Electronic version]. *Psychiatric Times, 25*. Retrieved October 20, 2008, from http://www.psychiatrictimes.com/display/article/10168/1285473

Benjamin, M. (2007, August 21). *Will psychologists still abet torture?* Retrieved September 5, 2007, from http://www.salon.com/news/feature/2007/08/21/psychologists/index.html?source=rss&aim=yahoo-salon

Benjamin, M., & Goodman, A. (2007, August 20). *American Psychological Association rejects blanket ban on participation in interrogation of U.S. detainees*. Retrieved August 21, 2007, from http://www.democracynow.org/2007/8/20/american_psychological_association_rejects_blanket_ban

Bloche, M. G., & Marks, J. H. (2005). Doctor's orders – Spill your guts [Electronic version]. *Los Angeles Times*. Retrieved December 10, 2008, from http://articles.latimes.com/2005/jan/09/opinion/op-brutality9

Bond, T. (2006, August 23). *APA confab whitewashes torture by shrinks: The American Psychological Association meets Dr. Mengele*. Retrieved August 23, 2006, from http://counterpunch.org/bond08232006.html

Bond, T. (2008a, August 7). Fixing Hell and curing obesity: The strange, post-Gitmo career of Col. Larry James. *CounterPunch*. Retrieved August 7, 2008, from http://counterpunch.org/bond08072008.html

Bond, T. (2008b, May 19). *If not now, when? An open letter to Dr. Stephen Behnke on psychologists engaged in torture*. Retrieved May 19, 2008, from http://counterpunch.org/bond05192008.html

Bradbury, S. (2005a, May 10). *Memorandum for John Rizzo Senior Deputy General Counsel of the Central Intelligence Agency Re: Application of 18 USC §§ 2340-2340A to certain techniques that may be used in the interrogation of a high value al Qaeda detainee*. Retrieved May 18, 2009, from http://stream.luxmedia501.com/?file=clients/aclu/olc_05102005_bradbury46pg.pdf&method=dl

Bradbury, S. (2005b, May 10). *Memorandum for John Rizzo Senior Deputy General Counsel of the Central Intelligence Agency Re: Application of 18 USC §§ 2340-2340A to the combined use of certain techniques in the interrogation of high value al Qaeda detainees.* Retrieved May 18, 2009, from http://stream.luxmedia501.com/?file=clients/aclu/olc_05102005_bradbury_20pg.pdf&method=dl

Bradbury, S. (2005c, May 30, 2005). *Memorandum for John Rizzo Senior Deputy General Counsel, Central Intelligence Agency Re: Application of United States Obligations Under Article 16 of the Convention Against Torture to certain techniques that may be used in the interrogation of high value al Qaeda detainees.* Retrieved May 18, 2009, from http://stream.luxmedia501.com/?file=clients/aclu/olc_05302005_bradbury.pdf&method=dl

Brewer, S. E., & Arrigo, J. M. (2009). Places that medical ethics can't find: Preliminary observations on why health professionals fail to stop torture in overseas counterterrorism operations. In R. Goodman & M. Roseman (Eds.), *Interrogations, forced feedings, and the role of health professionals: New perspectives on international human rights, humanitarian law and ethics* (pp. 1–19): Cambridge, MA: Human Rights Program at Harvard Law School.

Burnham, G., Lafta, R., Doocy, S., & Roberts, L. (2006, October 11). *Mortality after the 2003 invasion of Iraq: A cross-sectional cluster sample survey.* Retrieved October 11, 2006, from http://www.thelancet.com/webfiles/images/journals/lancet/s0140673606694919.pdf

Bybee, J. (2002, August 1). *Memorandum for John Rizzo, Acting General Counsel of the Central Intelligence Agency.* Retrieved April 18, 2009, from http://stream.luxmedia501.com/?file=clients/aclu/olc_08012002_bybee.pdf&method=dl

Central Intelligence Agency. (1983). *Human resource exploitation training manual.* Retrieved September 17, 2006, from http://www.gwu.edu/~nsarchiv/NSAEBB/NSAEBB122/#hre

Church, A. T., III (2005). *Review of Department of Defense detainee operations and detainee interrogation techniques* (previously redacted pp. 353–365). Retrieved May 19, 2008, from http://www.aclu.org/pdfs/safefree/church_353365_20080430.pdf

Coalition for an Ethical Psychology. (2008, January 16). *Analysis of the American Psychological Association's frequently asked questions regarding APA's policies and positions on the use of torture or cruel, inhuman or degrading treatment during interrogations.* Retrieved February 9, 2008, from http://psychoanalystsopposewar.org/blog/wp-content/uploads/2008/01/apa_faq_coalition_comments_v12c.pdf

Coalition for an Ethical Psychology. (2009, May 5). *Coalition for an Ethical Psychology calls for independent investigation of American Psychological Association.* Retrieved May 5, 2009, from http://www.zcommunications.org/znet/viewArticle/21384

Costanzo, M., Gerrity, E., & Lykes, M. B. (2006). The use of torture and other cruel, inhumane, or degrading treatment as interrogation devices [Electronic version]. *Analyses of Social Issues and Public Policy, 6,* 1–14. Retrieved June 21, 2006, from http://www.spssi.org/SPSSI_Statement_on_torture.pdf#search=%22apa%20zimbardo%20guantánamo%20%22

Council of Europe Committee on Legal Affairs and Human Rights. (2007, June 7).

Secret detentions and illegal transfers of detainees involving Council of Europe member states: Second report. Retrieved October 10, 2007, from http://news.bbc.co.uk/2/shared/bsp/hi/pdfs/marty_08_06_07.pdf

Danner, M. (2009a). The Red Cross Torture Report: What it means [Electronic version]. *New York Review of Books, 56.* Retrieved April 7, 2009 from http://www.nybooks.com/articles/22614

Danner, M. (2009b). US Torture: Voices from the Black Sites [Electronic version]. *New York Review of Books, 56.* Retrieved April 9, 2009 from http://www.nybooks.com/articles/22530

Davis, M. (Writer). (2008). *Interrogation psychologists: The making of a professional crisis.* In M. Davis (Producer): Focus Reframed.

Doran, A. P., Hoyt, G., & Morgan C. A., III. (2006). Survival, Evasion, Resistance, and Escape (SERE): Preparing military members for the demands of captivity. In C. H. Kennedy & E. A. Zillmer (Eds.), *Military psychology: Clinical and operational applications* (pp. 241–261). New York: Guilford.

Eban, K. (2007, July 17). *Rorschach and awe.* Retrieved September 5, 2007, from http://www.vanityfair.com/politics/features/2007/07/torture200707?printable=true&currentPage=all

Eban, K., Olson, B., & Goodman, A. (2007, July 30). *Rorschach and awe: As opposition grows over the APA's policy allowing psychologists to take part in military interrogations, vanity fair exposes how two psychologists shaped the CIA's torture methods.* Retrieved February 17, 2008, from http://www.democracynow.org/2007/7/30/rorschach_and_awe_as_opposition_grows

emptywheel. (2009, April 18). Khalid Sheikh Mohammed was waterboarded 183 times in one month. *Firedoglake.* Retrieved April 18, 2009, from http://emptywheel.firedoglake.com/

Ephron, D. (2008). The Biscuit Breaker: Psychologist Steven Reisner has embarked on a crusade to get his colleagues out of the business of interrogations [Electronic version]. *Newsweek.* Retrieved October 19, 2008, from http://www.newsweek.com/id/164497

Fink, S. (2009). Bush memos suggest abuse isn't torture if a doctor is there [Electronic version]. *ProPublica.* Retrieved April 17, 2009, from http://www.propublica.org/article/memos-suggest-abuse-isnt-torture-if-a-doctor-is-there-417

Flaherty, A. (2008, June 17). *Rumsfeld ignored torture warnings.* Retrieved June 17, 2008, from http://www.capitolhillblue.com/cont/node/8845

Frakt, D. J. R. (2009). Closing argument at Guantánamo: The torture of Mohammed Jawad [Electronic version]. *Harvard Human Rights Journal, 22,* 402–423. Retrieved May 6, 2009, from http://www.law.harvard.edu/news/2009/02/frakt-closing-argument.pdf

Freeze, C., & Akkad, O. E. (2008, July 9). Canada's secret documents on Khadr's treatment revealed [Electronic version]. *Globe and Mail.* Retrieved July 10, 2008, from http://www.theglobeandmail.com/servlet/story/RTGAM.20080709.wkhadr0709/BNStory/Front#

Glenn, D. (2009). Anthropologists adopt new language against secret research [Electronic version]. *Chronicle of Higher Education, 2009.* Retrieved February 19, 2009, from http://wikileaks.org/wiki/Anthropologists_Adopt_New_Language_Against_Secret_Research

Goodman, A. (2008, April 10). A torture debate among healers. Retrieved April 10, 2008, from http://www.alternet.org/story/81991/

Goodman, A., & Goodman, D. (2008). *Standing up to the madness: Ordinary heroes in extraordinary times.* New York: Hyperion.

Goodman, A., Koocher, G., Reisner, S., & Xenakis, S. (2006, June 16). *Calls grow within American Psychological Association for ban on participation in military interrogations: A debate.* Retrieved June 16, 2006, from http://www.democracynow.org/article.pl?sid=06/06/16/1355222

Goodman, D. (2008, March 1). The enablers: The psychology industry's long and shameful history with torture. Retrieved March 1, 2008, from http://www.motherjones.com/news/feature/2008/03/the-enablers.html

Greeenburg, J. C., Rosenberg, H. L., & Vogue, A. (2008a, April 11). *Bush aware of advisers' interrogation talks: President says he knew his senior advisers discussed tough interrogation methods.* Retrieved April 13, 2008, from http://abcnews.go.com/TheLaw/LawPolitics/Story?id=4635175&page=1

Greeenburg, J. C., Rosenberg, H. L., & Vogue, A. (2008b, April 13). *Sources: Top Bush advisors approved "enhanced interrogation."* Retrieved April 9, 2008, from http://abcnews.go.com/print?id=4583256

Greenberg, K. J., & Dratel, J. L. (Eds.). (2005). *The torture papers: The road to Abu Ghraib.* New York: Cambridge University Press.

Hersh, S. M. (2004a). *Chain of command: The road from 9/11 to Abu Ghraib.* New York: Harper Collins.

Hersh, S. M. (2004b, April 30). Torture at Abu Grhraib. Retrieved May 1, 2004, from http://www.newyorker.com/printable/?fact/040510fa_fact

Hersh, S. M. (2007, June 25). *The General's Report.* Retrieved June 16, 2007, from http://www.newyorker.com/reporting/2007/06/25/070625fa_fact_hersh?printable=true

Horton, S. (2007, November 18). *The psychologists and Gitmo.* Retrieved November 18, 2007, from http://harpers.org/archive/2007/11/hbc-90001695

Human Rights Watch. (2006, July). *"No blood, no foul": Soldiers' accounts of detainee abuse in Iraq.* Retrieved January 18, 2007, from http://hrw.org/reports/2006/us0706/

Human wrongs: Psychologists have no place assisting interrogations at places such as Guantánamo Bay. (2007, August 23). *Houston Chronicle.* Retrieved September 3, 2007, from http://www.chron.com/CDA/archives/archive.mpl?id=2007_4410052

International Committee of the Red Cross. (2004, February). *Report of the International Committee of the Red Cross (ICRC) on the treatment by the coalition forces of prisoners of war and other protected persons by the Geneva Conventions in Iraq during arrest, internment and interrogation.* Retrieved May 11, 2004, from http://www.antiwar.com/rep/red-cross-report.pdf

International Committee of the Red Cross. (2007, February). *ICRC report on the treatment of fourteen "high value detainees" in CIA custody.* Retrieved April 7, 2009, from http://www.nybooks.com/icrc-report.pdf

Iraq Family Health Survey Study Group. (2008). Violence-related mortality in Iraq from 2002 to 2006 [Electronic version]. *New England Journal of Medicine, 358,* 484–483. Retrieved January 15, 2009 from http://content.nejm.org/cgi/content/full/NEJMsa0707782

Jacobs, U. (2007a, December 5). *Farewell to the APA*. Retrieved April 9, 2008, from http://psychoanalystsopposewar.org/blog/2007/12/05/uwe-jacobs-a-major-moral-voice-in-psychology-leaves-the-apa/

Jacobs, U. (2007b, March 25). *Uwe Jacobs of Survivors International asks questions of Michael Gelles*. Retrieved March 26, 2007, from http://psychoanalystsopposewar.org/blog/2007/03/25/uwe-jacobs-of-survivors-international-asks-questions-of-michael-gelles/

Jaffer, J., & Singh, A. (2007). *Administration of torture: A documentary record from Washington to Abu Ghraib and beyond*. New York: Columbia University Press.

Järnefors, E. B. (2008). U.S. psychologists accused of participating in torture [in Swedish with English translation] [Electronic version]. *Swedish Journal of Psychology, 2008*, 8–13. Retrieved February 14, 2008, from http://www.psykologforbundet.se/www/sp/hemsida.nsf/objectsload/Tortyr_swe_eng/$file/Tortyr_swe_eng.pdf

Jaschik, S. (2007a, October 12). *Ethics rebellion in psychology*. Retrieved March 19, 2008, from http://insidehighered.com/news/2007/10/12/psych

Jaschik, S. (2007b, December 6). *Psychology protest grows*. Retrieved December 8, 2007, from http://insidehighered.com/news/2007/12/06/apa

Joint Task Force Guantánamo. (2003). *Camp Delta Standard Operating Procedures (SOP)*. Retrieved April 3, 2008, from http://wikileaks.org/wiki/Camp_Delta_Standard_Operating_Procedure

Kaye, J. S. (2009, January 7). *How the U.S. Army's Field Manual codified torture—and still does*. Retrieved January 7, 2009, from http://www.alternet.org/rights/117807/how_the_u.s._army%27s_field_manual_codified_torture_--_and_still_does/

Keiger, D. (2007). *The number*. Retrieved February 11, 2007, from http://www.jhu.edu/~jhumag/0207web/number.html

Kennedy, C. H., Malone, R. H., & Franks, M. J. (2009). Provision of mental health services at the detention hospital in Guantánamo Bay. *Psychological Services, 6*(1), 1–10.

Kiley, K. C. (2006, October 20). *Memorandum for Commanders, MEDCOM Major Subordinate Commands: Behavioral Science Consultation Policy*. Retrieved October 10, 2008, from http://wikileaks.org/wiki/Guantánmo_Bay_use_of_psychologists_for_interrogations_2006-2008

Lagouranis, T., & Mikaelian, A. (2007). *Fear up harsh: An Army interrogator's dark journey through Iraq*. New York: NAL Caliber.

Levin, C. (2008a, September 25). *Opening statement by Senator Carl Levin, Senate Armed Services Committee hearing on the authorization of SERE techniques for interrogations in Iraq: Part II of the committee's inquiry into the treatment of detainees in U.S. custody*. Retrieved September 25, 2008, from http://levin.senate.gov/newsroom/release.cfm?id=303575

Levin, C. (2008b, Jun 17). *Opening statement: Senate Armed Services Committee hearing: The origins of aggressive interrogation techniques*. Retrieved June 17, 2008, from http://levin.senate.gov/newsroom/release.cfm?id=299242

Levin, C. (2008c, December 11). *Statement of Senator Carl Levin on Senate Armed Services Committee report of its inquiry into the treatment of detainees in U.S. custody*. Retrieved December 11, 2008, from http://levin.senate.gov/newsroom/release.cfm?id=305734

Levine, A. (2007, January 8). *Collective unconscionable: How psychologists, the most liberal of professionals, abetted Bush's torture policy*. Retrieved January 8, 2007, from http://www.washingtonmonthly.com/features/2007/0701.levine.html

Lewis, N. A. (2004, November 30). *Red Cross finds detainee abuse in Guantánamo*. Retrieved November 30, 2004, from http://www.nytimes.com/2004/11/30/politics/30gitmo.html?oref=login&adxnnl=1&oref=login&adxnnlx=110183 1750-FbT+0bYfbchtnBvKJVZOBw&pagewanted=print&position=

Lewis, N. A. (2005, June 24). Interrogators cite doctors' aid at Guantánamo prison camp. Retrieved June 28, 2005, from http://www.nytimes.com/2005/06/24/politics/24gitmo.html?ei=5094&en=0bb87618febc3438&hp=&ex=1119585 600&partner=homepage&pagewanted=print

Lott, B. (2006). APA and the participation of psychologists in situations in which human rights are violated: Comment on "Psychologists and the use of torture in interrogations" [Electronic version]. *Analyses of Social Issues and Public Policy, 7*, 1–9. Retrieved June 19, 2007, from http://www.asap-spssi.org/pdf/0701Lott.pdf

Marks, J. H. (2005). Doctors of interrogation. *Hastings Center Report, 35*(4), 17–22.

Marks, J. H., & Bloche, M. G. (2008). The ethics of interrogation—The U.S. military's ongoing use of psychiatrists. *New England Journal of Medicine, 359*(11), 1090–1092.

Mayer, J. (2005, July 11). The experiment: Is the military devising new methods of interrogation at Guantánamo? *New Yorker*. Retrieved July 11, 2006, from http://www.newyorker.com/printables/fact/050711fa_fact4

Mayer, J. (2007, August 13). *The black sites*. Retrieved September 5, 2007, from http://www.newyorker.com/reporting/2007/08/13/070813fa_fact_mayer?printable=true

Mayer, J. (2008). *The dark side: The inside story of how the war on terror turned into a war on American ideals*. New York: Doubleday.

Mayer, J., & Goodman, A. (2007, August 8). *The black sites: A rare look inside the C.I.A.'s secret interrogation program*. Retrieved September 5, 2007, from http://www.democracynow.org/article.pl?sid=07/08/08/1338248

Mayer, J., & Sullivan, A. (2008, July 17). Mayer on Seligman. *Daily Dish*. Retrieved July 27, 2008, from http://andrewsullivan.theatlantic.com/the_daily_dish/2008/07/mayer-on-seligm.html

McCoy, A. W. (2004, September 9). *The hidden history of CIA torture: America's road to Abu Ghraib*. Retrieved September 9, 2004, from http://www.tomdispatch.com/index.mhtml?pid=1795

McCoy, A. W. (2006). *A question of torture: CIA interrogation, from the Cold War to the War on Terror*. New York: Metropolitan Books/Henry Holt and Co.

McCoy, A. W. (2009, June 7). *Confronting the CIA's mind maze*. Retrieved June 8, 2009, from http://tomdispatch.com/post/print/175080/Tomgram%253A%25 20%2520Alfred%2520McCoy%252C%2520Back%2520to%2520the%25 20Future%2520in%2520Torture%2520Policy

Mickum, B. (2007, January 8). *Guantánamo's lost souls*. Retrieved January 9, 2007, from http://commentisfree.guardian.co.uk/brent_mickum/2007/01/post_885.html

Miles, S. H. (2009). *Oath betrayed: America's torture doctors* (2nd ed.). Berkeley, CA: University of California Press.

Miller, G. D. (2003). *Assessment of DoD counterterrorism interrogation and detention operations in Iraq.* Retrieved November 19, 2008, from http://www1.umn.edu/humanrts/OathBetrayed/Taguba%20Annex%2020.pdf

Morlin, B. (2007, August 12). *Expert has stake in cryptic local firm.* Retrieved September 5, 2007, from http://www.spokesmanreview.com/tools/story_pf.asp?ID =204358

Office of the Inspector General of the Department of Defense. (2006, August 25). *Review of DoD-directed investigations of detainee abuse.* Retrieved June 1, 2007, from http://www.fas.org/irp/agency/dod/abuse.pdf

Office of the Inspector General of the Department of Justice. (2008, May 20). *A review of the FBI's involvement in and observations of detainee interrogations in Guantánamo Bay, Afghanistan, and Iraq.* Retrieved May 20, 2008, from http://graphics8.nytimes.com/packages/pdf/washington/20080521_DETAIN_report.pdf

Office of the United Nations High Commissioner for Human Rights. (1984). *Convention against Torture and Other Cruel, Inhuman or Degrading Treatment or Punishment.* Retrieved August 23, 2006, from http://www.unhchr.ch/html/menu3/b/h_cat39.htm

Olson, B. (2006). Human rights and the ethics of psychologist-involved interrogations. *Community Psychologist, 39*(4), 55–57.

Olson, B., & Soldz, S. (2007). Positive illusions and the necessity of a bright line forbidding psychologist involvement in detainee interrogations [Electronic version]. *Analyses of Social Issues and Public Policy, 7*, 1–10. Retrieved June 28 from http://www.asap-spssi.org/default.htm

Olson, B., Soldz, S., & Davis, M. (2008). The ethics of interrogation and the American Psychological Association: A critique of policy and process [Electronic version]. *Philosophy, Ethics, and Humanities in Medicine, 3.* Retrieved February 14, 2008, from http://www.peh-med.com/content/3/1/3

Opinion Research Business (ORB). (2008, January). *Update on Iraqi casualty data.* Retrieved February 18, 2008, from http://www.opinion.co.uk/Newsroom_details.aspx?NewsId=88

ORCON [Authoring agency classified by Originator Control]. (2003, January 11). *Secret Orcon interrogation log detainee 063.* Retrieved September 1, 2006, from http://www.time.com/time/2006/log/log.pdf

Otterman, M. (2007). *American torture: From the Cold War to Abu Ghraib and beyond.* Ann Arbor, MI: Pluto Press.

Physicians for Human Rights. (2008, June 18). *Broken laws, broken lives: Medical evidence of torture by US personnel and its impact.* Retrieved June 18, 2008, from http://brokenlives.info/?dl_id=5

Physicians for Human Rights. (2009, May 5). *PHR calls for investigation of American Psychological Association's ties to Pentagon.* Retrieved May 5, 2009, from http://www.prweb.com/releases/ethics/national_security/prweb2392144.htm

Pipher, M. (2007). *Why I've returned my award to the American Psychological Association—Because it sanctions torture.* Retrieved August 24, 2007, from http://www.opednews.com/articles/opedne_mary_pip_070824_why_i_ve_returned_my.htm

Pope, K. S. (2008, February 10). *Why I resigned from the American Psychological Association*. Retrieved February 10, 2008, from http://kspope.com/apa/index.php

Pope, K. S., & Gutheil, T. G. (2008). The American Psychological Association and Detainee Interrogations: Unanswered Questions [Electronic Version]. *Psychiatric Times, 25*. Retrieved October 10, 2008 from http://www.psychiatrictimes.com/display/article/10168/1166964.

Pope, K. S., & Gutheil, T. G. (2009). Psychologists abandon the Nuremberg ethic: Concerns for detainee interrogations. *International Journal of Law and Psychiatry, 32*, 161–166.

Price, D. (2008). The leaky ship of human terrain systems [Electronic version]. *CounterPunch*. Retrieved December 13, 2008, from http://counterpunch.org/price12122008.html

Psychological Ethics and National Security Task Force. (2009, May 5). *Email messages from the listserv of the American Psychological Association's Presidential Task Force on Psychological Ethics and National Security: April 22, 2005–June 26, 2006*. Retrieved May 5, 2009, from http://s3.amazonaws.com/propublica/assets/docs/pens_listserv.pdf

Psychologists and torture [Editorial]. (2008, August 30). *Boston Globe*. Retrieved August 30, 2008, from http://www.boston.com/bostonglobe/editorial_opinion/editorials/articles/2008/08/30/psychologists_and_torture/

Psychologists for an Ethical APA. (2008). *Psychologists for an Ethical APA*. Retrieved December 29, 2008, from http://ethicalapa.com/

Psychologists for an Ethical APA. (2009, May 2). *Psychologists for an Ethical APA calls for American Psychological Association investigation and resignations*. Retrieved May 17, 2009, from http://psychoanalystsopposewar.org/blog/2009/05/02/psychologists-for-an-ethical-apa-calls-for-american-psychological-association-investigation-and-resigations/

Psychologists for Social Responsibility. (2009a). *Psychologists for Social Responsibility*. Retrieved January 19, 2009, from http://psysr.org/torture

Psychologists for Social Responsibility. (2009b). PsySR Statement Urges Independent Torture Commission. Retrieved May 11, 2009, from http://www.psysr.org/about/pubs_resources/torture_commission.php

Reisner, S. (2007). *Ethical concerns about psychologists' participation in interrogation of detainees* [Electronic version]. *Psychologist-Psychoanalyst, 27*, 23–29. Retrieved April 25, 2007, from http://www.division39.org/pdfs/PPJan07web.pdf

Rejali, D. M. (2007). *Torture and democracy*. Princeton, NJ: Princeton University Press.

Roberts, L., Lafta, R., Garfield, R., Khudhairi, J., & Burnham, G. (2004, October 30). Mortality before and after the 2003 invasion of Iraq: cluster sample survey. Volume 364, Number 9445, Early Online Edition. Retrieved October 29, 2004, from http://www.zmag.org/lancet.pdf.

Sands, P. (2008). *Torture team: Rumsfeld's memo and the betrayal of American values* (1st ed.). New York: Palgrave Macmillan.

Schmitt, E., & Marshall, C. (2006, March 18). *Task Force 6-26: Before and after Abu Ghraib, a U.S. unit abused detainees*. Retrieved March 18, 2006, from http://www.nytimes.com/2006/03/19/international/middleeast/19abuse.html?ei=5088&en=e8755a4b031b64a1&ex=1300424400&partner=rssnyt&emc=rss&pagewanted=print

Seligman, M. (2008, July 14). *Former APA President Martin Seligman denies involvement in developing CIA tactics.* Retrieved July 25, 2008, from http://psychoanalystsopposewar.org/blog/2008/07/14/former-apa-president-martin-seligman-denies-involvement-in-developing-cia-tactics/

Senate Armed Services Committee. (2008a, June 17). *Documents released at the Senate Armed Services Committee hearing "The Origins of Aggressive Interrogation Techniques."* Retrieved June 17, 2008, from http://levin.senate.gov/newsroom/supporting/2008/Documents.SASC.061708.pdf

Senate Armed Services Committee. (2008b, December 11). *Senate Armed Services Committee inquiry into the treatment of detainees in U.S. custody: Executive summary and conclusions.* Retrieved December 11, 2008, from http://levin.senate.gov/newsroom/supporting/2008/Detainees.121108.pdf

Senate Armed Services Committee. (2008c). *Transcript of hearing on the origins of aggressive interrogation techniques: Panel I.* Federal News Service.

Senate Armed Services Committee. (2009, April 21). *Inquiry into the treatment of detainees in U.S. custody.* Retrieved April 21, 2009, from http://armed-services.senate.gov/Publications/Detainee%20Report%20Final_April%2022%202009.pdf

Shinn, M. (2006, April). Psychologists and Torture: APA, PENS, SPSSI, and DSJ. *Forward*, 1–2, 20.

Shinn, M. (2007, October 7). *Noted psychologist Beth Shinn resigns from American Psychological Association.* Retrieved April 7, 2008, from http://psychoanalystsopposewar.org/blog/2007/10/07/noted-psychologist-beth-shinn-resigns-from-american-psychological-association/

Slevin, P., & Stephens, J. (2004, June 10). *Detainees' medical files shared: Guantánamo interrogators' access criticized.* Retrieved September 20, 2007, from http://www.washingtonpost.com/ac2/wp-dyn/A29649-2004Jun9?language=printer

Society for the Study of Peace Conflict and Violence: Peace Psychology Division 48. (2005). *American Psychological Association Presidential Task Force on Psychological Ethics and National Security: 2003 members' biographical statements.* Retrieved April 9, 2007, from http://www.webster.edu/peacepsychology/tfpens.html

Soldz, S. (2004, November 5). *100,000 Iraqis dead: Should we believe it?* Retrieved February 17, 2006, from http://www.zmag.org/content/showarticle.cfm?SectionID=15&ItemID=6565

Soldz, S. (2006a, December 14). *Abusive interrogations: A defining difference between psychiatrists and psychologists.* Retrieved September 20, 2007, from http://www.dissidentvoice.org/Dec06/Soldz14.htm

Soldz, S. (2006b). *Protecting the torturers: Bad faith and distortions from the American Psychological Association.* Retrieved January 19, 2007, from http://www.counterpunch.org/soldz09062006.html

Soldz, S. (2006c, August 1). *Psychologists, Guantánamo and torture: A profession struggles to save its soul.* Retrieved January 19, 2007, from http://www.counterpunch.org/soldz08012006.html

Soldz, S. (2006d, September 13). *What the US reservations to UN Convention on Torture really means?* Retrieved January 10, 2008, from http://psychoanalystsopposewar.org/blog/2006/09/13/what-the-us-reservations-to-un-convention-on-torture-really-means/

Soldz, S. (2007a, August 22). *APA, torture, and the CIA*. Retrieved August 22, 2007, from http://psychoanalystsopposewar.org/blog/2007/08/22/apa-torture-and-the-cia/
Soldz, S. (2007b, August 25). *Mary Pipher returns award to American Psychological Association to protest torture stance*. Retrieved August 25, 2007, from http://www.zmag.org/content/showarticle.cfm?ItemID=13625
Soldz, S. (2008a, May 27). *American Psychological Association supports psychologist engagement in Bush regime interrogations: A critique of Stephen Behnke's letter to the ACLU*. Retrieved May 27, 2008, from http://www.counterpunch.org/soldz05272008.html
Soldz, S. (2008b, August 10). Ending the psychological mind games on detainees. *Boston Globe*. Retrieved, from http://www.boston.com/bostonglobe/editorial_opinion/oped/articles/2008/08/10/ending_the_psychological_mind_games_on_detainees/
Soldz, S. (2008c). Healers or interrogators: Psychology and the United States torture regime. *Psychoanalytic Dialogues, 18*, 592–613.
Soldz, S. (2008d). "That psychologists have prevented abuse against detainees is a fantasy" [Electronic Version]. *Psykologernastidning [Swedish Journal of Psychology], 2008*, 12–13. Retrieved February 15, 2008 from http://www.psykologforbundet.se/www/sp/hemsida.nsf/objectsload/Tortyr_swe_eng/$file/Tortyr_swe_eng.pdf.
Soldz, S. (2008e, June 26). The Torture Trainers and the American Psychological Association. Retrieved June 26, 2008, from http://www.zcommunications.org/znet/viewArticle/18002.
Soldz, S. (2009a). Closing eyes to atrocities: U.S. psychologists, detainee interrogations, and response of the American Psychological Association. In M. Roseman & R. Goodman (Eds.), *Interrogations, forced feedings, and the role of health professionals: New perspectives on international human rights, humanitarian law and ethics* (pp. 103–142). Cambridge, MA: Harvard Human Rights Program at Harvard Law School.
Soldz, S. (2009b). "Sleep deprivation": Euphemism and CIA torture of choice [Electronic version]. *OpEdNews*. Retrieved May 10, 2009, from http://www.opednews.com/articles/-Sleep-deprivation--Euph-by-Stephen-Soldz-090510-130.html
Soldz, S. (2009c). Psychologists and United States coercive interrogations. In J. M. Arrigo, R. Bennet & S. Soldz (Eds.), *Casebook on psychological ethics in national security settings*. Manuscript in preparation.
Soldz, S., & Assange, J. (2007, November 17). *Guantánamo document confirms psychological torture*. Retrieved November 18, 2007, from http://wikileaks.org/wiki/Guantánamo_document_confirms_psychological_torture
Soldz, S., & Barahnona, R. A. (2007, April 8). *Psicólogos, Guantánamo y tortura*. Retrieved April 8, 2007, from http://www.nacion.com/ln_ee/2007/abril/08/opinion1054680.html
Soldz, S., & Olson, B. (2008a, September 24). *Psychologists reject the dark side: American Psychological Association members reject participation in Bush detention centers*. Retrieved September 24, 2008, from http://www.zmag.org/znet/viewArticle/18906
Soldz, S., & Olson, B. (2008b). Psychologists, detainee interrogations, and torture: Varying perspectives on nonparticipation. In A. Ojeda (Ed.), *The trauma of psychological torture* (pp. 70–91). Westport, CT: Praeger.

Soldz, S., & Olson, B. (2008c, March 2). *A reaction to the APA vote on sealing up key loopholes in the 2007 resolution on interrogations*. Retrieved March 2, 2008, from http://www.zcommunications.org/znet/viewArticle/16711

Soldz, S., Olson, B., Reisner, S., Arrigo, J. M., & Welch, B. (2008, July 22). *Torture after dark: Torture and the strategic helplessness of the American Psychological Association*. Retrieved July 22, 2008, from http://www.counterpunch.org/soldz07232008.html

Soldz, S., Reisner, S., & Olson, B. (2007a, June 7). *A Q&A on psychologists and torture: The Pentagon's IG report contradicts what the APA has said about the involvement of psychologists in abusive interrogations*. Retrieved June 10, 2007, from http://www.counterpunch.org/soldz06072007.html

Soldz, S., Reisner, S., & Olson, B. (2007b, June 23). *Torture, psychologists and Colonel James: An open letter to Sharon Brehm, President of the American Psychological Association*. Retrieved February 10, 2008, from http://www.counterpunch.org/soldz06232007.html

Steele, K. D., & Morlin, B. (2007). Senate probe focuses on Spokane men [Electronic version]. *Spokesman Review*. Retrieved September 8, 2008, from http://www.spokesmanreview.com/tools/story_breakingnews_pf.asp?ID=10496

Stein, J. (2008, April 4). *Evidence grows of drug use on detainees*. Retrieved April 4, 2008, from http://public.cq.com/docs/hs/hsnews110-000002697912.html

Sullivan, A. (2008, September 19). *Know hope*. Retrieved September 19, 2008, from http://andrewsullivan.theatlantic.com/the_daily_dish/2008/09/know-hope.html

Summers, F. (2008). Making sense of the APA: A history of the relationship between psychology and the military. *Psychoanalytic Dialogues*, 18(5), 614–637.

Taguba, A. M. (2004a, May 4). *Article 15-6 Investigation of the 800th Military Police Brigade: Annex 40*. Retrieved December 10, 2008, from http://www1.umn.edu/humanrts/OathBetrayed/Taguba%20Annex%2040.pdf

Taguba, A. M. (2004b, May 4). *Article 15-6 Investigation of the 800th Military Police Brigade: Annex 46*. Retrieved December 10, 2008, from http://www1.umn.edu/humanrts/OathBetrayed/Taguba%20Annex%2046.pdf

United States Department of the Army. (2006, September 6). *Human intelligence collector operations*. Retrieved September 17, 2006, from http://www.army.mil/references/FM2-22.3.pdf

Valtin, (2006, June 7). Civil war in psychology over abetting U.S. torture. Retrieved June 14, 2006, from http://theyrereal.dailykos.com/storyonly/2006/6/8/0958/85979

Warrick, J. (2008, April 22). Detainees allege being drugged, questioned. Retrieved April 22, 2008, from http://www.washingtonpost.com/wp-dyn/content/article/2008/04/21/AR2008042103399.html?hpid=topnews

Warrick, J., & Finn, P. (2009, April 18). Psychologists Helped Guide Interrogations [Electronic version]. *Washington Post*. Retrieved April 18, 2009 from http://www.washingtonpost.com/wp-dyn/content/article/2009/04/17/AR2009041703690.html?hpid=moreheadlines

Welch, B. (2008, July 28). *Torture, political manipulation and the American Psychological Association: Why did they do it?* Retrieved July 28, 2008, from http://www.counterpunch.org/welch07282008.html

White, J. (2008, August 8). Tactic used after it was banned [Electronic version]. *Washington Post*. Retrieved January 26, 2009, from http://www.washingtonpost.com/wp-dyn/content/article/2008/08/07/AR2008080703004_pf.html

withholdapadues.com. (2008). *Welcome to the Withholdapadues website*. Retrieved April 25, 2008, from http://withholdapadues.com/

Woodward, B. (2009, January 14). Detainee tortured, says U.S. official. *Washington Post*. Retrieved January 14, 2009, from http://www.washingtonpost.com/wp-dyn/content/article/2009/01/13/AR2009011303372_pf.html

Woolf, L. M. (2007, September 1). *A sad day for psychologists: A major blow against human rights*. Retrieved September 1, 2007, from http://www.counterpunch.org/woolf09012007.html

Zagorin, A. (2005, February 7). *The Abu Ghraib scandal you don't know*. Retrieved December 10, 2008, from http://www.time.com/time/magazine/article/0,9171,1025139,00.html?iid=sphere-inline-sidebar

Zeller, S. (2007, September 17). *Torture issue ties up psychologists association*. Retrieved September 17, 2007, from http://public.cq.com/docs/cqw/weeklyreport110-000002585116.html

Chapter 6

From resistance to *resistance*
A narrative of psychoanalytic activism

Steven Reisner

> Suddenly, no, at last, at long last, I couldn't anymore, I couldn't go on. Someone said, you can't stay here. I couldn't stay there and I couldn't go on.
>
> Samuel Beckett
> *Stories and Texts for Nothing 1* (1967)

I plan to tell two intertwined stories. One is personal: the story of how I suddenly—no, at last, at long last—couldn't go on. I couldn't go on as simply an observer of the history of government abuses of detainees in the war on terror, but found myself unwittingly applying psychoanalytic skills to influence that history, and how I ultimately found that I had to act. It is not my story alone, but the story of how a small group of psychologists and psychoanalysts uncovered and changed the American Psychological Association's complicity in our country's military and intelligence torture programs, and perhaps, in the process interfered with its continued execution.

The second story, simultaneous with the first, is the story of the Bush administration's program of torture and abuse of detainees, and of the essential role psychologists played in that program.

Telling the two stories together and discovering what I have to tell continues to surprise me. I still find it unfathomable that pulling on the threads of the role of psychologists unraveled the Bush administration's covert torture program—its development, its execution, its legal justifications, and its dissemination—and unfathomable, too, that all of these involved psychologists and psychology. But the reason we pulled these threads in the first place is because, as it happens, the story we were presented with—from the Bush administration and from the American Psychological Association—about the role of psychologists and psychology in national security interrogations didn't make sense. It wasn't that the story was incoherent or inconsistent. It wasn't that it was ethically suspect, even though that played a role. It was that, for me at least, the story aroused skepticism and curiosity *as a psychoanalyst*. I found myself listening *clinically* to the material that the government and the APA were presenting, much as I listen to my patients' stories. I found myself listening for the hidden story—the story that was being obscured, precisely as the surface narrative was being perfected.

Of course, in one very essential way, listening to political or social material is different than clinical listening. The neutral stance whose aim is to mobilize and resolve resistance in the patient is out of place in political listening because the analyst is part of the events of the world. The resistance that is mobilized through psychoanalytic political listening is a very different form of resistance—more akin to the French resistance. As Badiou (2005) has put it: "When all is said and done, all resistance is a rupture in thought through the declaration of what the situation is, and the foundation of a practical possibility opened up through this declaration" (p. 8).

An implication in Badiou's statement is that psychoanalysis—the aim of which is to foster "a rupture in thought through the declaration of what the situation is"—makes possible, and perhaps makes necessary, the opening up of new practical possibilities: political resistance and activism. Badiou (2005) continues:

> This does not amount to believing that it is the risk, very serious indeed, which prevents a good many from resisting; it is, on the contrary, the non-thinking of the situation that prevents the risk or the examination of possibilities. Not to resist is not to think; not to think is not *to risk risking*. (p. 8, emphasis in original)

One of the two stories, then, is the story of how a small group of psychoanalysts used psychoanalytic methods to overcome resistances and join the resistance.

The other is the story that unfolded as our group began to research the backgrounds of the psychologists whom the APA leadership selected to decide its policy on psychological ethics and national security. We discovered that some of these psychologists and their colleagues were present at Guantánamo, Bagram, and Central Intelligence Agency (CIA) "black sites" when and where torture took place under their commands. Others organized "brainstorming" sessions on counterterrorism and interrogation techniques between psychologists and military and intelligence counterterrorist operatives, including, it turns out, torturers.

Following these links one degree of separation further, the threads unravel into intriguing and frightening associations—one former APA president was on the board of directors of Mitchell-Jessen Associates, the CIA's psychologist-consultants in torture (Marlin, 2007); another former APA president invited one of the two originators of the torture program to his home for a brainstorming session (Shane, 2009), and later gave a lecture on domination techniques to these and other CIA interrogations operatives (Mayer, 2008, p. 164). The psychological procedures that apparently came out of these sessions, combined with other psychological techniques developed elsewhere,* were approved by President Bush (Greenberg, Rosenberg, & De Vogue, 2008b) and directly overseen and guided by a White House team consisting of the vice president, the secretary of defense, the director of the CIA, the attorney general, the director of the National Security Agency, and the secretary of state for use in the U.S. torture program (Greenberg, Rosenberg, & De Vogue, 2008a). In other words, the story leads to a conspiracy at the highest levels of government to use psychologists and psychological methods to torture prisoners, implicating an entire administration in war crimes.

The extent to which these two stories are intertwined is the stuff of spy novels—not the usual ambit of a psychoanalyst. In the end, our small group of psychologists (calling itself the Coalition for an Ethical Psychology) played a small but significant role, not only in changing the policy of the APA and helping to restore its ethical obligations to "do no harm," but, more important and surprising, in exposing the government's psychological torture program. We were part of a network that included other psychologist-activists, investigative journalists, and a handful of dedicated human rights activists and investigators (particularly one, our "deep throat," who always managed to point us, the reporters, and eventually government investigative bodies in the right direction). Our network doggedly kept this issue on the Internet, in the news, and on the radar screen of the Senate Armed Services Committee.

Along the way, I came to believe that activism (personal and political) constitutes an important if underappreciated measure of the success of a psychoanalytic process. If psychoanalysis promotes rupture in thought, it promotes, too, the capacity to "risk risking."

APRIL 2007: THE HOLLYWOOD MOMENT

If this story was being presented as part of a thriller, it would begin in April 2007, with the following scene. I received a call from our deep throat (I'll

---

* Mitchell-Jessen Associates was responsible for the first reverse-engineered use of techniques taken from the military's SERE (Survival, Evasion, Resistance, Escape) program. SERE training is a rigorous training process wherein our own soldiers undergo torture techniques as a kind of "inoculation" program—preparing them to resist if they were to be captured and tortured by an enemy that doesn't observe the Geneva conventions. When our government decided to "take the gloves off" in interrogations, our nation became a nation that didn't observe the Geneva conventions. The very torture techniques that we aimed to protect our own soldiers and Special Forces from became the basis for our own interrogation procedures.

call him Morty). Morty was coming to New York to meet with another human rights investigator who had obtained a copy of a document each of them knew about but had never seen. It was called the "Brunswick memo" and laid out standard operating procedures for transforming Survival, Evasion, Resistance, and Escape (SERE) techniques into interrogation techniques at Guantánamo. Up until this point, the government had claimed that any abusive interrogation activities at Abu Ghraib and at Guantánamo were carried out by a few "bad apples." APA psychologists, like Col. Larry James, who had been the chief of psychological interrogation activity at Guantánamo, had made the case that such activity was exceptional and had been cleaned up.* According to James (2005), "since Jan. 2003, where ever we have had psychologists no abuses have been reported." The Brunswick memo revealed a very different story.

Brunswick was a classified memorandum apparently put together by a psychologist from Navy SERE school in Brunswick, Maine. The memorandum described SERE techniques in use at Brunswick that could be reverse-engineered for use at Guantánamo: "The premise behind this is that ... these tactics and techniques are used at SERE school to 'break' SERE detainees. The same tactics and techniques can be used to break real detainees during interrogation operations" (Moss, 2002). This "GTMO 'SERE' standard operating procedure," dated 10 December 2002, followed a request, made two weeks earlier, by William Haynes, general counsel, to Defense Secretary Rumsfeld recommending that the "SECDEF approve the USSOUTHCOM Commanders' use of those counter-resistance techniques ... during the interrogation of prisoners at Guantánamo Bay" (Haynes, 2002).† Accompanying e-mails, from as late as the end of December, described conversations between a psychologist at Brunswick and one at Guantánamo, addressing the great care necessary to institute these procedures without "going ... too far." The procedures detailed in the Brunswick standard operating procedure (SOP) included slapping, stress positions, hooding, manhandling, and walling.‡

Morty and I, and our three colleagues, had spent the previous 6 months trying to interest government and the press in the story that we had uncovered of the intimate role of psychologists, with the support of the APA, in abusive interrogations. We had decided the previous September to

---

* In his book *Fixing Hell* (2008), James takes personal credit for insuring that no abusive techniques were being used and that he personally instituted protocols to guarantee that the abuse stopped: "My role was to teach rapport and relationship-building approaches between the detainee and the interrogator without the abuse" (p. 55).
† The copy of the memo released by the Senate Armed Services Committee bears Rumsfeld's signature, giving his approval.
‡ "Ensure only the broad part of the shoulders contact the surface of the wall. Grip the detainee's clothing firmly enough so the collar acts as a restrictive constraint to preclude the detainee's head from contacting the wall does this. If the detainee's head inadvertently touches the wall, walling will be ceased immediately" (Moss, 2002).

offer one reporter, writing for *Vanity Fair*, exclusive rights to the story and funneled everything through her. But the more we uncovered, the further up the chain of command the story seemed to go. We had learned about two psychologists, former SERE instructors for the Army now under contract with the CIA, who had tortured Abu Zubaydah at a CIA black site in Thailand using SERE techniques. The *Vanity Fair* reporter, using our contacts and others, had verified the information and was going to publish the story.

I suggested we meet in Chinatown at the Oriental Garden, a Cantonese seafood restaurant. From New York, Morty would be heading to Washington to confer with the Senate Armed Services Committee, which had begun an investigation into abusive interrogation practices; hopefully, he would find a way to share the information we were getting with them. He had stopped in New York to arrange a complicated passing of information.

Morty had a casual, Midwestern gait and look. He spoke in an easy-going manner, creating aphorisms and analogies that were simultaneously ridiculous and spot on. "Look, if the American program of torture is malaria, psychologists are the tsetse flies," he explained to me as we entered the restaurant. "They aren't the ones responsible for the disease—that's a virus emanating from the top echelons of the system. But they are the carriers; they're how it's spread."

The waiter came over and before I had a chance to say a word, Morty began speaking rapidly in Mandarin and ordered, off the menu, for both of us. Morty explained to me that if the story of the two contractors, from the Department of Defense (DoD) working for the CIA, was able to see the light of day, it might not bring down the APA leadership, but it could bring down the government. He explained what was, to me, a new concept: *jointness*. If both the CIA and the Department of Defense were sharing the same two psychologists, and these psychologists were using heretofore illegal interrogation techniques, then the approval for the use of these techniques must have come from a source higher up the chain of command than either the top echelons of the CIA or the DoD. A joint operation such as this, violating international and domestic law, would require executive branch approval. That meant that, were this to be made public, some major players in the Bush administration would be subject to prosecution.*

Morty went straight from the restaurant to Penn Station to catch a train for his meeting with the Senate Armed Services Committee. As he was

---

* As reported in the recently released CIA Inspector General's report: In accordance with the [Torture] Convention, the United States criminalized acts of torture in 18 U.S.C. 2340A(a), which provides as follows: "Whoever outside the United States commits or attempts to commit torture shall be fined under this title or imprisoned not more than 20 years, or both, and if death results to any person from conduct prohibited by this subsection, shall be punished by death or imprisoned for any term of years or for life."

leaving, he gave me a package. In it was a t-shirt of Dustin Hoffman and Robert Redford from *All the President's Men*.

2003: THE BEGINNINGS

It began for me in March and April of 2003. I had begun to comb the Internet for different perspectives of the news, because it had become clear since the attack on the World Trade Center and since the government's hijacking of public sentiment after that attack, that the American press, apparently traumatized by its own coverage of the World Trade Center attack, was reporting the war news superficially and in a manner that was heavily influenced by the government. To get some sense of what was really going on, I would habitually read multiple versions of the same story, not only from the American press, but, thanks to the newly introduced Google News and the Internet, from the European, Asian, and Arab press as well.

In March, I read a story in the British Press about the capture of Khalid Shaikh Mohammed, the alleged mastermind of September 11. Mentioned in the article was the fact that months earlier the CIA had kidnapped Mohammed's two young sons, ages 7 and 9, and continued to keep them in custody to pressure their father. According to the report, the CIA interrogators in charge of the boys didn't see anything wrong with kidnapping children to achieve their ends. The article continued with the kind of statement I could imagine from a Pinochet henchman, but it was a CIA "official" who stated, "His sons are important to him. The promise of their release and their return to Pakistan may be the psychological lever we need to break him" (Craig, 2003).

The American press presented the story from a wildly different perspective than the European press, reflecting a kind of journalistic "machismo" analogous in tone to that of the new journalists of the '60s. Except that where Hunter S. Thompson, Norman Mailer, and others were part of an anti-establishment journalistic rebel-machismo, these "war on terror" reporters piggybacked on the government-sanctioned bravado of national security operations and abuse of power. Like their neoconservative counterparts in government and in think tanks, these journalists had learned to co-opt the 1960s-style radical practice in support of neoconservative ideology:

> Military interrogators say their prisoners can be lied to, screamed at ... stripped, forcibly shaved and deprived of religious items and toiletries ... [A]s long as the pain and suffering aren't "severe," it's permissible to use physical force and to cause "discomfort," as some U.S. interrogators euphemistically put it. Among the techniques: making captives wear black hoods, forcing them to stand in painful "stress positions"

> for a long time and subjecting them to interrogation sessions lasting as long as 20 hours. U.S. officials overseeing interrogations of captured al Qaeda forces at Bagram and Guantánamo Bay Naval Base in Cuba can even authorize "a little bit of smacky-face," a U.S. intelligence official says. "Some al Qaeda just need some extra encouragement," the official says ...
>
> Initially, interrogators will aim to disorient Mr. Mohammed. "You deprive him of food, water and sleep. You make morning night, and you make hot cold ..." U.S. authorities have an additional inducement to make Mr. Mohammed talk, even if he shares the suicidal commitment of the Sept. 11 hijackers: The Americans have access to two of his elementary-school-age children. (Bravin & Fields, 2003)

In many articles written at the time in the mainstream American press, there was not only the tendency to accept military and intelligence propaganda at face value, and not only the tendency to celebrate the machismo-revenge attitude offered by interrogators and spokespeople, but there was simultaneously a hint of something even more insidious: the role of health professionals in creating a strategic language for describing the effects of imprisonment and torture. The military directed reporters to pay attention, not to international standards of human rights or the Geneva Conventions to describe conditions of confinement and their effects, but to the psychological conditions and symptoms of the detainees. Health professionals appeared to be colluding with the military to describe imprisonment as a kind of mental health treatment, complete with group therapy sessions to treat depression. In this context, suicide attempts among juvenile prisoners at Guantánamo, for example, were described as stemming from previous conditions and not a result of confinement conditions.

> Cmdr. Brian Grady, the staff psychiatrist at the camp's medical facility, said in a recent interview that most prisoners suffering from depression brought their symptoms with them to Cuba ... Officials at Guantánamo have generally dismissed the notion that the confinement and uncertainty about the future are specifically to blame. "I would not particularly say these circumstances are a factor," Commander Grady said. (Gall & Lewis, 2003)

I found myself reading such news in much the same way as I listened to my patients' associations; I flagged certain claims that seemed to indicate something hidden, some "unthought known" and made a mental note of it.

I suspected that health professionals had begun to use psychological knowledge, including diagnoses and treatment, as a military tactic, even while presenting them as healing processes. I suspected a cynical misuse of diagnostic jargon, and I intuited that there was a sinister purpose embedded in the peculiarly American denial of responsibility contained in the phrase "preexisting condition." Perhaps, just as psychologists and psychiatrists on the payroll of insurance companies used this phrase to deny needed treatment, the psychologists and psychiatrists at Guantánamo were denying the effects of torture.

2004–2005: BSCTS AND PENS

It was more than a year later that the *New York Times* published segments from a leaked report submitted to the White House by the International Committee of the Red Cross (ICRC). Although it had been widely reported in the press that our government was torturing detainees at Guantánamo, at Bagram Air Force Base in Afghanistan, and at secret black sites around the world, the ICRC report revealed a striking, and in retrospect, essential part of the record of the United States and torture:

> The [ICRC] team of humanitarian workers, which included experienced medical personnel, also asserted that some doctors and other medical workers at Guantánamo were participating in planning for interrogations, in what the report called "a flagrant violation of medical ethics." Doctors and medical personnel conveyed information about prisoners' mental health and vulnerabilities to interrogators, the report said, sometimes directly, but usually through a group called the Behavioral Science Consultation Team, or BSCT. The team, known informally as Biscuit, is composed of psychologists and psychological workers who advise the interrogators, the report said. (Lewis, 2004)

There was a huge uproar in the medical community in response to this obvious breach of medical ethics. The fact that health professionals were peering into the medical records of detainees in order to exploit vulnerabilities seemed beyond the pale to most professional organizations. But hidden in the exposé was not only the first public mention of the BSCTs, but a vague reference to bringing in "outside doctors": "The report said that sometimes 'outside doctors' are brought in to help interrogators plan their strategy of interviewing detainees" (Lewis, 2004).

In January 2005, the *New York Times* published a follow-up article in which interrogators from Guantánamo confirmed the Red Cross report on the role of health professionals in interrogations:

> The interrogators also discussed another factor in the Red Cross report, the use of a Behavioral Science Consultation Team, known as Biscuit,

comprising a psychologist or psychiatrist and psychiatric workers. The team was used to suggest ways to make prisoners more cooperative in interrogations ... "They were supposed to help us break them down," one said. (Lewis, 2005a)

Rather quickly, physicians and medical associations around the world condemned any role for medical professionals in interrogations in general, and specifically condemned the use of confidential medical records as part of detainee exploitation. But the American Psychological Association remained silent on the issue.

To be honest, disturbing as this news was, it didn't interfere with my going on. I didn't really expect much from the American Psychological Association. I was a member for two reasons: First, because it enabled my involvement with the Division of Psychoanalysis (39). And second, it provided malpractice insurance. It never seemed to me that Division 39 was really a part of the national organization. Rather, Division 39 seemed to me to be in relation to the APA in much the same way that Spaulding Gray had described the relationship of Manhattan to the rest of the United States: "I moved to Manhattan [because] I wanted to move to an island off the coast of America" (Gray, 1985). For me, psychoanalysis was an island off the coast of the APA.

I found the literature I received from the APA embarrassing. The *Monitor* invariably presented psychology as an arm of the American government and industry—trumpeting the role of psychologists as facilitating symptom reduction and restoration of productivity. On the other hand, as an analyst, I saw my role as working to promote change by mobilizing my patients to face the immanent meanings of so-called symptoms and promoting their courage to live richer, more effective, and loving lives. I saw the APA's emphasis on symptom, on the other hand, as a commercialization and simultaneously a depoliticization of the healing arts. The *Monitor's* coverage of military psychology was uniformly supportive of the U.S. military and government position, and was similarly symptom focused and depoliticized.*

---

* For example, the *Monitor's* article on the Joint Personnel Recovery Agency (JPRA) and repatriation (http://www.apa.org/monitor/feb04/helping.html) focused primarily on psychologists' role in "decompression" and "normalization" of repatriated American detainees who had been captured and tortured in places such as Vietnam. In the article, SERE training was described as an "inoculation—providing folks with the information, skills and, most importantly, the confidence to survive the captive experience." The article continued, "JPRA psychologists provide an unstructured, unsupervised environment to help 'normalize' detainees' experience, says Lieutenant Colonel Debra Dunivin, PhD, deputy chief and director of residency training at Walter Reed Army Medical Center and a newly trained SERE psychologist. They give a range of psychosocial interventions to those who need it, she says." Of course there is no mention of the fact that Dunivin, at the time the article was written was a BSCT psychologist at Guantánamo Bay.

Accordingly, my first response to the news of psychologists' involvement in interrogations had nothing to do with the APA; I rededicated my efforts to opposing the increasingly oppressive policies and actions of the Bush administration. I marched against the war in Iraq. I continued to have monthly dinners with a group of leftist-activists who originally had gotten together to protest the CIA's activities in the "dirty wars" in Latin America in the '70s and '80s. I attended a meeting of political psychoanalysts to analyze the susceptibility of the American public to what we saw as cynical and repressive policies of the Bush administration. But none of this prevented me from "going on." I could go on. We all could pretty much go on.

In June 2005, the *Times* published another piece on the specific role of psychologists in interrogations. The *Times* article mentioned an APA Task Force that was addressing the issue.* By this point, other health professional associations, including the American Medical Association, the American Psychiatric Association, and the American Nursing Association, had all made it clear that there was no place for their members in coercive interrogations. I was naively confident that the APA would do the same. I was relieved to discover that a psychoanalyst colleague, a member of our psychoanalyst-activist listserv, was a part of the task force. Good, I thought. One of our analysts is on it. She has been an advocate against torture. I don't have to act.

But within days, a second piece in the *Times* challenged my complacency:

> The report by a group convened to study the ethical boundaries for psychologists at places like the detainment center at Guantánamo Bay, Cuba, concluded that it was acceptable to act as behavioral consultants to interrogators of the prisoners from Afghanistan who are held there ... The report said that psychologists may not engage in torture or cruel, inhuman and degrading treatment. But in seeming to refer to the situations reported at Guantánamo, which might fall short of torture or cruel treatment, it said only that they "require special ethical consideration." (Lewis, 2005b)

I found myself, as I often do, discovering my views as I debated this issue on the listserv. It was becoming clearer that something was amiss with the APA's response, and I felt the APA was contributing to, rather than opposing, abuse. And, looking back, I can see that I was becoming increasingly dissatisfied with debate as a response to institutional collusion. I wrote:

> When it comes to "coercive interrogation" there is only one side for an ethical psychologist to take. Anything else is a negotiating process

---

* The Task Force, created by then APA President Ron Levant, was called the APA Presidential Task Force on Psychological Ethics and National Security (PENS).

which only supports the "incremental" deterioration of certain internationally recognized standards ... And while I believe there is a value in working "within the system," there comes a point, where one is losing one's voice and one's very presence gives support to the other side, where one must confront that system.

One result of this discussion was that one of our members, Neil Altman, mobilized the APA's Divisions of Social Justice to try to stem the damage from the PENS report by urging the APA Council of Representatives, the democratically elected body of APA Division representatives, to add resolutions that would require ethical behavior, even if involvement in interrogations wasn't itself proscribed. The idea was to prohibit psychologists from any action that violated "basic human rights." At the August 2005 council meeting, when the council members were presented with the APA's PENS Task Force report on the ethics of interrogations, Altman and other psychologists from Divisions of Social Justice successfully passed a series of such resolutions.

2006: THE TURNING POINT

For the ensuing months, we all turned to other issues, confident that we had won a victory and that psychologists would join other health professionals in refusing to participate in interrogations that violate human rights, like those going on at Guantánamo. But in December, there were renewed reports in the press of the APA leadership extolling the important role psychologists played in such interrogations. I began to understand that the APA leadership was engaged in a duplicitous process—of representing the association as condemning torture and abuse, while crafting policy that permitted psychologists to continue to contribute to those very efforts.* For instance, Stephen Behnke, the director of the APA Ethics Office, said,

> [When] we talk about words like "isolation" and "sleep deprivation," we need to be careful. If one talks about isolation about a very few minutes, say, five minutes, I don't think anyone would argue seriously that isolating someone for five minutes rises to the level of torture or cruel, inhuman, or degrading treatment.

---

* As of this writing, the APA is continuing this practice. In 2008, against the opposition of the APA leadership, the membership passed a referendum that precludes psychologists' participation in operations that violate international law. In public, the leadership cites the referendum as exemplifying progressive APA policy, but in practice the same leadership refuses to implement the referendum, even in conditions that have been certified by the UN Special Rapporteur on Torture as violating its provisions.

Behnke went on to argue that "[P]sychologists have an obligation to take part in prisoner interrogations—in an ethical manner, when doing so can help protect Americans from terrorists and other dangerous criminals" (Behnke, 2005).

During this period, while the dominant U.S. press remained silent on continuing reports of abuse, rendition, and torture, the few journalists and ethicists who continued to publish exposés on these issues invariably described the role of psychologists and other health professionals. Jane Mayer (2006) published a piece in the *New Yorker* on Rumsfeld's memo approving abusive psychological techniques as standard operating procedure at Guantánamo Bay; Alfred McCoy's *A Question of Torture* (2006) highlighted the history of the CIA's psychological research used to develop techniques of torture that left no marks; and the United Nations' "report on the situation of detainees at Guantánamo" (United Nations Commission on Human Rights, 2006) explicitly stated that

> health professionals in Guantánamo Bay have systematically violated widely accepted ethical standards set out in the United Nations Principles of Medical Ethics and the Declaration of Tokyo … In sum, reports indicate that some health professionals have been complicit in abusive treatment of detainees detrimental to their health.

On June 7, 2006, the *New York Times* published an article that pushed a group of us over the line to action (Lewis, 2006). Up until that moment, we continued to discuss strategies. But on June 7, Neil Lewis's article made clear that psychologists in the military weren't simply a part of the problem of abuse and torture—psychologists were the essential overseers of the abuse and torture. The article reported that Pentagon officials had announced that they would try to use *only* psychologists, and not psychiatrists, to help interrogators devise strategies to get information from detainees at places like Guantánamo Bay, Cuba, and that the new policy favoring the use of psychologists over psychiatrists was a recognition of differing positions taken by their respective professional groups.

Three psychoanalysts on the listserv took direct steps to challenge the APA's position. Stephen Soldz created a petition on the Internet, specifically protesting the unique role psychologists now played in military interrogations. Within 2 weeks, there were over 1,000 signatures. Ghislaine Boulanger called for withholding APA dues until the policy was changed. In short order, hundreds of members began withholding their dues or resigned outright. And I wrote public protest letters to APA's CEO on June 9, and to the APA president, Gerald Koocher, on June 11:

> I am embarrassed that the American Psychological Association has not been willing or able to combat the publicly held view that

psychologists are more willing than psychiatrists to participate in coercive, possibly abusive, interrogations of prisoners at Guantánamo Bay. And worse, I am horrified that the APA and officials at the Pentagon believe that psychologists' participation in BSCT teams is acceptable; whereas both the American Psychiatric Association and the World Medical Association have stated unequivocally that such participation violates the Hippocratic Oath and is unacceptable. ... I request that you, or another authorized spokesperson, issue a statement which makes clear unequivocally that psychologists are prohibited from participation in cruel and inhuman treatment of anyone held against their will in general, and in centers of abuse such as Guantánamo, in particular. I would suggest further that the statement make it clear that psychologists are prohibited from advising in coercive interrogations, and in environments where their participation gives the impression that psychologists approve or assist in such techniques. ...

Koocher responded the same day:

The APA Board of Directors understands and appreciates that its members have strong opinions about psychologists' involvement in interrogations, and that their opinions are not uniform. Please recognize that interrogation does not equate to torture and that many civilian and military contexts exist in which psychologists ethically participate in information gathering in the public interest without harming anyone or violating our ethical code. Please also examine press reports with healthy skepticism and seek facts, rather than reflexively engaging in letter-writing campaigns predicated on inadequate access to the data.

Many others, from the listserv and beyond, also posted letters of protest. Koocher began responding angrily. "You are dead wrong!" he wrote to one member. To another, who wrote a second time because he hadn't received a response to his first e-mail, Koocher wrote, "Don't hold your breath!" Soon he took a different approach and began to address us as analysts: "Would you offer an interpretation to an analytic patient without carefully assessing all the facts? I doubt it. So please do try to get the facts straight and ask your Division 39 colleagues to do likewise." And "Unnamed sources and unnamed alleged perpetrators do not constitute valid data in my view, and when members of Division 39 start beating this drum they demonstrate no concern for the truth or protection of the innocent. Instead they smear all, just as when years ago our psychiatric colleagues claimed that psychologists were unqualified to practice analysis." In the end he was deluged with protest letters, and assumed that "an orchestrated campaign" was behind what was a genuine expression of widespread outrage:

> In the past 48 hours I have received 300 email messages clearly part of an orchestrated campaign that is ill informed and conflates very appropriate anger at administration policies with incorrect assumptions about APA policies and actual behavior of APA members. If you don't like my tone, consider what it feels like to get such messages and do a better job of education [sic] yourself to the facts.

Although it was not widely known, Koocher had played a significant role on the PENS Task Force on the ethics of such interrogations. He came on to the Task Force, ostensibly, as a "second" liaison for the APA board of directors. It is noteworthy that no other APA task force has ever had a second liaison and that Koocher (at that time, APA president-elect) played a dominant and guiding role.

A day later, a member of the listserv forwarded my letter and Koocher's response to Amy Goodman, who produces the daily independent television and radio show, *Democracy Now!* Before another 48 hours had passed, I was contacted by a producer at *Democracy Now!*, and by the end of the week, on June 16, I was on the air debating Koocher on the ethics of psychologists' involvement in abusive and coercive interrogations at Guantánamo Bay.

NEW ORLEANS, AUGUST 2006: THE (FIRST) HOLLYWOOD MOMENT; JEAN MARIA ARRIGO BREAKS PENS' CONFIDENTIALITY

I was at the American Psychological Association's Annual Convention in New Orleans, attending a wine and cheese party held by Psychologists for Social Responsibility (not to be confused with Psychoanalysts and Social Responsibility or Psychotherapists for Social Responsibility), with Morty and Brad Olson, incoming president of the Divisions for Social Justice. I had spoken to both of them during the heady days before the convention, but this was the first time I had actually met either of them. I was particularly interested in speaking with Jean Maria Arrigo, one of the three nonmilitary members of the PENS Task Force. I wanted to speak with Arrigo, because it was becoming clearer and clearer that the PENS Task force was key to the APA's relentless commitment to keeping psychologists in the interrogation business at all costs.

The APA had kept the PENS proceedings confidential, but Arrigo, as an oral historian, could not on principle accept the notion of confidentiality in perpetuity and had decided on her own initiative to deposit all the PENS materials in the *Intelligence Ethics Collection* of the Archives of the

Hoover Institution on War, Revolution, and Peace at Stanford University, to be made public after 10 years. I wondered if Arrigo might be willing to divulge something of her experience on the task force.

We began our conversation by sharing stories about how we became activists. Arrigo explained to me that her father had been an unrepentant interrogator for the CIA, including participating in the nefarious MKULTRA research, where unsuspecting subjects were given mind-altering drugs. Arrigo believed that interrogations could be done ethically and had become an oral historian of intelligence-gathering practices.

I told her about my parents, survivors of the Holocaust. My mother had been in Auschwitz as a teenager, where she lost every member of her family. My father, who lost his mother, father, and younger brother in the Warsaw Ghetto, had fled to Russia when the war began. He was arrested by the NKVD (the Soviet secret police that later became KGB), interrogated, and sent to prison in Siberia, accused of spying for the Germans. When Hitler dissolved his pact with Stalin and attacked the Soviet Union, Polish prisoners were freed and my father joined the First Polish Division of the Soviet Army. He told me stories of being pressed to follow Soviet orders during horrendous battles and frontline activities, including killing prisoners of war, actions that today would be considered war crimes. I explained to Arrigo that I grew up with a personal family history of stories of both victimization and perpetration of gross political violence.

I had come to New Orleans (my first APA convention in nearly 20 years) because, after the debate with Gerald Koocher on *Democracy Now!*, psychologists' role in military interrogations and the APA's support for that role had become national news. Koocher had become the APA spokesperson in favor of psychologists' participation and I had become a public voice against it. It became clear that if the APA was going to change its policy, pressure had to be applied at the convention. Since nothing had come of the resolutions passed by the council the year before, Neil Altman and others from the Divisions of Social Justice put forth a second, more explicit proposal prohibiting psychologists' involvement.

A few days after the debate, I had received a call from Mark Benjamin, a reporter with *Salon.com*. He had been trying to get the names of the members of the PENS Task Force from the APA leadership, but the APA refused, citing confidentiality of the proceedings. Finally a contact on Capitol Hill had forwarded him the names of the members and he wanted to know if these names meant anything to me. I plugged them all into a Google search and discovered something that neither the APA leadership, nor members of the press seemed to know—the names and biographies of the members of the PENS Task Force had been published on the Web site of APA's Division 48 (Peace Psychology), after they had been distributed to the APA council representatives. Biographical information on the PENS members, meant for a small group of APA governance, had been available for all, if one only knew where to look.* Benjamin (2006) published the names and biographies and drew the inescapable conclusion: 6 of the 10 psychologists on the task force had close ties to the military; and the majority of these were involved in military interrogation practices taking place at Guantánamo, Afghanistan, Iraq, and elsewhere.

I learned from Benjamin that Koocher had invited the Army's primary BSCT apologist, Lieutenant General Kevin C. Kiley, as the sole speaker to address the APA council during the deliberation process when the council would be debating the ethics of the BSCTs and Altman's new resolution. I had already learned the power of writing protest letters to Koocher and wrote another requesting that Koocher invite a second speaker to present the other point of view: "It would show that the Administration of the APA is interested in a full and fair debate of these important issues, rather than what has thus far appeared to be a rubber stamping of the position of the current United States administration and it's military services." When Koocher didn't respond, I wrote again saying that colleagues had asked me to release my letter to the press. This time I received a rather speedy response: "I am very interested in providing a balanced platform for Council's discussion of the issues surrounding psychologists' role in national security interrogations. Toward that end, I would like to invite you to make a presentation to the Council meeting …"

The morning of the council meeting, Kiley offered an ode to the value added to the military by its psychologists. I spoke immediately after lunch. I had done some serious research, because I was trying to understand why

---

* The APA leadership has since denied that the names of the members of the task force had ever been kept secret. Both Stephen Behnke, director of the APA Ethics Office, and Olivia Moorehead-Slaughter, chair of PENS and former chair of the Ethics Committee, have publicly denied any such secrecy, and the APA press office released a statement stating that the notion that the names of the task force members were kept secret "is totally false. In reality, the names and composition of the Task Force is public information. The names and biographical statements of each of the Task Force members are, and have been for some time, available through the APA website." But this assertion is belied by evidence from the PENS listserv itself. A post from August 22, 2005, by a military/intelligence member of the task force stated: "I wanted to leave a short note regarding the ethics in national security panel presentation at the APA conference on Friday. While this was not related to the task force, there were many questions and comments regarding the task force report posed to Dr. Steve Behnke who chaired the panel. I was once again impressed with how Dr. Behnke eloquently represented our work and insured the confidentiality of the panel, despite pressure to reveal the identities of the task force members and the process that unfolded during the task force meetings. Steve was respectful, gracious and polite in response to some very direct and provocative questions and comments" (PENS Listserv, 2005, p. 169). Moorehead-Slaughter responded to this e-mail, also on August 22, 2005: "I have no doubts that Steve [Behnke] was respectful and masterful in preserving the integrity of our Task Force process" (PENS Listserv, 2005, p. 170).

it seemed so important to the APA to keep psychologists in what seemed to me to be the indefensible position of participating in the purposeful abuse of prisoners, many of whom were obviously innocent. I made the case to the council that since the other health professions refused to take part, currently only psychologists were overseeing these interrogations:

> You have a rare opportunity to make a significant difference for good in the world. If I am right, and the US government believes it can only legally justify these shameful techniques by claiming medical supervision, then the refusal of psychologists to participate may finally put an end to these practices. At the very least, it is time you stopped psychologists from being associated with them.

The next evening, when I was speaking with Arrigo, I told her how increasingly surprised I was to discover more and more high-level connections between APA leadership and the military. I mentioned that, in an apparent attempt to ingratiate himself with the psychologist opposition, General Kiley had told Brad Olson, incoming president of the Divisions of Social Justice that Debra Dunivin, the wife of the chair of the APA Practice Directorate, was a BSCT psychologist at Guantánamo.

Arrigo was taken aback. Not simply because a leader of the APA was married to a Guantánamo BSCT, but, she explained, because that particular leader, Russ Newman, was present during the PENS deliberations. She said:

> No one knows this, because of the confidentiality agreement, but Russ Newman was there at the meeting. Not just present; he took a leadership role, right alongside Gerald Koocher. He never said anything about having a connection to the military. He kept saying that our job as a Task Force was to "put out fires" and he kept emphasizing how important psychologists' contributions were at Guantánamo, how appropriate their actions were, how vital a role they were playing.

I asked Arrigo if Morty and Olson might join the conversation. She agreed and went on to explain to us that during the meeting Koocher and Stephen Behnke, chair of the APA Ethics Office, worked with Newman and two of the military psychologists, Morgan Banks and Larry James, to guide the conclusions of the task force, almost from the start. Koocher shot down the voices of dissent among the nonmilitary members, and Behnke created drafts of the report that prioritized the views of the military members and simply ignored other positions.

Arrigo said that there were even more observers to the task force than she had mentioned. She had since done some research and there had never been so many observers to a task force before, and furthermore no task force report was ever issued unsigned before. She said she had taken notes when the observers introduced themselves and that most of them seemed to be involved in lobbying for APA's interest with Congressional military committees, as well as directly with the Department of Defense (DoD) and various intelligence agencies. However, a few hours into the meeting she was told to stop taking notes; that was the first clue that she had that something was amiss. How could the members of the task force contribute anything to the language of the group product if they were prohibited from taking notes?

She said that, at the last moment of the meeting, when the report had been finalized by Behnke (the only one permitted to take notes) after only two and a half days of discussion, the military folks in the room were just ecstatic. They began talking about sending it to Rumsfeld, and meeting with Kiley. Arrigo said that it suddenly dawned on her that the group was not simply working to offer the APA's position on interrogations—that something more important was going on that only the military members, the APA officials, and the observers understood. The three nonmilitary members were all in the dark. "It's ironic," she said, "the whole time on the listserv, I was trying to organize an agenda for that meeting, and I kept getting shot down by Koocher. I didn't realize that there already was an agenda and we weren't privy to it."

I asked her about the listserv. This was the first any of us had heard about it. She explained:

> There was a listserv for months before and months after the meeting. If you want to see how Russ Newman was invited to the Task Force, you can see how the military members pushed for him. I didn't understand what PENS had to do with private practice, but that's the argument James and Banks used to get him in there. It's all in the listserv.

Morty began to speak, assuming a much more serious tone than earlier; a sharp contrast with his Middle American folksy humor. He explained that the more we learned about the APA's role in determining what psychologists could and could not do in military and intelligence interrogations, the clearer it became that this was not simply a matter of a task force botching its assigned task of establishing appropriate ethical boundaries, nor was it simply a matter of conflicting interests. More and more, the role of the task force appeared to be a premeditated collusion between the DoD, the APA, and perhaps certain intelligence agencies. Morty said that for some reason that we didn't yet understand, it seems it had become extremely important for the military and the CIA to ensure that no military or intelligence psychologist involved in interrogations ran afoul of APA ethics. He asked Arrigo if there was any way we might see the documents she had archived from the PENS Task Force. From what she had said, the PENS listserv seemed especially important.

Arrigo paused and made a decision then and there. She decided that because of the duplicity and hidden agendas of the APA leadership—especially the fact that she didn't know that Newman's wife was a Guantánamo BSCT—her obligation under the confidentiality agreement was nullified. She said that she would therefore be willing to give copies of the material to me. Something in the history we had shared gave her the impression she could trust me. She said that she didn't mind if Morty or Olson saw the material, but that I would be the only one who could keep a copy.

It would take 4 weeks for us to arrange to transport Arrigo and her trove of documents to New York, so that she, Morty, and I could go through them. While waiting, I continued to hear Arrigo's words: "It's all in the listserv." Like Alexander Butterfield's revelation that led the Senate Committee to the discovery of the Nixon tapes: "There's a recording system in the White House."

Over the following months, the research that Arrigo, Morty, Olson, and I did began to expose the threads that connected psychologists and the American Psychological Association with just about every facet of the Bush administration's "enhanced interrogation" program, which, outside of the American government's and the APA's jargon, is called by its right name: torture. This research led us to two distressing, but unavoidable, conclusions: *psychologists had been instrumental in operationalizing the Bush administration's program of abuse and torture. And some members of the APA leadership had been intimately involved in ensuring that no APA policy or ethical standard interfered with psychologists continuing to play these roles.*

THE PENS PROCESS: THE APA RUBBER STAMPS DOD BSCT POLICY

What became painfully clear in our initial review of the PENS papers was that these notes and documents strongly supported Arrigo's suspicions that the task force followed a covert agenda, shared by the military members, the APA leadership, and certain of the observers, but not known by the nonmilitary members of the task force. This was easily deduced by attending to the sequence of comments in the listserv and in Arrigo's contemporaneous notes (taken before the military members and APA leaders voted against note taking), and by closely observing the role Behnke played in selecting which aspects of the group discussions found their way into the draft reports and which were left out. Included almost invariably was language consistent with BSCT policy, whereas topics that challenged that policy, offered by the non-military psychologists, fell by the wayside. Most apparently damning was the fact that the first draft of the PENS report, produced by Behnke within 4 hours of the start of the meeting, reflected the major points contained in a document provided to the task force by Colonel Banks, supervisor of all Army BSCT Psychologists (Banks, 2005), and described by him in the listserv as "the written instructions I give my [BSCT] psychologists" (PENS listserv, 2005, p. 17).

Each draft contained the statement, introduced into the proceedings by Russ Newman and Larry James: "Psychologists have a valuable and ethical role to assist in gathering information that can be used in our nation's defense" (Arrigo, 2006) and another, introduced out of the blue after the first break, by Morehead-Slaughter, "safe, legal, ethical, effective."* Like Banks's instruction manual, which stated unequivocally, "The ethics code is subordinate to the law and regulations", each draft of the PENS report contained a reference to following law and military regulations. The final draft directly referenced APA ethics standard 1.02, with its infamous "Nuremberg clause" that, in cases of conflict between ethics and law, military orders or regulation, permits following the law or military regulations.

The listserv revealed that Banks, James, and Debra Dunivin, met with Kiley about a month after the PENS weekend to "establish the doctrinal guidelines and training model for psychologists performing this [BSCT] job" (PENS listserv, 2005, p. 167). Following this meeting, BSCT standard operating procedures were rewritten, integrating the PENS conclusions, which in turn had been derived, in part, from earlier BSCT materials. The difference was that now, when the BSCT protocols described the BSCT psychologist's ethical priorities, it did so with the imprimatur of the American Psychological Association: "The ethics code does not supersede applicable US and international law, regulations or DOD policy" (Kiley, 2006, p. 17). A copy of the PENS report was attached.

THE COALITION IS FORMED

At this point, I was becoming a committed activist. It seemed to me, reading the PENS listserv and minutes, and the presentations offered at the council meeting, that the APA leadership simply thought that the membership would take what they said at face value. If they said that psychologists do not torture, that would be the end of it; if they invoked September 11 and spoke about looking at terrorists face-to-face, the rest of us would simply grant them the authority to protect the nation in whatever way they saw fit. And certain members of the PENS Task force and the APA leadership seemed simply to think that they were smarter than everyone else, and no one would catch on to what they were doing.

Arrigo, Olson, Soldz, and I (with Morty always coming in at the right time to give us a lead or to validate or correct our wilder theories) quickly became an extraordinary team. Our different strengths were

---

* The first documented use of this phrase is in a BSCT standard operating procedure from 2004 (DoD Operational Policy Memorandum #14, p. 1). It is described as the mission of the BSCT in both Kiley's investigation into Guantánamo and Abu Ghraib abuses in early 2005 (Kiley, 2005) and in Banks's instructions (Banks, 2005).

complimentary, and enabled us to respond quickly and powerfully to the APA's propaganda, the press, and government agencies. We called ourselves the Coalition for an Ethical Psychology, a purposefully vague moniker that we knew would be confused with other similarly named groups. We never let on how many members the coalition had, but purposely encouraged the notion that we spoke for multitudes.* Our small band posted exposés on the role of the PENS members, we took on the APA leadership when they either prevaricated in public or attacked the integrity of their opponents. We were a clearinghouse of information, some classified, most divined after extraordinary efforts at research, for a host of reporters, for the Senate Armed Services Committee, and for other government agencies. And, most important, we, along with a small coterie of other psychologists and psychoanalysts-turned-activists, kept the pressure on the APA by exposing collusion and advocating for policy changes to prohibit psychologists from participating in these atrocities.†

What ensued was an extraordinary chess game, with the APA leadership working to look like it opposed torture and abusive Bush-era tactics, without actually committing itself or any psychologists to a position that wasn't fully consistent with administration policy, and our working to both expose the APA's collusion and simultaneously to hold it to the assertions of the APA public relations campaign. We don't know who, exactly, ensured that the language introduced into all APA resolutions and policies was so carefully worded as to be consistent with Yoo and Bybee's doctrines, as well as other Bush-era DoD and intelligence agency policy, but we had no doubt that someone was overseeing this effort. Every attempt our side made to pass policy against torture was undermined by an eleventh-hour rewording from the other side to take the teeth out of the resolution. Every one of their last-minute maneuvers to undermine the resolutions was met on our side by a publicity campaign exposing APA's duplicity.

2006 RESOLUTION AGAINST TORTURE

Our first inkling of what we were up against came during the council's August debate of the 2006 Resolution Against Torture (right after I addressed the council). I was new at the game and wasn't paying attention

---

* Since those beginnings we added Bryant Welch and Ted Strauss to the coalition.
† I want to reiterate that the coalition was one, small, focused part of a large group of activists. Ghislaine Boulanger and the WithholdAPAdues group organized a powerful political arm called Psychologists for an Ethical APA. Neil Altman, Laurie Wagner, and other activists within APA governance led the battle to pass resolution after resolution, and, as we gained momentum, other human rights organizations, including the American Civil Liberties Union (ACLU), the Center for Constitutional Rights, and Amnesty International, joined Physicians for Human Rights (PHR) in supporting our efforts.

to the nuances of the discussion. A member of council wanted to add the word *knowingly* to the resolution so that it read, "psychologists shall not *knowingly* engage in, tolerate, direct, support, advise, or offer training in torture or other cruel, inhuman, or degrading treatment or cruel, inhuman, or degrading punishment." I wasn't paying attention because it occurred to me that the weak point of the resolution was that, although there was a definition of torture, there was no definition of "cruel, inhuman, or degrading treatment." I quickly wrote out a "friendly amendment" to add a definition, derived from the United Nations Convention Against Torture (UNCAT), which frequently addressed these issues in its history of jurisprudence. I brought it to members of the Peace Division, who assured me that they understood the problem and would try to address it. By the time of the vote, however, *knowingly* had been added, but nothing was said about the issue I had raised. A new version of the resolution was suddenly produced, which no one had a chance to read, the vote was called, and the resolution passed overwhelmingly.

It was only on the flight home that I had a chance to read the revised resolution. I discovered that a definition of cruel, inhuman, or degrading treatment had, in fact, been added. The definition was not aligned with the UNCAT, as I had hoped, nor was it left vague, as in the original document. Instead, the resolution now stated that "the term 'cruel, inhuman, or degrading treatment or punishment' means treatment or punishment by any psychologist of a kind that ... would be prohibited by the Fifth, Eighth, and Fourteenth Amendments to the Constitution of the United States, as defined in the United States Reservations ... to the United Nations Convention Against Torture." So, during the break, when the APA leadership inserted *knowingly* into the text, it also managed to slip in a clause that aligned the resolution with the definition of torture, cruel, inhuman or degrading treatment contained in the in famous Yoo–Bybee "torture memos."*

2007: REAFFIRMATION OF THE RESOLUTION AGAINST TORTURE

Once we were able to publicize the meaning of the words that had been inserted into the Resolution Against Torture, Neil Altman began another

---

* "As we explain in Part III, U.S. obligations under international law are limited to the prevention of conduct that would constitute cruel, unusual or inhuman treatment prohibited by the Fifth, Eighth, and Fourteenth Amendments" (Yoo, 2003, p. 2f). Knowingly also makes the text consistent with the Yoo torture memo, in that Yoo argues that an interrogator is not guilty of torture unless he knows that the effect of his tactics will be "severe" and "lasting": "We believe that if an interrogator acts with the honest belief that the interrogation methods used on a particular detainee do not present a serious risk to the detainee's health or safety, he will not have acted with deliberate indifference" (p. 65).

attempt to change APA policy from within. Altman's argument was that since the Resolution Against Torture stated that "should torture or other cruel, inhuman, or degrading treatment ... evolve during a procedure where a psychologist is present, the psychologist shall attempt to intervene to stop such behavior, and failing that exit the procedure," and since the UN Rapporteur on Torture and the UN Committee Against Torture had agreed that the conditions at Guantánamo themselves were cruel, inhuman, and degrading treatment, psychologists had an ethical duty, under the resolution to "exit the procedure" and leave Guantánamo. Altman and a group of like-minded council members offered a resolution to establish "a moratorium on all psychologist involvement, either direct or indirect, in any interrogations at U.S. detention centers for foreign detainees. This moratorium is necessary as detainees may be currently denied protections outlined under the Geneva Conventions and interrogations techniques in violation of the 2006 APA Resolution against Torture." The APA Council scheduled a vote on this second resolution for the next APA convention to be held in August 2007 in San Francisco.

Meanwhile, the Office of the Inspector General of the Department of Defense (2006) released a report that made clear that SERE tactics had been taught to interrogators at Guantánamo, and that psychologists from Banks' command and James' command were implicated.

This time, the opposition was well prepared. What had been a small group of dedicated activists, working to change APA policy and thereby interfere with the Bush administration's torture program had grown into a large, well-organized protest movement. Damning articles on psychologists' role in torture had been published widely, including in *Vanity Fair* (Eban, 2007), *Salon.com* (Benjamin, 2007), and on the Internet.*

The APA leadership had joined with the Divisions on Social Justice to organize a "miniconvention" on interrogations, where the issues were to be discussed with representatives from both sides. The idea, as frequently articulated by Behnke, representing the APA, was that "reasonable people may disagree." But when the mini-convention convened, the APA leadership seemed overwhelmed by the strength and size of the opposition, and by the fact that when it came to opposing psychologists' presence in a program of torture and abuse, reasonable people were relatively united. Every panel was filled beyond capacity, with the majority in the audience opposed to the APA's position. There was a public protest that gathered hundreds of psychologists and San Franciscans to the convention center. Journalists and TV news covered the story in print, on the Internet, and on television news broadcast around the world.

---

* Stephen Soldz and Jeffrey Kaye especially have been tireless in their integration, analyses, and exposure of psychologist and APA complicity. See Soldz' blog, www.psychoanalystsopposewar.org, and Kaye's blog, http://valtinsblog.blogspot.com/.

Nonetheless, the APA leadership used their mastery over APA governance processes to ensure that no real change in APA policy was possible. First, it recast the resolution as a "reaffirmation" of the 2006 Resolution Against Torture, specifically adding a prohibition against 20 specific torture and abusive detention tactics (this seemed like a genuine change in APA's position, and, since it prohibited outright involvement in most of the known abuses, we all supported the new wording wholeheartedly). Second, the APA leadership convinced the negotiating group that if the resolution contained the "moratorium" amendment, it was likely to fail, and the whole resolution condemning these tactics would fail. The group agreed to separate the resolution prohibiting the tactics and the moratorium for separate votes. While this seemed reasonable, the purpose behind it only became clear when new language of the first resolution was released, just hours before the vote, changing the applicability of the resolution considerably.

Although, even in the new wording, the most egregious tactics, including waterboarding, rape, and the use of drugs were still banned outright, seven of the techniques—"hooding, forced nakedness, stress positions, the use of dogs to threaten or intimidate, physical assault including slapping or shaking, exposure to extreme heat or cold, threats of harm or death"—were only prohibited if "used for the purposes of eliciting information in an interrogation process" (i.e., they were not prohibited if used as preparation for interrogation). But most disturbing was the change in the prohibition against four of the most widely used psychological techniques: "isolation, sensory deprivation and over-stimulation and/or sleep deprivation" were only prohibited if "used in a manner that represents significant pain or suffering or in a manner that a reasonable person would judge to cause lasting harm" (American Psychological Association, 2007).

In other words, once again someone in the APA leadership changed the resolution language to be consistent with the legal standards of the Yoo–Bybee torture memos (which Rumsfeld's group cited approvingly to permit the use of these techniques at Guantánamo: "For purely mental pain or suffering to amount to torture under [the federal torture statute], it must result in significant psychological harm of significant duration" [Bybee, 2002, p. 1]).

This resolution passed overwhelmingly, as, once again, the council did not have the opportunity to learn the meaning of the new language. And once again, the APA leadership, working with the Department of Defense, brought in a surprise speaker to address the council while it was debating the moratorium resolution. Colonel Larry James came from his deployment at Guantánamo to speak out against the resolution. Even though James had not been elected to the council, he was permitted to speak because he had been selected by APA leadership to replace the Division 38 representative, who had resigned her position the night before the meeting. Appearing in uniform, James spoke *against* removing psychologists from sites that violated the Geneva Conventions, repeating the argument that psychologists serve as "safety officers." He added a melodramatic coda: "When you don't have psychologists involved in the day-to-day activity, bad things are going to happen, innocent people are going to die" (Goodman, 2008). The moratorium resolution failed.

That afternoon the APA held a town meeting for the membership to address these same issues. Again the hall overflowed, far beyond the expectations of the APA leadership. When I spoke I asked those present to repeat the vote on the moratorium resolution. I read the resolution aloud and asked how many were in favor and how many opposed. The vote was overwhelmingly in favor. I said, "I want to know why the Council of representatives is so different in how it votes from the members of the American Psychological Association" (Goodman, 2007).

Fortunately, the town meeting was carried on public television, and journalists not only covered James's statement at the council meeting, but Laurie Wagner's (council rep of the Division of Psychoanalysis) response: "If psychologists have to be there in order to keep detainees from being killed, then those conditions are so horrendous that the only moral and ethical thing to do is to protest by leaving" (e.g., Alfano, 2007).

Soon afterward, the APA put out an explanatory position paper, which was such a gross example of the increasingly distorted mechanisms the APA used to claim opposition to coercive and abusive interrogations while, in practice, maintaining support, that it permitted the Coalition for an Ethical Psychology to offer our most comprehensive history of the APA's duplicity (Coalition for an Ethical Psychology, 2008).

2008: AMENDMENT TO THE "REAFFIRMATION," "REISNER FOR APA PRESIDENT," REFERENDUM

The public and membership outcry after the 2007 resolution vote spurred the council to forego the usual rules, and pass an amendment to the resolution, banning all the "techniques" outright, without exception, the following February. Again, however, at the last moment, the promilitary side of the group that negotiated the language changed the resolution, without the council or its negotiating partners understanding the import of the change. The original resolution applied the prohibitions to "interrogations or other detainee-related operations." That was why the exceptions were so odious—they undermined the purpose of the resolution by exempting conditions of confinement and requiring significant and lasting harm. When the council was pressed by the membership to amend the resolution and ban all tactics without exception, it removed the clause on "other detainee-related operations" altogether, and replaced it with "An absolute prohibition against *the following techniques* …" (American Psychological Association, 2008a). The APA has since interpreted this as applying only to techniques used in an interrogation: "APA has specifically prohibited 19 interrogation techniques as torture" (American Psychological Association, 2008b). Thus, the leadership managed to undo, in the guise of repairing, the far-reaching ban on psychologists participation on all modes of detainee abuse that the council believed it was enacting.

It was clear to many of us that the APA leadership would continue to maintain its ties to the military and the CIA, and to these abusive interrogations and conditions unless it was somehow forced to do otherwise. And that the only way to change that policy would be for us to find a way to take the power away from the leadership and put it back in the hands of the membership. Our group decided to try a major strategy shift and run a candidate for president of the APA. I was drafted to be the candidate and hastily put together a 50-word statement to be included with the nomination ballot:

> I am running for President to restore APA as a voice for human rights and social justice. The APA currently supports psychologists who facilitate detainee and enemy-combatant interrogations, even if these interrogations violate human rights and the Geneva conventions. Let's change this policy. Please read my full statement at www.ethicalapa.com.

Simultaneously, WithholdAPAdues member Dan Aalbers discovered an obscure APA bylaw that required a vote of the membership on any policy change proposal that was petitioned by 1% of the membership. Within a few months, I had been nominated for the presidency of the APA and received the most nominating votes of any candidate, and a petition had garnered enough votes to force a referendum on the moratorium issue by the entire membership.

From then on, we all began to believe that victory was near. We knew we would be helped by the upcoming report of the Senate Armed Services Committee investigation. We knew, too, that the CIA Inspector General had written a report that exposed the role of psychologists, and hoped that it would either be leaked or declassified. The days of the Bush administration were numbered, and we hoped that all these would combine to give us success with our campaigns to turn the APA away from complicity and back toward human rights.*

In the end, the referendum won overwhelmingly and I lost. I believe that the two were related, in that the referendum vote preceded the election, allowing members to believe that the interrogations issue was finally settled

---

* Our cause was further aided by a book by Amy and David Goodman, who wrote about the Coalition for an Ethical Psychology in *Standing Up to the Madness: Ordinary People in Extraordinary Times* (2008) and by an article in *Newsweek* about the campaign (Ephron, 2008).

and that they could vote for the president based on other issues. To be frank, I was relieved—I don't think I could have handled 2 more years (as president-elect and then as president) devoted to APA politics and policy. But with the referendum, we truly believed we finally had inserted a wedge between the psychologists and abuse.

2009: THE BATTLE CONTINUES

After the passage of the referendum, 2008 APA President Alan Kazdin did not wait for the board or the council's approval and immediately sent letters to the president and the heads of all government agencies to inform them of the new APA policy: "The effect of this new policy is to prohibit psychologists from any involvement in interrogations or any other operational procedures at detention sites that are in violation of the U.S. Constitution or international law" (Kazdin, 2008). Kazdin appointed a working group to draft a policy for implementation of the referendum. The group, composed of many of the original group who negotiated the antitorture resolutions, unanimously produced an extraordinary document, which determined that "relevant information about whether a specific site operates outside of, or in violation of, international law can be accessed by contacting ... the Special Rapporteur Against Torture" (American Psychological Association, 2008c). It went on to state that a "determination of whether a particular detention setting is 'in violation of international law' ... may include a ... denial of access to the site and to detainees by U.N. monitors" (pp. 4–5).

But when the council met in February to vote on the referendum, the referendum was approved unanimously,* but the implementation policy was deferred until the August meeting. At the August council meeting it was not put on the agenda. The result was that while the APA information office publicly touted the referendum as evidence of the APA's absolute prohibition against participation in detainee abuse,† the leadership privately

---

* Council had no choice but to approve it, since the bylaws mandated that the results of the vote would become APA policy.
† APA President James Bray wrote, for example, in a letter to editor of *The New York Times* on August 13, 2009:
  The allegations against Drs. Jessen and Mitchell are extremely serious. While they are not members of the APA and are therefore out of the reach of our ethics process, it is the positions of the APA that any psychologist involved in the abuse of national security detainees should be held accountable for their actions. It is also noteworthy that a recent APA policy referendum states that psychologists may not work in settings where "persons are held outside of, or in violation of, either International Law (e.g., the UN Convention Against Torture and the Geneva Conventions) or the U.S. Constitution (where appropriate), unless they are working directly for the persons being detained or for an independent third party working to protect human rights" or providing treatment to military personnel.

communicated that the referendum was not binding (e.g., Garrison, 2009) and that the referendum cannot be implemented because there is no way to determine whether a site is in violation of international law.

To combat this last example of APA duplicity and complicity, I wrote Manfred Nowak, the United Nations Special Rapporteur on Torture, and apprised him of the language of the referendum, the text of the advisory committee report, and the APA's argument that there was no way to determine whether a site was in violation of international law.

On August 7, 2009, Nowak wrote a letter to 2009 APA President James Bray, which simply stated:

> I certainly conclude that the overall conditions of detention at Guantánamo Bay constitutes to be "outside, or in violation of, international law" … Thus, in keeping with both the APA's own policy and relevant international law and ethical guidelines, I request that you do all that is necessary to invoke the referendum and immediately request that the Obama administration, the Department of Defense, and the US intelligence agencies remove psychologists from Guantánamo and any other sites where international law is being violated or where inspectors are prohibited from assessing that conditions are in compliance with international law.

The APA has simply made no public response, or acknowledgment, of the letter from the Special Rapporteur. The referendum is on the books, Guantánamo conditions still violate international law, and psychologists still may participate with impunity.

CONCLUDING THOUGHTS

I began this history with the idea that my particular way of listening as a psychoanalyst influenced my understanding of what was being presented by the Bush administration and by the APA about the roles of health professionals in national security interrogations. And further, how this mode of listening pushed me over the edge, from analyst-observer to analyst-activist. As an analyst, my instinct, training, and curiosity directed me to those spots where ethical violations were being obscured by a narrative woven of heroism and ethical responsibility.* And the more I looked into it, the clearer it became to me that the psychologists involved had created a subterfuge of denial and disavowal—a cover story to disguise something hidden and more sinister.

---

* Freud (1900) used the metaphor of "the embroidered mark on Siegfried's cloak" (p. 515), to point to where the defensive structure itself reveals the point of vulnerability.

I found myself, in Badiou's (2005) terms, compelled to "risk risking" and join the resistance. But what surprised me as much as the psychologists' subterfuge was the extent to which our continued exposé encountered, among the vast majority of psychologists, that other resistance, a psychological resistance not only to action but, at times, to thought. To revisit Badiou's conceptualization: "the non-thinking of the situation that prevents the risk or the examination of possibilities" (p. 8). How could that small coterie of psychologists supporting the Bush interrogation policies continually enforce their hold on APA policy, in spite of the fact that their stories were debunked and their complicity exposed; how were the majority of members of the APA who opposed those policies so easily manipulated, in exactly the same way, over and over? The same questions could be asked of our nation. How could it be that government spokespeople could, early on in the war, brag to reporters about using enhanced techniques, and then later deny that the techniques had ever been used, without the reporters or the public calling them to account?

I believe that during the Bush administration our nation's psychological relationship to clandestine domestic and foreign policy activity changed dramatically. The activity itself has not changed; the CIA, for example, has researched and supervised torture for at least 50 years. Psychologists and psychiatrists have been part of that research (see, for example, McCoy, 2006). What has changed, I believe, are the psychodynamics of the representation of that activity.

To understand these psychodynamics, I propose we view the nation's intelligence and national security agencies as a kind of tacit unconscious force. It is unconscious insofar as these agencies exist to carry out aims and means of governmental policy that cannot be acknowledged publicly. They are tacit in that, although they are known to exist, by public agreement their operations are unacknowledged.

Historically, the citizens' relationship to the "tacit political unconscious" functioned in a manner akin to the psychodynamics of neurotic individual and family processes. The government, in this sense, is experienced as the parental agency, responsible for law, order, and maintaining conditions of well-being. The clandestine agencies are permitted to deviate from those standards, so long as they (a) work abroad, and (b) are seen to be fulfilling national security operations without the gross breach of moral and legal standards that would cause them to be noticed (i.e., brought into awareness). In cases where clandestine activities breached those standards and were brought into public awareness, the "public" (Congress, the press, public opinion) was forced to address the breach. The results are not so different than when an individual is exposed for having secretly acted upon forbidden and repressed desires: shame, guilt, redress, and the institution of new restrictions.

The Bush administration changed this essentially neurotic process of repression, breach, and repair into something else altogether. That administration's relation to clandestine activity and its public face shifted away from the neurotic paradigm to a narcissistic/sociopathic one. The fundamental principle that all players were beholden to legal and ethical standards was replaced by a philosophy of unchecked executive power. The clandestine agencies ceased their role as enactors of tacitly unconscious political aims, still beholden to a higher standard of law and the constitution, and began functioning as an arm of the self-justifying executive, no longer beholden to the law.

Unfortunately, the American public was primed, prior to the advent of the Bush administration to take a corollary narcissistic-dependent position, ready to grant exceptional power to the government in exchange for a promise of exceptional protection.

I have previously written (Reisner, 2003) about the cultural shift toward narcissism, describing the elevation of trauma as one of its central symptoms:

> These elements, taken together—the exceptional, privileged status of the traumatized, the assertion that the needs of the traumatized are responded to as if they reflect life-or-death urgency, the location of the threat to well-being as coming exclusively from the outside ... the aim of the effort toward restoration of a former idealized state ... reveal that, to use psychoanalytic terms, trauma has become the venue in our culture, and in our treatment, where narcissism is permitted to prevail. (pp. 407–408)

September 11 solidified what was already a move in this country toward a culture of exceptionalism. After September 11, the Bush administration emphasized aspects of that terrible event to promote a fear-based, narcissistic response in the populace (much as malignant narcissism in a parent often promotes dependent narcissism in the child). Americans were encouraged to employ primitive defenses, such as splitting good and evil, locating the evil exclusively outside, and idealizing a national identity that had been victimized ("Why do they hate us?") in order to justify an "any means necessary" approach to restoring the wounded ideal. The manipulation of fear and narcissistic regression resulted in a population willing to perceive every threat as life and death, and because of this activated, regressed state, the old rules of following ethical and legal standards and precedents no longer applied.

More disturbing to the social psyche, the Bush administration made clear that it would determine the legality of its tactics, without regard to any other historic or political authority. The executive branch simply denied the applicability of precedent, or moral or legal traditions. In claiming its own law and violating heretofore agreed upon treaties and traditions, especially with regard to gross violence and abuses of power, the Bush administration institutionalized a sociopathic model of authority—permitting no recourse to external standards of morality. The relation of the public to the clandestine agencies changed dramatically, eschewing the public check on exposed illegal behavior in favor of a culture of disavowal and blind support, because, in a sense, without external referent for what is legal or illegal the (parental) authority can only be seen as protective or frightening—there is no middle ground.

The APA leadership crudely invoked the power dynamics of the Bush administration, counting on the fact that, if treated with a combination of fear-inducing rhetoric and the aggressive assertion of the right to abrogate any law, treaty, or ethical standard in the name of self-protection, the APA Council would disavow any evidence that its leadership was supporting and, in some cases, participating in abuses. This proved to be an accurate assessment.

Although we may not have been conscious of what we were doing, our group's aim could be seen as an attempt to restore the neurotic position, vis-à-vis government and the APA. Our aim was to reintroduce a sense of guilt and responsibility, by reinvoking the obligation to legal and ethical standards outside of the leadership. We tried to apply international law and the Geneva Conventions to psychologists' role in national security activity, and simultaneously to restore the independence of the APA ethics code. As we discovered that the APA leadership, like the Bush administration, was functioning according to a sociopathic model (acting as if it were in complete control of association rules and needing only to promote a fear-based manipulation of the membership), we realized that we had to go around the leadership and reinstate a process where the exposure of tacit unconscious processes that broke the law and violated ethics might once again be received with guilt, redress, and the imposition of renewed standards. We bypassed the leadership and brought a referendum to a successful vote. Now it is a matter of forcing the leadership to comply with what is now APA policy.

LOVE, WORK, RESIST, ACT

Should we continue the struggle? After all, the Bush administration is history and the Obama administration has vowed to close Guantánamo and asserts that its actions will comply with the Geneva Conventions. What does it really matter, if the APA wants to deny its complicity and maintain its economic and political ties to the Defense Department and the CIA?

For me the problem remains a serious one in the following ways. First, in spite of the new administration's assertions, indefinite detention, isolation, force-feeding, and other abuses continue at Guantánamo and other detention centers. Second, the BSCTs remain a part of intelligence operations, and the same psychologists who created their operating procedures during the times of terrible abuses are determining their future operations. Third, not a single psychologist has been held to account for these abuses, by the APA or any ethics or licensing board. Fourth, APA policy on national security has been dictated and continues to be dictated not only by psychologists who were part of the operations of abuse, but by others who profit privately from military and intelligence contracts.

But at the same time I have been heartened by colleagues around the world who have watched what we've been doing and applaud our willingness to stand up to both the APA and the U.S. government, and to have had a hand in exposing and changing policies of abuse and torture. And I've been part of a generation of analysts making use of our analytic skills to change society for the better.

I've come to believe that the final stage of psychoanalysis is action. Analysis, when all is said and done, is a process of progressive rupture: internal rupture, in that habitual resistance to full experience and cognizance is progressively undermined; and external rupture, in that the analysand becomes progressively adept at resisting the manipulations of personal and political power narratives. Psychoanalysis can be seen as the progressive undermining of internal resistance in order to foster external (or political) resistance. To Freud's dictum that psychoanalysis frees the analysand to love and to work, I would add that to complete the process, it is also necessary to find the freedom to resist and act.

REFERENCES

Alfano, S. (2007, August 20). Psychologists back off ban on detainee aid. *CBSNews.com*. Retrieved October 20, 2009, from http://www.cbsnews.com/stories/2007/08/20/terror/main3184002.shtml

American Psychological Association. (2007, August 19). *Reaffirmation of the American Psychological Association position against torture and other cruel, inhuman, or degrading treatment or punishment and its application to individuals defined in the United States Code as "enemy combatants."* Retrieved August 24, 2007, from http://www.apa.org/governance/resolutions/councilres0807.html

American Psychological Association. (2008a, February 22). *Amendment to the reaffirmation of the American Psychological Association position against torture and other cruel, inhuman, or degrading treatment or punishment and its application to individuals defined in the United States code as "enemy combatants."* Retrieved March 27, 2008, from http://www.apa.org/governance/resolutions/amend022208.html

American Psychological Association. (2008b). *Statement of the American Psychological Association*. Retrieved October 20, 2009, from http://www.apa.org/releases/statement.html

American Psychological Association. (2008c, December) *Report of the APA Presidential Advisory Group on the implementation of the petition resolution*. Retrieved on October 20 from http://www.apa.org/ethics/advisory-group-final.pdf

Arrigo, J. M. (2006). *Unofficial records of the American Psychological Association Task Force on Psychological Ethics and National Security, June 25–28, 2005*. Stanford, CA: Intelligence Ethics Collection, Archives of the Hoover Institution on War, Revolution, and Peace, at Stanford University.

Badiou, A. (2005). *Metapolitics* (J. Barker, Trans.). London: Verso.

Banks, M. (2005). Providing psychological support for interrogations. *Unofficial records of the American Psychological Association Task Force on Psychological Ethics and National Security, June 25–28, 2005*. Stanford, CA: Archives of the Hoover Institution on War, Revolution, and Peace, at Stanford University

Beckett, S. (1967). *Stories and texts for nothing*. New York: Grove Press.

Behnke, S. (2005, December 12). Medical experts debate ethics of military interrogations. *VOA News*. Retrieved October 18, 2009, from http://www.voanews.com/english/archive/2005-12/2005-12-12-voa54.cfm

Benjamin, M. (2006, July 26). Psychological warfare. *Salon.com*. Retrieved October 20, 2009, from http://www.salon.com/news/feature/2006/07/26/interrogation/index.html

Benjamin, M. (2007, June 21). The CIA's torture teachers. *Salon.com*. Retrieved October 19, 2009, from http://www.salon.com/news/feature/2007/06/21/cia_sere/print.html

Bravin, J., & Fields, G. (2003, March 4). *How do interrogators make terrorists talk?* Retrieved October 18, 2009, from http://www.teamdelta.net/WSJ-Interrogation.htm

Bray, J. (2009, August 13). Unpublished letter to the *New York Times*.

Bybee, J. C. (2002, August 1). *Memorandum for Alberto R. Gonzales, Council to the President re: Standards of conduct for interrogation under 18 U.S.C. §§ 2340-2340A*. Retrieved October 19, 2009, from http://www.tomjoad.org/bybeememo.htm

Coalition for an Ethical Psychology. (2008, January 16). *Analysis of the American Psychological Association's frequently asked questions regarding APA's policies and positions on the use of torture or cruel, inhuman or degrading treatment during interrogations*. Retrieved February 9, 2008, from http://psychoanalystsopposewar.org/blog/wp-content/uploads/2008/01/apa_faq_coalition_comments_v12c.pdf

Craig, O. (2003, March 9). *CIA holds young sons of captured Al-Qaeda chief*. Retrieved October 18, 2009, from http://www.nogw.com/download/_07_ksm_young_children_held.pdf

Department of Defense. (2004, December 10). Operational Policy Memorandum #14. Behavioral Science Consultation Team. Retrieved October 18, 2009, from http://humanrights.ucdavis.edu/projects/the-guantánamo-testimonials-project/testimonies/testimonies-of-standard-operating-procedures/bsct_sop_2004.pdf

Eban, K. (2007, July 17). Rorschach and awe. *Vanity Fair*. Retrieved September 5, 2007, from http://www.vanityfair.com/politics/features/2007/07/torture200707?printable=true&currentPage=all

Ephron, D. (2008, October 18). The Biscuit breaker: Psychologist Steven Reisner has embarked on a crusade to get his colleagues out of the business of interrogations. *Newsweek*. Retrieved on October 20, 2009, from http://www.newsweek.com/id/164497

Freud, S. (1900). *The interpretation of dreams*. In J. Strachey (Ed. & Trans.), *The standard edition of the complete psychological works of Sigmund Freud* (Vols. 4–5). London: Hogarth Press.

Gall, C., & Lewis, N. (2003, June 17). Threats and responses: Tales of despair from Guantánamo. *New York Times*. Retrieved October 18, 2009 from http://www.nytimes.com/2003/06/17/international/asia/17PRIS.html?pagewanted=all

Garrison, E. (2009). *Letter to the Director of the Association of State and Provincial Licensing Boards*. Retrieved October 20, 2009, from http://www.apa.org/releases/073109-letter-demers.pdf

Goodman, A, (2007). *APA members hold fiery town hall meeting*. Retrieved October 19, 2009, from http://www.democracynow.org/2007/8/20/apa_members_hold_fiery_town_hall

Goodman, A. (2008, July 18). *The dark side: Jane Mayer on the inside story of how the war on terror turned into a war on American ideals*. Retrieved October 19, 2009, from http://www.democracynow.org/2008/7/18/the_dark_side_ jane_mayer_on

Goodman, A., & Goodman, D. (2008). *Standing up to the madness: Ordinary heroes in extraordinary times*. New York: Hyperion.

Gray, S. (1985). *Swimming to Cambodia*. New York: TCG Group.

Greenberg, J. C., Rosenberg, H. L., & De Vogue, A. (2008a, April 9). Sources: Top Bush advisors approved "enhanced interrogation": Detailed discussions were held about techniques to use on al Qaeda suspects. *ABC News.com*. Retrieved September 30, 2009, from http://abcnews.go.com/print?id=4583256

Greenburg, J. C., Rosenberg, H. L., & De Vogue, A. (2008b, April 11). Bush aware of advisers' interrogation talks: President says he knew his senior advisers discussed tough interrogation methods. *ABC News.com*. Retrieved April 13, 2008, from http://abcnews.go.com/TheLaw/LawPolitics/Story?id= 4635175&page=1

Haynes, W. J. (2002, November 27). *Letter for Secretary of Defense: Counter-resistance techniques*. Retrieved October 20, 2009, from http://news.findlaw.com/hdocs/docs/dod/gcrums1127120202mem.pdf

James, L. (2005, May 23). *APA Presidential Task Force on Psychological Ethics and National Security (PENS) listserv* (p. 47). Retrieved September 30, 2009, from Propublica.org: http://documents.propublica.org/e-mails-from-the-american-psychological-association-s-task-force-on-ethics-and-national-security/page/47#p=47

James, L. (2008). *Fixing hell*. New York: Grand Central Publishing.

Kazdin, A. (2008, October 6). *Letter to General Michael V. Hayden, USAF, Director, Central Intelligence Agency*. Retrieved October 20, 2009, from http://www.apa.org/releases/CIA%20Letter%20on%20New%20APA%20Policy.pdf

Kiley, K. C. (2005, April 13). *Assessment of detainee medical operations for OEF, GTMO, and OIF*. Retrieved October 18, 2009, from http://www.globalsecurity.org/military/library/report/2005/detmedopsrpt_13apr2005.pdf

Kiley, K. C. (2006, October 20). *Memorandum for Commanders, MEDCOM Major Subordinate Commands: Behavioral Science Consultation Policy*. Retrieved October 18, 2009, from http://content.nejm.org/cgi/data/359/11/ 1090/DC1/1

Lewis, N. (2004, November 30). Red Cross finds detainee abuse in Guantánamo. *New York Times*. Retrieved October 17, 2009, from http://query.nytimes.com/gst/fullpage.html?res=9C03E1DE113EF933A05752C1A9629C8B63&sec=&spon=&pagewanted=all

Lewis, N. (2005a, January 1). Fresh details emerge on harsh methods at Guantánamo. *New York Times.* Retrieved October 17, 2009, from http://query.nytimes.com/gst/fullpage.html?res=9D00E0DD1539F932A35752C0A9639C8B63&sec=&spon=&pagewanted=3

Lewis, N. (2005b, July 6). Psychologists warned on role in detentions. *New York Times.* Retrieved October 17, 2009, from http://www.nytimes.com/2005/07/06/politics/06gitmo.html?scp=1&sq=Psychologists%20Warned%20on%20Role%20in%20Detentions%20&st=cse

Lewis, N. (2006, June 7). Military alters the makeup of interrogation advisers. *New York Times.* Retrieved October 17, 2009, from http://www.nytimes.com/2006/06/07/washington/07detain.html?_r=1&scp=1&sq=alters%20the%20makeup%20of%20interrogation%20advisers&st=cse

Marlin, B. (2007, August 12). Expert has stake in cryptic local firm: Consultations tied to CIA interrogations. *Spokesman-Review.* Retrieved September 30, 2009, from http://www.spokesmanreview.com/tools/story_pf.asp?ID=204358

Mayer, J. (2006, February 27). The memo. *The New Yorker,* 32–41.

Mayer, J. (2008) *The dark side: The inside story of how the war on terror turned into a war on American ideals.* New York: Doubleday.

McCoy, A. (2006) *A question of torture: CIA interrogation, from the Cold War to the War on Terror.* New York: Henry Holt and Company.

Moss, T. K. (2002, December 10). *JTF GTMO "SERE" interrogation standard operating procedure.* Retrieved October 20, 2009, from http://wikileaks.org/wiki/Guantánamo%27s_SERE_Standard_Operating_Procedures

Nowak, M. (2009). *Letter to James Bray.* Retrieved October 20, 2009, from http://psychoanalystsopposewar.org/blog/2009/08/12/un-special-rapporteur-on-torture-on-torture-calls-on-american-psychological-association-to-withdraw-psychologists-from-guantánamo-detention-camp/

Office of the Inspector General of the Department of Defense. (2006, August 25). *Review of DoD-directed investigations of detainee abuse.* Retrieved October 19, 2009, from http://www.fas.org/irp/agency/dod/abuse.pdf

PENS Listserv. (2005). *E-mails from the American Psychological Association's task force on psychological ethics and national security.* Retrieved October 18, 2009, from http://documents.propublica.org/e-mails-from-the-american-psychological-association-s-task-force-on-ethics-and-national-security

Reisner, S. (2003). Trauma: The seductive hypothesis. *Journal of the American Psychoanalytic Association, 51,* 381–414.

Shane, S. (2009, August 11). 2 U.S. architects of harsh tactics in 9/11's wake. *New York Times.* Retrieved September 30, 2009, from http://www.nytimes.com/2009/08/12/us/12psychs.html?_r=4&hp=&pagewanted=all

United Nations Commission on Human Rights. (2006, February 15). *The situation of detainees at Guantánamo Bay. Report of the Working Group on Arbitrary Detention.* Retrieved October 19, 2009, from http://www.globalsecurity.org/security/library/report/2006/guantánamo-detainees-report_un_060216.htm

Yoo, J. C. (2003, March 14). *Memorandum for William J. Haynes II, General Counsel of the Department of Defense; Re: Military interrogation of alien unlawful combatants held outside the United States.* Retrieved October 19, 2009, from http://www.aclu.org/pdfs/safefree/yoo_army_torture_memo.pdf

Chapter 7

Torture and the American Psychological Association

A one-person play

Neil Altman

The U.S. response to the attacks of September 11, 2001, initiated a seismic shift in national self-image. Driven by fear and a sense of vulnerability rarely felt on these protected shores, there evolved a sense of entitlement to take whatever action promised to protect the United States from further attacks, leading to support for preemptive war in Iraq and the opening of detention centers outside the framework of previously accepted international law. It was only in 2004, when reports of torture at Abu Ghraib prison, complete with pictures, hit the media that there arose a general public feeling that perhaps this had gone too far. Around the same time, there were reports in the media of possible involvement by psychologists in the development of abusive interrogation techniques (Lewis, 2006; Mayer, 2005) at Guantánamo Bay and elsewhere, accompanied by disavowal by Bush administration officials of the applicability of the Geneva Conventions to the treatment of "illegal enemy combatants." Those placed in this category were held outside the United States to avoid the constraints of the U.S. judicial system as well. Of particular interest to psychologists were reports, in the same articles, that psychologists had been involved in training interrogators at the Guantánamo Bay detention center in psychological methods of torture. For many psychologists, these reports brought into sharp focus the fact that psychologists had made central contributions to modern methods of torture in the form of research in sensory deprivation, sensory overload, the exploitation of psychological vulnerability, and so on, and that we were considered experts in interrogation, including abusive methods of interrogation, by those who organized and ran the new detention centers for illegal enemy combatants. Suddenly, for some of us, this was not just a question of taking a stand on a major social and political issue, such as the American Psychological Association (APA) had done previously in supporting gay rights in the military, for example. This issue of torture implicated psychologists, and our professional organization, as major players.

From its side, the president of the APA, Ron Levant, responded by convening a task force on Psychological Ethics and National Security (PENS). This task force endorsed the notion that there is a valid and ethical role for psychologists in relation to national security interrogations, while maintaining that psychologists may not have any role, including a supporting role, in interrogations that involve torture or cruel, inhuman, or degrading treatment. The report of this task force was approved by a subgroup of the Council of Representatives (COR), the board of directors, shortly before COR was due to meet, so that the members of COR did not have an opportunity to consider the contents of the report before it was made APA policy. At the same time, questions arose about two provisions of APA's ethics code, principles 1.02 and 1.03, which stated that if there is conflict between the ethics code and law, regulations, orders, or organizational requirements, psychologists may follow law, regulations, orders, or organizational requirements. It seemed to some of us that in the light of this loophole, nothing APA said about psychologist participation in interrogations had any teeth. In response to these concerns, COR passed a resolution at its meeting immediately following the adoption of the PENS report, requesting that the Ethics Committee reconsider ethical principles 1.02 and 1.03, and directing the Ethics Committee to produce a "casebook" that would specify more precisely what behaviors were considered ethical and unethical in relation to interrogations.

At the time of the 2006 COR meeting, the Ethics Committee had made no recommendation with respect to ethical principles 1.02 and 1.03 and there was no sign of a casebook. A resolution had been prepared, however, jointly by the Divisions of Military Psychology and the Peace Psychology Division, that reaffirmed a 1986 APA resolution against torture while pointing out that the APA had meanwhile become a nongovernmental organization accredited to the United Nations (UN), bound to support and promote UN conventions and resolutions, such as the Geneva Conventions and those related to the Universal Declaration of Human Rights. The makers of this resolution from Peace Psychology thought that this resolution would form the basis for further, more specific resolutions, related to psychologist participation in interrogations, in line with international law as outlined in the Geneva Conventions and other documents. It seemed to me a slam dunk to think that the next step should be a moratorium on psychologist participation in interrogations in settings where there was no due process for the detainees (i.e., where Geneva Convention protections were absent, as was the case at Guantánamo Bay).

It was thus a shock when, at a meeting of the Divisions for Social Justice (DSJ) at the 2006 COR meeting, the representatives decided not to pursue a resolution calling for a moratorium. One representative said that she felt that a moratorium would be unpopular and DSJ had other issues to which it wanted to devote its efforts, rather than into an issue that appeared to be a losing cause. I was quite surprised and shocked at this assessment; I wondered if the DSJ representatives themselves were opposed to a moratorium and if so, why? Why would these generally socially progressive people be opposed to putting distance between the APA and the detention of people with no due process? The American Psychiatric Association had taken a similar position. What was I missing?

My sense of shock only increased and deepened when I could get only 16 members of COR, less than 10% of the total membership, to cosponsor the resolution, and when people with whom I had been allies on a variety of other issues declined to lend their support to the resolution. Months later, at a meeting of the various boards and committees that consider COR legislation, I ran into strong opposition, to the point of hostility, from members of the committee on Psychology and the Law, who stated with considerable vehemence that it was not the APA's place to take a stand on a legal or political issue (why, then, was there a committee and a division on Psychology and the Law?). The Division of Military Psychology (less surprisingly) decided not to support the resolution and not to offer any suggestions as to modifications to the resolution that would enable the division to support it. At this point, I heard through the grapevine that I was considered "intransigent." Finally, the Ethics Committee of the APA declined to support the resolution. Shortly before the COR meeting at which a vote was to be taken on the moratorium resolution, the board of directors offered a substitute resolution that specified certain prohibited interrogation techniques, and acknowledging that the "conditions of confinement" could constitute cruel, inhuman, and degrading treatment, but without mentioning a moratorium.

The persistence of my sense of shock (Adrienne Harris, in a March 2008 personal communication, points out that I used this word three times in this account) highlights the question as to what, in this process, I was perhaps refusing to recognize and accept. I was preoccupied with the suffering of the detainees and a sense of distress and guilt about the violations of human rights and suffering imposed by my country, in my name. I kept running up against resistance to the moratorium proposal that reflected a different set of priorities that I couldn't, or didn't want to, identify with. Following Freud's (1919/2003) sense of the "uncanny" as something once familiar that has been repressed or, I might prefer to say, disowned, I took the opportunity of being asked to contribute a chapter to this book to see if I could articulate the side of myself that *could* identify with those who opposed my efforts. This effort also meant coming to terms with the anxiety and pain behind such striking disavowal. Thus, emerged the following "dialogue with myself."

NEIL 1: [*Shouting*] This is absolutely outrageous! I feel like I don't know these people anymore! I don't often feel that there's something going on behind the scenes, a conspiracy, but clearly there's some investment in being a part of this "national security" industry that's developing, including Guantánamo Bay. This is the same American Psychological Association that took a stand against the Vietnam War? These are the same people who line up to fight for social justice, against racial and ethnic discrimination inside and outside the APA, and with whom I worked so well on the task force to resolve differences about the APA's participation at the UN's World Congress Against Racism? Then as soon as they become part of the APA's power structure, suddenly they're buying the party line that psychologists need to be part of the Guantánamo Bay interrogation project to [*mockingly*] "protect the detainees"?

NEIL 2: Hold your horses! Listen to your tone of voice! You've got to be careful not to get swept up into a war of your own here. You of all people, doesn't your psychoanalysis teach you that polarized people tend to become mirror images of each other, unconsciously? Do you want to perpetuate a war within the APA?

NEIL 1: [*Taken aback, and briefly speechless*] Well—OK, but what room is there not to be totally outraged here?

NEIL 2: Is it possible that you're not the only one with ethical principles here? Is it possible that these so-called conspirators, in bed with the torturers, are actually following ethical principles that you're overlooking? After all, if so many people that you like and respect are coming down on the other side of an issue, wouldn't you want to get off your high horse, and stop and think about what you're missing? Aren't you getting a little self-righteous here?

NEIL 1: [*Stung*] OK, OK, tell me what I'm missing.

NEIL 2: Ummm—let me think about that for a second. OK—suppose they really believe that psychologists do serve a useful function keeping an eye on things; after all, Michael Gelles, a psychologist, did blow the whistle on the torture of Mohammad Al-Qatani, leading to changes in what were considered permissible interrogation techniques at Guantánamo Bay. And isn't there a point to the idea that the APA can work toward change better from within than by backing out? Wouldn't that amount to an abandonment of the detainees? But even more fundamentally, can't you make room for the possibility that some of these "conspirators" actually believe that they're doing the right thing, and not just the expedient thing?

NEIL 1: Let me think about that for a second ...

NEIL 2: Plus, and I hate to say this, but you sound to me a little like one of those White liberals who could afford not to have to consider a military career to pay your way through college, who can afford to sit in your cafés in the sophisticated big cities, drive your Priuses and leave it to the less privileged to fight so you can have your comforts, while you denigrate the dirty work they do on your behalf!

NEIL 1: [*Defensively*] Now, hold on a second ... [*stumbling over his words*] but ... we're talking about very misguided efforts to protect the United States, aren't we? [*Regaining his footing*] What's going on at Guantánamo

Bay isn't helping to protect this country—most of the people were just picked up by anti-Taliban people in Afghanistan for the bounty money, and there's no due process in place even to separate those who might be truly dangerous from those who are not. Meanwhile, as the U.S. loses its moral standing in the world, and enrages people with its use of violence, are we not just playing into the hands of those who want to damage this country? Stop for a minute and think about the lives already ruined among those being held without charges at Guantánamo. How much are we helping those people by helping to plan [*sarcastically*] "humane" interrogations? Much better that we should take a forceful position that the whole thing is misguided and morally wrong and inconsistent with the basic ethical principles of the APA. And nothing I'm saying would be against psychologists being there as health care providers; they could still be alert to torture from that vantage point.

NEIL 2: All very true on a theoretical basis (your preferred basis, I must say), but we're also talking about emotional realities here, aren't we Mr. Psychoanalyst, oh excuse me, Dr. Psychoanalyst?

NEIL 1: Now who's getting nasty? And, by the way, you put down the Prius drivers, but what kind of car do you drive? We haven't even gotten to the complacency that's going to leave the legacy of global warming to our children and grandchildren. Wars for water, massive displacements of people, etc. We'll get to the issue of complacency here when you're done with your particular rant.

NEIL 2: OK, fine. [*Taken aback*] I'm just saying that you can't simply disassociate yourself from the brutality of Guantánamo Bay. You weren't personally affected by 9/11, but don't you remember how impressed you were at the time with the incredible and mindless destructiveness of the people who organized and carried that out? You may be right about the counterproductiveness of the U.S. response to 9/11, but can you dissociate yourself from the emotional response? Don't you want to be in dialogue with that emotional response, rather than simply wash your hands of it?

NEIL 1: That sounds right. So then how do we, if I can speak for both of us here, take a strong stand without getting self-righteous? It seems to me that recognizing that but for the grace of God I could be a Guantánamo interrogator or military policeman means that I'm no less obligated to take a stand against that part of myself and that part of other people. As a matter of fact I'm more obligated to take a stand if I recognize that I have the potential to go the wrong way on an issue of such importance. And make no mistake about how important this is. Speaking of "that but for the grace of God"—that but for the grace of God, you and I are picked up by the Northern Alliance in Afghanistan and held at Guantánamo Bay indefinitely. Can you even begin to imagine how unbearable life then becomes? How unbearable even one second of such a life becomes? And I'm not even talking about torture yet. I'm just talking about being held indefinitely by people who assume you're some kind of monster. And our government is doing this! Your government and mine!

NEIL 2: I didn't vote for Bush!

NEIL 1: But you're paying your taxes and some of that money is paying for what's going on at Guantánamo Bay, not to mention the CIA black sites. Doesn't that make you feel like you have to get up and take a stand?

NEIL 2: Yes, but what you don't seem to be getting is that there's a difference between getting up and taking a stand to make yourself feel better because you feel guilty, and doing something that's actually going to make a difference in getting the APA to move toward a stronger position. You forget I'm actually basically on the same side as you. If people feel you're being self-righteous or moralistic, they're not going to listen to you. Nobody wants to be made to feel guilty. Sometimes I have trouble listening to you myself. Didn't someone once say that the first rule of effective political behavior is not to argue your positions on a moral basis? You can't set yourself up as morally superior, or people will shut down.

NEIL 1: You know, I might agree with you if I thought that everyone on the council were well meaning people who were really struggling with the humanitarian issues. But, to tell you the truth, there are some people like that, and then there are others who care only about their self-interest and the interests of their particular guild group. Have you noticed how on the council listserv, so many of the messages have to do with whether the fitness room in the health club will be available to council members at the meetings, or whether there will be outlets for their laptops at the meeting? The APA makes everyone so comfortable by putting them up in fancy hotels, feeding them, paying for business class on Amtrak Acela trains, etc. They make you so comfortable, you feel so well cared for by APA, it's very seductive. Nothing is conducive to making you feel that there is a moral crisis at hand, or that the APA might be anything other than a benevolent and generous parent.

NEIL 2: So given what you're saying, what does that imply for your strategy? Does that mean it's going to be effective to pound the table and lecture about human rights?

NEIL 1: No, but it means that a certain number of people don't want to hear what you're saying no matter how you say it. They will follow the APA leadership wherever the leadership wants to take them.

NEIL 2: And how do you propose to influence the leadership?

NEIL 1: Ahhh—now you're getting to the tough part, and I think the part where you're going to make me admit that I don't have the stomach for real political change. I think there are strong forces in the APA that have a vested interest in developing a "national security" specialization in psychology. Take a look at the paper by Shumate (2004) that describes the "contributions" such specialized psychologists could make. I think these people have been a strong influence on the APA leadership, and that the leadership is sympathetic to such government collaborations to begin with. They will move away from the present position if and when the political winds blow differently in Washington and it begins to look like there's no future in supporting the work at Guantánamo Bay.

NEIL 2: You make it sound as if nothing you or your fellow travelers do can make a difference, then, it's all about larger political currents.

NEIL 1: I think there's something to that. But we need to recognize that the "national security" lobby, a subset of the politically conservative group that also supports collaboration with the corporate establishment (e.g., on managed care and "evidence-based" [in the narrow sense] treatments), has done its homework and constitutes a majority of the council members. If we really want to see change in the APA, we need to organize people who agree with us, and then get them to nominate and vote for people who share our values. That's what it takes. It's a matter of hard work over a period of years. And to tell you the truth, I don't have the temperament for that. I get too angry and I want to walk away. I want to hang out with people who already share my values, rather than work hard and over a very long term to change institutions that have a lot of inertia and vested interests at work in them.

NEIL 2: So maybe, in the end, people recognize that you aren't all that committed to the APA, and that's why they won't listen to you.

NEIL 1: Maybe so—but hey, wait a minute, what do you call commitment, if not that I served for 6 years on the Council of Representatives and the task force that resolved a highly fraught controversy over the APA's participation at the UN's World Congress Against Racism, a controversy that threatened to precipitate a Black–Jewish split within the council?

NEIL 2: OK, OK, maybe I didn't say that right. Maybe what I mean is that you don't have the right temperament to hang in there past a certain point with people who disagree with you.

NEIL 1: Excuse me—it's not simply a matter of disagreement! I can deal with disagreement, but when there's a moral imperative involved, like not colluding with torture, I do run out of patience at a certain point, especially when I feel there's self-interest driving the collusion.

NEIL 2: Even if you know that, strategically speaking, you might make some progress, not total progress, but some progress, by hanging in there with those people and by virtue of your continuing to agitate for your position, over time you extract concessions, and even turn some people around.

NEIL 1: You know what I realize—this discussion replicates what is at issue about the APA and Guantánamo Bay: the APA leadership says stay there so we can continue to press for changes, maybe not a complete turnaround, but by virtue of continuing to show up, changes can be made, whereas I'm saying "what's going on here is egregious, I want nothing to do with it," hoping thereby both to cleanse myself morally and to put maximum pressure on the APA. If I give the maximum benefit of the doubt to the APA leadership, which I'm not sure I do, I'd say we've gotten to the heart of the difference in approach here.

NEIL 2: Maybe we can agree that there may be some elements in the APA leadership with a more cynical and self-serving motivation, while others truly buy the argument that psychologists are there to protect the detainees.

NEIL 1: I guess I'll join you there.

[CURTAIN.]

POSTSCRIPT AND COMMENTARY

As mentioned above, shortly before the meeting of the Council of Representatives in August 2007, the APA Board of Directors proposed a substitute motion meant to preempt the moratorium I had proposed. This substitute motion did not address the question of a moratorium but instead listed specific interrogation techniques that would be prohibited in terms of psychologist participation. While the substitute motion acknowledged that the conditions of confinement themselves could be abusive, it did not take the further step of stipulating that psychologists should not support the purpose for which these detention centers existed. Some of the cosponsors of the moratorium resolution engaged in negotiations with the board of directors to reinsert a proposal for a moratorium that would be an amendment to the board of directors' substitute motion. In the end the substitute motion was approved and the amendment was defeated.

I now recount one series of events that took place in the course of this process that was particularly meaningful to me. Some of those who supported the moratorium resolution at one point made allegations against Larry James, an African American colonel who had been chief psychologist at Guantánamo Bay at one point and who was transferred to Abu Ghraib after the revelations of abuse there to repair the situation. Some of the ethnic minority members of the council reacted with outrage to these allegations, marshalling evidence that they were unfounded. This incident seemed to bring to the fore an ethnic, racial, and social class dimension to the discussions around the moratorium proposal, on which I will elaborate shortly.

On the other hand, the Division of Ethnic Minority Psychologists had voted to support the moratorium resolution, and the council representatives of that division were among the cosponsors. The event on which I want to focus took place at the meeting of the "Ethnic Minority Caucus" at the August 2007 council meeting. I made a brief presentation in support of the moratorium proposal. The chairperson of the caucus then asked me: "How would a moratorium affect ethnic minority psychologists?" I asked "Are you asking that because ethnic minorities are overrepresented in the military and would thus be disproportionately put in a position of conflict if the APA takes this stand?" He answered in the affirmative. I then responded that it should be kept in mind that the detainees were also largely people of color, and that historically the human rights of people of color were violated disproportionately. Subsequent to this meeting, the ethnic minority members of COR voted as a group to support the moratorium amendment.

This incident captures some of the complexity and multidimensionality of the discussions around the moratorium proposal. People from poverty or working class backgrounds disproportionately use military service as a way to fund vocational training and higher education, and to find employment. Thus, opposition to military activity of various kinds can easily seem to be the prerogative of those of higher class who can find employment or fund their own education without the assistance of the military. In the course of advocating for the moratorium resolution there were people who had to remind me that the plight of those who would be caught in the middle between the APA and the military were disproportionately of working class and ethnic minority background.

Neil 2 brings out a social class dimension to the discussion when he accuses Neil 1 of being a privileged White liberal "Prius owner." On one hand, the allegation here is that White people and people of economic privilege benefit from the protection of the police and the military, then try to have it both ways by turning against them with accusations of violence and human rights violations. The poor and working class fight wars for the privileged, and are disowned and criticized for their trouble, in an exercise of White liberal guilt expungement. Neil 1 struggles to acknowledge the truth in this characterization, and the way it feeds opposition to his position, while maintaining a space for principled opposition to human rights violations. Neil 1 must also find a way to advocate for his position while acknowledging the way it can be seen as out of touch with the socioeconomic realities of the poor, the working class, and people of color. I was able to do so in the Ethnic Minority Caucus, as discussed earlier, by acknowledging the concern of the meeting's chair and by pointing out how the sword of being a member of an ethnic minority group cuts both ways when it comes to torture.

Having engaged in a "therapeutic dissociation" (Davies, 1999), pulling out the various voices competing to be heard within me, I am now faced with the task of pulling myself together somehow. Having listened to both voices, I am more securely in a state of conflict rather than dissociation (Bromberg, 1998), in a position to make a choice as to which voice to listen to at a particular moment. Does it help to recognize that the dilemma as to whether to quit in protest or to keep on working from the inside is perpetuating itself, that a kind of enactment is organizing my dilemma? I think perhaps it helps me to realize that to the extent that I am open to the situation around me, I am bound to be conflicted in the way that I have described. As I write, the choice that presents itself is whether to continue withholding my dues to press the APA to take a stronger stand against abusive, cruel, and inhuman conditions of confinement, or whether to quit altogether, giving up on any effort to change the APA from the inside. I can see that those who are working from the inside are making progress, strengthening language that makes it clear that so-called enhanced interrogation techniques are unacceptable to the APA. For what it's worth, however, I feel myself most powerfully responding to a sense of alienation from the guild mentality that I observed on the APA Council of Representatives. For a period of time, I could accept that self-interest was what organized most of the discussions there, because there was a large enough forum for discussion and action around social justice and related issues. I could also recognize, and do continue to recognize, that I count on the economic privileges that organized psychology has won. However, when a collision developed between guild interest and making a statement against the cruel and inhuman treatment of people, and when the APA leadership took a stand that seemed to me to be so clearly and one-sidedly motivated by guild interest, I feel a wave of alienation that makes me feel this is not my organization.

[FROM THE WINGS] SO YOU'RE GOING TO ACT FROM YOUR GUT—AFTER ALL THAT THINKING WE DID TOGETHER?
[STAGE LIGHTS GO OUT.]

REFERENCES

Bromberg, P. (1998). *Standing in the spaces: Essays on clinical process, trauma, and dissociation*. Hillsdale, NJ: The Analytic Press.

Davies, J. (1999). Getting cold feet, defining "safe-enough" borders: Dissociation, multiplicity and integration in the analyst's experience. *Psychoanalytic Quarterly, 68*, 184–208.

Freud, S. (2003). *The uncanny* (A. Phillips, Ed.). London: Penguin Books. (Original work published 1919)

Lewis, N. (2006, June 7). Military alters the makeup of interrogation advisers. *New York Times*. Retrieved from http://www.nytimes.com/2006/06/07/washington/07detain.html

Mayer, J. (2005, July 11). The experiment: Is the military devising new methods of interrogation at Guantánamo? *The New Yorker*. Retrieved from http://www.newyorker.com/archive/2005/07/11/050711fa_fact4

Shumate, R .S. (2004). Psychologists' evolving role in federal counterterrorism activity. Paper presented at American Psychology and Law Conference, Scottsdale, AZ, March 5, 2004.

Chapter 8

Violence in American foreign policy
A psychoanalytic approach

Frank Summers

The American public has in the past few years accepted two programs of violence that lack legal, political, and ethical justification. The responsibility for America's unjustifiable transgression of Iraq's borders and the transformation of the United States into a torture society lies not only with the Bush White House, but also the American people who supported those policies of violence for which the administration proposed flimsy and often transparently false rationalizations. The question addressed by this essay is: How can we understand the acquiescence of the citizenry in these blatant violations of international law and universal ethical values?

The Iraq War was justified on two grounds: Saddam Hussein's alleged possession of weapons of mass destruction and the bombing of the World Trade Center. When the Bush administration began a propaganda campaign to "sell" an invasion of Iraq (e.g., Rich, 2007), the public quickly and overwhelmingly supported the invasion. Despite the fact that the 9/11 commission announced in 2004 that it found no evidence of any relationship between Hussein's regime and the Islamic terrorist group, as recently as 2006, 64% of the American public held the view that there was a "strong connection" between Hussein and al-Qaeda. Moreover, no weapons of mass destruction were ever found, and the United Nations (UN) inspectors concluded none had been there for years before the invasion.

Furthermore, detention centers, such as the sites at Guantánamo Bay and Bagram Air Base, in which people are held indefinitely without being charged with crimes and denied due process rights, violate U.S. and international law as well as agreements to which the United States is a signatory. Abundant evidence has been made public showing that torture techniques were used at these sights and routinely at Guantánamo Bay (e.g., Human Rights Center, 2008; Lewis, 2004; Office of Inspector General, 2004; UN Commission on Human Rights, 2006). So, less than 2 years after the World Trade Center attacks, the United States violated the territorial integrity of a nation that was no threat to it, set up illegal detention camps, and established the use of techniques defined as torture by international agreements.

Why were the American people so easily deceived into supporting a war with a clearly baseless rationale and so willing to accept the use of illegal detention centers? Whatever factors led the Bush administration to invade Iraq under false pretenses, its success in controlling American public opinion has to be understood in terms of the emotional state of the nation. The public was so willing to go along with a call to action that avowed values and principles were overwhelmingly ignored and violated with little objection. Here a psychoanalytic approach to the state of the nation is called for. When an individual acts in violation of long-held ethical principles, psychoanalysts search for current precipitants and unacknowledged character traits to make sense of the behavior. Similarly, in the case of a nation that transgresses principles it claims to uphold, we must understand that behavior from the viewpoint of both its current state and disavowed national character traits. It is the purpose of this essay to use psychoanalytic inquiry to understand both the state of susceptibility of the American public after the World Trade Center attacks and the long-standing national characteristics that gave rise to this state.

RESPONSES TO HELPLESSNESS: UNDERSTANDING THE AMERICAN PSYCHE

In the aftermath of 9/11, the country experienced a sense of vulnerability that was perhaps unprecedented. The ability of an adversary with primitive resources to penetrate the boundaries of a nation that long regarded itself as impenetrable was a humiliating injury as well as a devastating loss of life and national resources. Without a clearly identified enemy toward whom to direct revenge, a sense of helplessness pervaded the national response along with shock, grief, anger, and mourning. The attack on Iraq and the establishment of detention centers reflected a need to *do* something in retaliation for the 9/11 attacks. The invasions of Iraq and Afghanistan received wide public support, and the nation did not differentiate between the two wars, despite the fact that the Taliban regime in Afghanistan was a haven for training al-Qaeda terrorists and Iraq had no role in the 9/11 assault. Shortly after the invasion, polls found that 72% of Americans supported the Iraq War despite the lack of evidence of chemical or biological weapons and 79% thought the war was justified with or without conclusive evidence of illegal weapons (Milbank & VanderHei, 2003). The fact that these two invasions received similarly overwhelming public approval indicates widespread backing for the use of force however irrational it might be. Military action transformed national helplessness into a sense of control, a renewed belief that the country was taking charge of the situation.

The same analysis can be applied to the introduction of torture techniques as an interrogation tool. The establishment of detention centers outside of all international and civil law began shortly after 9/11 without protest from the population at large. Former Attorney General Ramsey Clark's (2003) appeal to maintain the rule of international law was barely noted by the American media and the polity. The rounding up of citizens of Afghanistan and other countries and sending them to indefinite detention without due process rights provided a sense of relief that action was being taken. The fact that there was no evidence of wrongdoing for the majority of the detainees (Denbeaux, 2006) was disregarded in favor of the feeling of potency. Holding prisoners indefinitely and even the use of torture enjoyed overwhelming public support two years after the establishment of detention centers (Stephens, 2003). Even after it was clear these camps did little good and, in fact, isolated the United States and damaged its international reputation, as recently as 2007 Americans were evenly split in their views of Guantánamo (Moore, 2007). The fact that there was a drop in public support following the negative consequences of these camps does not diminish the reality that they enjoyed broad support for many years.

Moreover, the detainees were subjected to illegal so-called enhanced interrogation methods, such as prolonged isolation, sensory deprivation, excessive heat and light, waterboarding, and physical intimidation and violence, which constitute torture under international law (Human Rights Center, 2008; Office of the Inspector General, 2004; Sands, 2008; UN Commission Report, 2006). While rationalized as methods to extract information from unwilling prisoners, the research is clear that such torture does not provide accurate information (Fein, Lehner, & Vossekuil, 2006). The most reliable information is obtained through rapport building, as all professional interrogators know (e.g., Alexander & Bruning, 2008; Kleinman, 2008). Furthermore, as has been well established by the photographs from Abu Ghraib, torture inflames hatred of the torture nation and erodes the ethical standing of the country (Arrigo, 2003).

The purpose of torture is not to extract information, but to subjugate the victim and obtain complete control. When General Ricardo Sanchez decided to "Gitmoize" Abu Ghraib, General Janis Karpinski said he put it this way, "You have to treat the prisoners like dogs. If you treat them, or if they believe that they're any different than dogs, you have effectively lost control of your interrogation from the very start" (Danchev, 2006). This attitude toward prisoners is not about obtaining information or keeping the country safe, but about the exhilarating experience of strength that comes from being in complete control of others. Lieutenant Colonel Diane Beaver, a lawyer at Guantánamo, observed that support for the most egregious techniques was a sign of "toughness," and she was struck by the excitement of the officials at Guantánamo as they discussed ideas for torture techniques such as smothering and waterboarding, "You could almost see their dicks getting hard," she noted (Sands, 2008, pp. 62–63). Helplessness disappears in the excitement of subjecting the victim to one's will. Beaver's comment on the specifically male representation of strength confirms that potency was most coveted in the aftermath of 9/11, and torture provided it.

The fact that the sense of vulnerability evoked by the attack was intolerable and had to be replaced by a show of strength indicates that the nation had suffered not only a loss of lives and resources, but a blow to the way it views itself. What is this self-representation that suffered so greatly from a terrorist assault that it produced a violent, irrational, and ultimately counterproductive response? Just as individuals can be understood only via their history and character, to answer this question, we must review, however briefly, the history and character of the American self-representation.

THE AMERICAN SELF-REPRESENTATION

The image the Puritans possessed of themselves in their mission to the New World was depicted by minister John Winthrop in his famous 1630 sermon delivered before the ship landed: "We are as a shining city on a hill. The whole world is watching." Winthrop's belief that the Puritan purpose was a mission from God that affected people everywhere was continued in the establishment of the republic. The major political figures of the American Revolution saw their mission as spiritual as well as political. They saw the democracy they created as a "glorious task assigned to us by Providence" for the liberty of all people. Jonathan Edwards, the most famous preacher of the 18th century, preached that America would usher in the Second Coming. Ezra Stiles (1783), the president of Yale, expressed in eloquent terms a common view of the revolutionaries when he called the young nation "God's American Israel" that would lead the march toward perfectibility by "spreading the seeds of liberty through the habitable world." Even the committed scientist Benjamin Franklin shared the belief that the establishment of the republic was a task assigned by Providence for the benefit of all humanity. This notion of America as specially chosen by Providence to create and sustain liberty for humankind was central to the self-representation of America as it became a nation. Often labeled "American exceptionalism," it tethered American prosperity, virtue, and strength to the fate of humanity.

As the nation expanded, the term American exceptionalism came to embrace potency as much as special goodness. In 1807 when the British brazenly attacked American ships including the *Chesapeake* just 12 miles off the American shore, the young country was enraged at what Henry Clay called the "shameful degradation" and felt obliged to defend its honor. John Calhoun warned the British that the citizenry "not only inherited the liberty which our fathers gave us, but also the will and power to maintain it" (Watt, 1987, p. 266). The success of the Battle of New Orleans in the War of 1812 further emboldened the image of America's strength and superior virtue as reflected in Francis Scott Key's paean to the nation's endurance. Furthermore, military ability increasingly became the definition of national strength. Tocqueville (1835) noted that the national reverence for Andrew Jackson showed the power of military prowess over the nation's spirit. Newspapers described the victory in the Battle of New Orleans as "virtue over vice" and the salvation of liberty (Watt, 1987). Such encomiums reflect the nationwide belief that America's military successes constituted the triumph of good over evil, the spread of virtue for all. Indeed, one of Tocqueville's (1835) strongest impressions of the American people was their self-regard: "They believe they are the only religious, enlightened, and free people. ... They have an immensely high opinion of themselves and are not far from believing that they form a species apart from the rest of the human race" (p. 374). Even more poignantly, Tocqueville (1835) said of America "the least reproach offends it, and the slightest sting of truth turns it fierce; and one must praise everything. ... Hence the majority lives in a state of perpetual self-adoration" (p. 317).

What many historians call the "myth of America," a narrative of moral superiority, omnipotence, and a destiny of prosperity, has been used historically to justify international intervention. Fittingly, in the 19th century, the "shining city on a hill" was transformed into the Monroe Doctrine, which, despite its putatively protective purpose, implied American superiority in the Americas and was used by John Quincy Adams as a rationale for westward expansion. It was a short conceptual step to Manifest Destiny, the doctrine used to justify the acquisition of all the Western territory to the Pacific Ocean on the basis of America's "goodness" (Stephanson, 1995). That policy, in turn, was "inverted" by Theodore Roosevelt's famous 1904 corollary to become a basis for American intervention in the Western Hemisphere. For our purpose, Roosevelt's rationale is telling: "A chronic wrongdoing or impotence which results in a general loosening of the ties of civilized society, may in America, as elsewhere, ultimately require intervention by some civilized nation and ... may force the United States ... to the exercise of police power." By implying that the United States was the only "civilized society" in this part of the world, Theodore Roosevelt not only brazenly declared American superiority, but also arrogated to the United States the status of policeman of the Western Hemisphere.

Neither the Monroe Doctrine nor Roosevelt's Corollary were enforceable until the United States became a major military power after the victory in World War I. President Woodrow Wilson adopted a foreign policy rooted in the principles of self-determination and international cooperation, both of which he hoped would be guided by the newly formed League of Nations. Nonetheless, he was not hesitant to use unilateral military force to intervene in Nicaragua, Haiti, the Dominican Republic, Mexico, and Panama even if that meant overthrowing democratic governments. By assuming American superiority and purity of intention, Wilson felt justified in taking action in blatant contradiction of his principles. As with Theodore Roosevelt, the assumption that the superiority of the United States entitles it to decide what is acceptable in other countries was not questioned by either the champion of "self-determination" or the citizens he led.

The policy of overt or covert deposing of regimes to ensure governments of the United States' choosing became an increasingly common feature of American foreign policy throughout the 20th century. From the end of World War II to the close of the 20th century, the United States attempted to depose 40 foreign governments unilaterally and on 30 other occasions tried to suppress nationalist movements organized against dictatorial regimes (Kinzer, 2006). Examples of the U.S. overthrow of democratic governments in favor of American backed dictatorships in the post-World War II era include the 1953 Central Intelligence Agency (CIA) coup ousting of the only democratically elected leader in Iranian history, Mohammed Mossadegh, the Arbenz regime in Guatemala in 1954, the Goulart government in Brazil in 1964, and the CIA-backed coup of Salvador Allende in Chile in 1973. None of these violations of territorial integrity drew significant opposition from the American people, and none of the regimes responsible lost any public support due to their intervention. Less successful efforts include the Bay of Pigs invasion and the wars in Vietnam and Iraq both fought on false pretenses.

Wilson's contradictory foreign policy stands as an iconic reflection of the character of the United States not just for Wilson's time but for much of the nation's history. Interference in the sovereignty of other nations has coexisted alongside American claims of upholding the principle of self-determination for all peoples at least since Wilson's presidency. This willingness of the American public to support this long history of interventionist exceptionalism with little significant opposition demonstrates that a sense of superiority and entitlement is deeply ingrained in the self-representation of the United States. Widespread opposition to aggression against sovereign nations has tended to occur only after the cost has become great in lives and resources, as in Vietnam and Iraq. This stark reality indicates that limitations in America's sense of entitlement to interfere with other countries have largely been imposed by harsh consequences rather than motivated by violation of ethical principle.

GRANDIOSITY AND AMERICAN IDENTITY

The self-representation of the United States as a special nation that by virtue of its superiority can determine events and reality for the world is equivalent to that of an empire. Nonetheless, the United States sees itself as a republic, not an empire. Whereas other imperial powers such as Persia, Rome, Napoleonic France, or the Soviet Union conquered other ethnic groups and absorbed them into their territorial boundaries, the United States ceased such territorial expansionism at the end of the nineteenth century while continuing to extend its empire. Called an "informal empire" or "invisible empire," the American empire, rather than being demarcated by territorial boundaries, operates by controlling the governments of nations that maintain their identity. In fact, one of the key characteristics of the American self-representation as "all good" is that it is not an empire, but a powerful nation that uses its resources to promote freedom throughout the world. This denial of imperial status allows the United States to operate at least 750 military facilities in 159 countries and territories while claiming to be a beacon for self-determination for the peoples of the world (Hoff, 2008). This narrative puts the rationale for violence and covert invasion on a putatively ethical basis. That is to say, the national self-representation is split between the conscious image of ethical purity and a disavowed omnipotence that believes in the U.S. capacity and right to control the world and events as it sees fit. Both are images of national superiority, a grandiose view of the United States as special and above the rules that apply to other nations.

That exceptional capabilities of the United States are a critical part of the American ethos is demonstrated in the emphasis it receives in national political campaigns by candidates of both parties. The continuity and power of the shining hill image has been explicitly affirmed by two modern presidents, John Kennedy and Ronald Reagan, as well as 2008 presidential candidate John McCain. Whether liberal or conservative, no politician runs a campaign on the basis of the United States as a nation with strengths and weaknesses. As the grandiosity of U.S. identity has been increasingly tied to military strength, politicians live in fear of being labeled "weak." Lyndon Johnson kept expanding the Vietnam War because he feared that if the Communist side were victorious, he would be seen as a coward and the United States as weak. He believed that the American people "will forgive you for anything except being weak" (R. Dean, 2001). Similarly, Richard Nixon's (1970) rationale for invading Cambodia was to insure that America not act like a "pitiable, helpless giant." Fritz Kraemer, the patriarch of the neoconservative movement, told Donald Rumsfeld on the eve of the Iraq invasion, "No provocative weakness, Mr. Secretary" (Hoff, 2008, p. 128). Tellingly, "weakness" is equated with reluctance to use military force.

As Hannah Arendt (1969) demonstrated long ago, violence is not power but its opposite. Where leaders lack authority from the citizenry, power is absent, and violence is used to obtain control. The United States, in its efforts to control other nations, resorts to violence because it lacks the power to influence. The extraordinary American capability for violence has stimulated a grandiosity that is confused with power by the American people and their leaders.

Public figures who simply question the ethic of a particular action, such as the invasion of Iraq or Vietnam, or the adoption of torture, without questioning the goodness or capability of the country, have become targets of vituperous assaults for "tearing it down." One can only imagine the invective that would be directed at any who would set into question the image of the country as holding a special place in the family of nations. While every nation has its narrative of national pride, what differentiates the cultural identity of the United States is the self-concept of omnipotence and moral purity historically imbued with religious connotations that conflate the interests of the United States with the liberation of the people of the world (Hixson, 2008).

The belief in American omnipotence requires a disavowal of failure in our ability to control events. A component of the American illusion of grandiosity is that we have never lost a war. Referring to possible withdrawal from Vietnam, Richard Nixon famously said he would not be the "first American president to preside over the loss of a war." Not only did this statement conveniently ignore the War of 1812, it also presaged our withdrawal from Vietnam and the defeat in that war. Ronald Reagan revised history by claiming that, in fact, the United States had won the war in Vietnam. This gross reversal of reality was challenged neither by politicians nor citizens. No public figure was willing to risk his career by a disagreement that would have acknowledged an American defeat. Elections can be won by trumpeting our extraordinary capabilities, never on a realistic assessment of strengths and weaknesses. The appeal to the American public of the rhetoric of national self-aggrandizement is demonstrated in political rhetoric.

Those politicians who have dared to question the grandiose assumptions of American foreign policy reside in the dustbin of failed political careers. In his 1972 presidential campaign, George McGovern tried to highlight the ethical malaise in which the country was ensconced by saying that we had "lost our ideals." Regarded by the American electorate as "weak" and "soft on communism," McGovern lost all but one state to Richard Nixon who emphasized America's strength and promoted a virtually unlimited image of our nation's capabilities.

Jimmy Carter shared McGovern's belief that the country had lost its moral compass, but his campaign rhetoric did not tarnish the image of goodness or strength of the American people. Carter campaigned as an outsider to Washington on the idea that the government must be "as good as the American people." This rhetoric applauding America's virtues while demonizing politicians who did not uphold them worked well enough for an electoral victory. However, when, after assuming the presidency, Carter shifted his discourse by inveighing against American consumer society and the resulting meaninglessness of American life, he lost significant popularity even before the Iranian takeover of the American embassy. His subsequent inability to free the hostages was regarded as a weakness that resulted in a national humiliation. Ronald Reagan (1979) beat Carter decisively by railing against "those who would have us believe that the United States, like other great civilizations of the past, has reached the zenith of its power." In announcing his candidacy, Reagan referred to Winthrop's depiction of America as a "shining city on a hill." His winning campaign was built on the proposition that America would gain ever greater power in the world and increasingly bountiful prosperity at home. That message of unblemished American capability resonated sufficiently with the American people for an overwhelming electoral victory.

To be sure, attempts to conduct U.S. foreign policy on a more realistic basis than unlimited U.S. power have emerged at times when events have limited America's ability to impose its will on other countries. George Kennan's "containment" policy of the Soviet Union to Richard Nixon and Henry Kissinger's détente were sober efforts to accept restrictions on America's ability to control the world, however each such effort has elicited a strong counterreaction asserting American entitlement. The neoconservative movement began to swell in response to dissatisfaction among some Americans with Kissinger's détente policy, which they regarded as an unnecessary accommodation to America's enemies. Dick Cheney's Defense Planning Guidance of 1992 promoted unilateral, preemptive military action with no basis other than U.S. military capability (Hoffman, 2005).

The Project for the New American Century (PNAC) was founded during the Clinton years with a plan for American domination of the world including unilateral military intervention and preemption before the bombings of the World Trade Center (Wolfowitz, 2000). Eighteen of its members, in an open letter to President Clinton, insisted that regime change in Iraq should be the aim of American foreign policy. Ten of the 18 signatories joined the Bush White House in key positions. In 2000 PNAC called for preemptive action against Iraq, the construction of several bases there, and the transfer of nationalized Iraqi oil industry to Western companies. The fact that these documents were issued well before 9/11 indicates that the Iraq invasion was a product of American entitlement rather than a mistaken effort at self-defense. Proposals based on frustrated grandiosity found adherents but could not be adopted as U.S. policy until 9/11 because the failure in Vietnam had still relegated American grandiosity to a latent state. The attack provided the narcissistic blow that brought broad support, or at least acquiescence, to the preemptive use of military force. The alacrity with which the American people were willing to go forward with an invasion of a country innocent of the attack, illegal detention, and even torture, indicates that the assault on the World Trade Center evoked a previously split off grandiosity and its associated sense of entitlement. While some Americans were vocal in their opposition to military action, polls showed that at the onset of the Iraq War, an overwhelming majority supported the invasion (Benedetto, 2003). The Bush Doctrine based on Cheney's "masterwork" written 9 years before

9/11, which openly violated the UN charter in its unparalleled assertion of unilateralism, was simply the next step in this process. This policy was new not in content but in the "blatant arrogance and self-righteous indifference to international law cloaked in evangelical religiosity" (Hoff, 2008, p. 169).

To understand this type of response to the introduction of reality adaptation in American foreign policy, it is necessary to distinguish between the relinquishment of grandiosity and its disavowal. The former is a decisive change in self-representation to a realistic appreciation of strengths and weaknesses; and the latter, a denial of grandiosity that remains in an unconscious state vulnerable to injury. Awareness that the United States could not free the Soviet satellite states, like the eventual acceptance that it could not impose its will on Vietnam, was a reality constraint on American grandiosity that led to its disavowal, not its abandonment. Arms control and other international agreements along with détente not only fueled the neoconservative movement but evoked a generally strong reaction to any constraints on American military might. Congress rejected the Kyoto Protocol, the ABM Treaty, the landmines, and comprehensive test ban treaties, and the authority of the International Criminal Court has been ridiculed. All of these refusals to cooperate internationally were justified on the basis that America holds a special position in the world order that exempts it from agreements to which other countries are subject (Reisman, 1999–2000). American exceptionalism was now American "exemptionalism."

The Puritan belief that America is "uniquely pleasing to God" has been central to American self-representation in the post-World War II era and beyond. Roosevelt invoked God as an ally in the war against Hitler. After the explosion of the atomic bomb, the White House and religious institutions conveyed the belief that God had blessed the United States uniquely with nuclear capability. In the 1950s, the words "under God" were added to the Pledge of Allegiance, and "In God We Trust" to our coins. Ronald Reagan sounded a popular theme when in 1985 he asked rhetorically, "Can we doubt that only a Divine Providence placed this land, this island of freedom here as a refuge for all those people who yearn to breathe freely?" President George W. Bush was even more definitive in his avowal that the United States was created with a special historical and divine mission to ensure freedom throughout the world and that his invasions of Iraq and Afghanistan were "missions from God" (McAskill, 2005).

Fearful of being identified with any limitations on American superiority, politicians raised not a whisper of opposition to the belief in America's divine mission. The evidence is that politicians are judging the American electorate correctly. A 2003 poll found that 71% of Evangelical Christians, 40% of mainline Christians, and 39% of Catholics believe that our country has the special protection of God. These figures indicate that the view of the United States as possessing divine providence is widely held by a citizenry that is all too willing to believe the leader who calls on God to justify policy. The belief in American moral superiority is reflected in the abundant use of moral and quasireligious terms in national self-descriptions (e.g., McAskill, 2005; Stam, 2003; Wills, 2006). As neoconservative writer Robert Kagan (2003) put it: "America did not change on 9/11, it only became more itself." Those who doubt that grandiosity plays a critical role in the image that the American people have of themselves need only observe the fates of two men who challenged that view: Jimmy Carter and George McGovern.

In a moment of candor, one Bush administration official told Ron Suskind (2004) that people who believe that solutions emerge from a judicious study of reality are "what we call the reality-based community. ... We're an empire now, and when we act, we create our own reality. And while you are studying that reality we act again, creating other new realities" (p. 48). This official was proclaiming that America does not have to accept a distinction between its wishes and reality. This vision of creating reality goes beyond the imperial to the delusional. The notion that America creates reality for the world brings to an ultimate conclusion the grandiosity that began with the mission of the "city on the hill" and continued through the entitlement used to justify a history of military intervention. American grandiosity is an ingrained self-representation that finds justification in a sense of providential destiny, ties American self-interest to the fate of humankind, and rests ultimately on the use of military force to impose its will on other nations.

GREED OR GRANDIOSITY?

Critics of American interventionism frequently claim that the United States interferes with the affairs of other nations not to promote peace, justice, and democracy, but to pursue greed and economic self-interest disguised by ethical rhetoric (e.g., Mills, 1958). This view sees greed as the primary motivating force of American foreign policy with the political strategy secondary to economic self-interest. Even if it is granted that such endless wealth acquisition is, in fact, the cornerstone of U.S. foreign policy, this pursuit must say something about the American psyche. More deeply, it is questionable if greed is the rock bottom of U.S. international motivation. One may ask: If profit is the motivator, why does an affluent nation continue to spend blood and treasure for the acquisition of wealth for which it has little need? Why do the captains of industry pursue treacherous policies of intervention and war that cost dearly in lives and resources to make more money they have little use for? In fact, after World War II the United States pursued "free trade" agreements that accepted protectionist policies of other countries while cutting tariffs at home. This policy hurt much American industry, such as steel, that could not compete on this uneven playing surface. It was this very trade policy, according to economic historian John M. Culberston (1989), that was the single greatest source of the economic decline of the United States. This economically damaging strategy was followed in order to influence the nations that benefited, such as Japan, Europe, Korea, Taiwan, Mexico, and Brazil (Hoff, 2008). The political objective of expanding the "silent empire" and maintaining a foothold of control over titularly independent nations trumped economic well-being.

While much of U.S. foreign policy is motivated by efforts to increase the profits of private corporations, this motive tends to be tethered to and is frequently overridden by the political objective of controlling the world. As we have seen, the Iraq war is a current example. There are windfall profits for certain companies, such as Halliburton, to be sure, but the U.S. economy has suffered from the loss of resources and the spectacular rise in borrowing to finance the military invasion. Use of military and political might to gain control over parts of the world tends to win out over economic self-interest.

SELF-INTEREST OR SHORTSIGHTEDNESS?

So, while it is true that the United States intervenes irrespective of an ethical basis for the imposition, this fact only evokes a deeper question: Is this policy in the self-interest of the United States? Whatever the outcome of the Iraq war, the invasion has been a windfall in funding, propaganda, and recruits for al-Qaeda and given the organization a foothold in Iraq (Emerson, 2008). The danger to Westerners from radical Islamic violence is far greater now than before the invasion (Military Balance, 2004). As John Dean (2005) has observed, there has been no better recruitment tool for al-Qaeda than American troops killing Muslims in an Islamic nation. For the foreseeable future, United States citizens may well have to be fearful of attacks in retaliation for what much of the Muslim world sees as an effort by America to control and dominate, or even eliminate, their way of life. Even more fundamentally, by invading Iraq, the United States squandered the worldwide good will it enjoyed after the 9/11 attacks. From an object of world sympathy, it became the target of vicious hatred and embroiled in an endless war that cost dearly in blood, treasure, and world standing (Stiglitz & Bilmes, 2008).

This shortsightedness has a history in American Mideast policy. The country paid for its 1953 overthrow of the democratically elected Mossadegh government in Iran with the takeover of the American embassy in 1979 and continues more than half a century later with an anti-American regime (Kinzer, 2006). In the short term, continuous military takeovers and the establishment of hundreds of permanent bases create either a tax burden on the citizens or an onerous debt. In the long term, robbing other nations of their self-determination elicits enmity toward the United States for which it pays with hostile governments, as can also be seen today in Iran and the growing number of anti-American regimes in South America.

A self-interested approach to the bombing of the World Trade Center would have been to build on the worldwide goodwill and the support for the United States it evoked by rallying the world around ideals of social justice and freedom while refusing to reduce ourselves to the violent methods of indiscriminate killers. In that way, the United States could have positioned itself as a moral leader with great influence in spreading its message around the globe, instead of making itself a focus of hatred. However, any such suggestion would have been ridiculed as soft on terrorism or, as Karl Rove (2005) put it, "giving therapy to the terrorists." The success of such intimidation is demonstrated in the fear the overwhelming majority of politicians displayed in refusing to question the invasion of a sovereign nation. The relief of helplessness and sense of potency provided by the invasion trumped economic and political self-interest.

LESSONS LEARNED?

Many social scientists and commentators have suggested that the important lesson to be learned from the protracted nature of the ongoing struggle in Iraq is that the United States has to accept the limits of its power (e.g., Bacevich, 2008; Hoffman, 2005). These critics implore the United States to adopt a more realistic foreign policy based on the strengths and limits of American capability. A highly popular, devastating critique of the Bush approach to terrorism concludes "the longer policy makers nurture the pretense that this country can organize the world to its liking, the more precipitous will be its slide when the bills finally come due" (Bacevich, 2008, p. 174). To stay the American decline, this insightful and courageous author advocates a policy of "containment" of terrorism, strategic use of allies, negotiation, and diplomacy, all of which acknowledge the inability of the United States to control world events by itself.

The problem with this advice is that it has been given after each major post-World War II war save the Gulf War only to be ignored as the country launched the next effort to control an area of the world. (The Gulf War is an exception precisely because the United States recognized the limits of what it could do and recognized international authority by refusing to go beyond the UN mandate to overthrow the Iraqi government.) After the compromise agreement that ended the Korean War, conventional wisdom was that the United States had to accept the limits of its power (e.g., Osgood, 1965). Nonetheless, 12 years after the armistice was signed ending that conflict, President Lyndon Johnson dramatically escalated the war in Vietnam; eight years and 58,193 American, more than 3 million Vietnamese, and almost 2 million Laotian lives were lost after U.S. troops were withdrawn, Vietnam was united as a communist nation, and the pundits assured the public that we had learned the limitation of our power to influence other nations. President Jimmy Carter summed up the national mood in the years after the Vietnam debacle: "We have learned that never again should our country become militarily involved in the internal affairs of another nation unless there is a direct and obvious threat to the security of the United States or its people." If that lesson was ever learned, it was forgotten by the time of the 2003 invasion of Iraq. Now political analysts and commentators are concluding that the quagmire in Iraq has taught us the limits of our ability to dominate another nation militarily and impose our political will on its citizens. These pundits are apparently oblivious to the fact that they are repeating a "conclusion" the country claimed to have reached at least twice before in the post-World War II era.

Critics who believe that the Iraq war shows that the United States has to accept the limits of its power are not incorrect, but they tend to avoid asking: After we had supposedly absorbed that lesson after the Korean and Vietnam Wars, why did we make that mistake yet again? And more fundamentally: What entitles the United States to believe it has the right to impose governments on other nations?

These questions cannot be answered by historical and political analyses alone. Such deliberations see the assumptions of omnipotence and exceptionalism that underlie American foreign policy, but they cannot explain why the United States refuses to accept constraints on its ability to control world events (e.g., Bacevich, 2008; Hoff, 2008). Political scientists and historians are limited to imploring the nation to accept the limitations of its power, to yield the belief in its omnipotence and superior goodness (e.g., Bacevich, 2008; Ignatieff, 2005; Kinzer, 2006; Ruggie, 2005). The history of the country suggests that such warnings go unheeded. Political and historical understanding by themselves cannot explain the intransigent nature of the U.S. illusion of its omnipotence and moral superiority. This is where psychoanalytic understanding can add a necessary component to historical and political thinking.

THE NATIONAL PROPENSITY FOR VIOLENCE: A PSYCHOANALYTIC APPROACH

To understand the stubbornness of the American insistence on attempting to dominate nations that it cannot control and of which it has little knowledge is to appreciate the resilience of grandiosity. For a narcissistic patient, there is little sense of self other than grandiosity (e.g., Kernberg, 1984; Kohut, 1971). The grandiose self is erected to protect against a sense of weakness and inadequacy that cannot be admitted to consciousness. The emotional investment in the self is concentrated in the grandiose self.

Any slight to the elevated self-image evokes shame and helplessness and threatens the very sense of self. If the grandiose self is assaulted so that it cannot be successfully protected, the very sense of self is threatened, resulting in what Kohut (1984) called "disintegration anxiety." To protect the very sense of self, defenses are employed, such as denial of all mistakes, failures, and limitations. One common way to achieve this defensive posture is to devalue the other and even reduce the other to helplessness where that is possible. Although such behavior is ultimately self-defeating because it alienates others rather than evokes the admiration the patient seeks, the narcissistically organized individual opts for immediate narcissistic gratification rather than long-term self-interest.

The same dynamic applies to the grandiose nation. Tocqueville's (1835) observation that American citizens were easily offended by any suggestion of the country's limitations was a comment on American grandiosity. Because the dominant view of the U.S. populace denies the nation's flaws, the loss of wars and invasions of other countries for perceived self-interest is disavowed. As Tocqueville noted, most Americans are offended by the association of "failure" with the United States. That is why the lessons learned from failed wars such as Vietnam tend to be limited to self-compliments such as "we do too much for others," and it is doubtful that even that putative "lesson" has been learned. After a confrontation with real-world limitations of its capabilities, the United States will attempt to restore the belief in its omnipotence by disavowing any defeat and eventually seeking victory or domination in another conflict. History shows that when the United States claims to have learned from its military interventions, it disavows grandiosity rather than relinquishes it. As long as America is able to convince itself that it has learned the lesson without becoming conscious of its grandiosity, the sense of entitlement will remain and render the nation vulnerable to narcissistic injury and ultimately further use of violence.

Now conventional wisdom once again asserts that America has learned the limits of its power. However, the very fact that there is so little recognition of America's need to hold to an omnipotent self-image while the country believes it has learned the lessons of the Iraq War gives a basis for pessimism that the grandiosity will be relinquished.

The reality constraints of the conflict in Iraq and the economic crisis are now pushing American grandiosity into a less active, almost hibernating state except for groups who cannot tolerate even a temporary loss of omnipotence, primarily given voice in the conservative media. Delimiting American omnipotence tends to build a gradual resentment among the citizens and political leaders with a consequent need for assertion of potency. In the post-World War II era, each time America has embarked on what appears to be a "realist" foreign policy, the narcissistic blow to American omnipotence gave rise to an unconscious bitterness that eventually fueled a new outbreak of violence. The lack of a national discourse on what went wrong in Iraq and how we became a society that practices torture, the refusal to confront the limits on American military force, and the smoldering resentment of the war's failure all suggest that America is once again undergoing a temporary reality constraint on its grandiosity, rather than a relinquishment of it.

Both the colonies and the republic were established with a sense of specialness, a mission for all humankind, that became the very identity of the nation. However laudable the intent, this self-aggrandizement has become the basis for superior military strength and violence. To be sure, national pride is characteristic of any nation and expressions of patriotism are part of the campaign recipe for politicians everywhere, but there is a critical difference between national pride and the grandiosity captured in the sense of providential mission embedded in American exceptionalism (Lockhart, 2003). Such grandiosity is reserved for empires.

Much as the narcissistic patient gives little consideration to the subjectivity of others, the experience of other countries is either ignored or regarded as inferior if it conflicts with U.S. aims. Needless to say, the objects of American arrogance take offense, and the United States makes enemies of many, even offending potential allies. Just as the narcissistic individual attempts to maintain her grandiosity by behavior that ultimately undermines its maintenance, the United States, by its attitude of superiority and nativism, offends the pride of Third World nations. That U.S. foreign policy gives precious little consideration to cultural differences is seen by the obliviousness to 1,400 years of Islamic history, Iraqi culture, and ethnic groups with which the United States crossed the Iraqi border with armored tanks and more than 130,000 troops. Fourteen hundred years of Islamic history were disregarded by the hubris of an American government that assumed its invasion would be welcomed by a culture of which it had little knowledge. Those who did raise concerns about the lack of sensitivity to Islamic culture were ridiculed as being soft for daring to question the absolute right of the United States to intervene where it sees fit. To take into account the experience of the other culture is to concede there is a reality to which the United States must adapt, and such a concession undermines narcissistic entitlement.

Similarly, the establishment of detention centers in which citizens of Third World countries were picked up with little or no evidence, stripped of all rights, subjected to degrading and inhumane conditions, and often tortured provides a sense of potency in being able to subjugate others to American will. The grandiosity of the nation was reestablished by this show of strength, but the cost to the United States in ill will and even hatred from the countries whose citizens were subjected to this humiliation will be long term. Like the narcissistic patient, the United States did not believe it needed to consider the impact on world opinion of its mistreatment of innocent individuals. Further, any American who suggested that the government should consider the long-term consequences of its abusive treatment of other nations was subject to ridicule. This belief that one need not consider the impact on others is rooted for the individual as for the nation in the primacy of maintaining grandiosity. Concern for the effect of the nation's behavior on other countries offends the self-sufficiency of a grandiose self-image. The photos of Abu Ghraib have already done and will continue to do damage to United States' reputation throughout the world with the inevitable impact on its ability to conduct foreign policy irrespective of the outcome of the Iraq War. Nonetheless, at the time the torture policy was approved, no thought was given to its long-term consequences beyond the need to "show them we're in charge" (Sands, 2008).

The surprise Americans felt in response to the strong Iraqi insurgency and virulent anti-Americanism of much of the population is testimony to the lack of consideration given to the Iraqi experience by the invaders. Ultimately, the political, economic, and military cost of the invasion is a major defeat for the United States. This outcome is analogous to the narcissistic patient's effort to overcome humiliation via arrogant and abusive behavior toward others that incurs enmity instead of the desired admiration. Just as rage outbursts ultimately hurt the narcissistic patient, the thoughtless use of force in Iraq has resulted in an exponential increase in enemies, such as al-Qaeda, and ill will throughout the Islamic world in addition to the waste of national resources.

Once American exceptionalism is seen as an ingrained grandiosity that forms the very self-image of the country, the resilience of the pattern of intervention and belief in omnipotence becomes understandable. To suggest that the United States should relinquish its illusions of omnipotence and superiority is analogous to telling a narcissistic patient to give up her grandiosity. Informing the patient that her self-image is exaggerated and needs to be more realistic would be futile because the patient does not believe it, disintegration anxiety threatens, and the only alternative the patient sees is humiliation and helplessness. Any deflation of American grandiosity would be an abandonment of national identity, an identity that has and continues to provide irresistible gratification in the form of an illusion of invincibility and moral purity. No appeal to "reality" or "common sense" can convince the narcissistically organized individual to give up the giddy gratification of exalted beliefs about herself. So, at the national level, it is not a simple matter of telling the U.S. citizenry that the country has an exaggerated perception of itself and that it must be realistic in its expectations. If those admonitions could work, they would have after the Korean and Vietnam Wars. The problem has to be attacked at its roots: the grandiose self-image of the United States as a superior nation with a providential mission that was born with the establishment of the colonies approximately 400 years ago.

CONCLUSION: THE REPAIR OF AMERICAN GRANDIOSITY

If the admonishment to the United States to accept the limitations of its power does not work, what does? Clearly, there is no simple or definitive answer to this very difficult question, but if the American propensity for violence is to be reduced, its grandiosity must be relinquished, rather than disavowed. Again, a clue can be found in the experience of the narcissistic patient who does not relinquish her grandiosity unless an alternative form of narcissistic gratification can be found. Kohut (1971) pointed out that grandiosity can be given up only if the analytic process provides a certain amount of gratification along with the frustration of narcissistic disappointment. Kernberg (1975, 1976) added that normal narcissism consists of a realistic appreciation of faults and strengths. Narcissism becomes abnormal only when the need for self-appreciation is grandiose. In Kohut's view, in the successful psychoanalysis of the narcissistically organized individual, the grandiose self is converted into realistic ambitions and ideals (Kohut, 1971, 1984). The relevance for national identity is that in a successful outcome, grandiosity does not disappear; it is transformed into a more realistic sense of self that nonetheless provides narcissistic gratification (Kohut, 1966, 1971). The successfully treated grandiose patient finds meaning in life from the fulfillment of realistic ambitions that substitute for the illusionary grandiosity (Kohut, 1971, 1984).

This clinical finding suggests that American grandiosity will only be relinquished if more realistic ambitions bring gratification. That in turn can only happen if the United States is willing to form an identity based on realistic strengths and limitations. It may seem quixotic to expect a nation to change an identity that has endured longer than the republic itself. Nonetheless, the fragility of that grandiose identity contains the roots of American violence. Consequently, only a transformation can hope to diminish the American propensity to react violently to events it cannot control. As long as the country defines itself in grandiose terms, it will be vulnerable to humiliation and the use of violence to bring immediate narcissistic relief and restore the grandiose self-image. Nothing short of a decisive change in national identity can hope to effect a significant, long-term change in America's ability to accept its limitations without a sense of insult.

The excitement in such an identity lies in the ability to achieve meaningful goals in concert with one's ideals. Such achievement provides a realistic form of pride that endures, rather than the temporary relief resulting from gratification of grandiosity. In the United States, national pride from achievements in accordance with its ideals of justice and liberty offer that opportunity.

Obviously, such a task is difficult, but it is a challenge that has been avoided to the detriment of the country and the world. The blood and treasure lost to protect American grandiosity is almost inestimable. We are at a historical juncture in which the exorbitant price in money and lives can be seen by many, if not all, Americans. That sobering and even depressing fact can be a source of further shame leading to a renewed motive to use force and coercion, thus continuing the historical cycle. Or, it can be used to demonstrate that we must redefine who we are. The right kind of delicate and forceful leadership is required if such a transformation is to have a chance to seize the popular imagination.

Presidents who endorse the grandiose definition of America will never lead a national dialogue in this direction. A president who sees the problem has a chance to show the nation that grandiosity has led the country into unnecessary and self-destructive conflicts, but must be careful to avoid the pitfalls into which George McGovern and Jimmy Carter fell by attacking the American people or the nation's spirit. The lesson of those two well-intentioned political figures is that telling the American people they suffer from a malaise only offends the public rather than inspires them to meet challenges. Leadership that wishes to help move the country toward a new way of viewing itself must promote national pride from a realistic appreciation of America's strengths and limitations while drawing attention to the long-term consequences of pursuing grandiose ambitions. The fact is that when the United States has provided moral leadership, it reaps benefits for itself as well as the world. Leadership of politicians, scholars, community leaders, and the business community is required to convince an inevitably skeptical public that the long-term national self-interest lies in pursing realistic goals that issue from an appreciation of our talents and limitations and that take into account the values, aspirations, and needs of others. It is imperative that the public sees that the mistakes in Vietnam and Iraq were not simply blunders or excessive zeal, but the symptoms of a temporarily gratifying yet dangerous view of ourselves that leads to self-defeating ends. The gratification of peaceful cooperation and international understanding can then become apparent.

No doubt, these suggestions will be opposed as a program for a weak America by those who remain adhesively attached to American grandiosity. That argument must be confronted at its root: the assumption of American superiority and its inevitable consequence of offending other nations, creating international conflict, expending national resources for little gain, and continuing the cycle of violence. The current expenditure of lives, money, and resources for questionable goals in Iraq provides the opportunity to challenge the Iraq War as not simply a mistake, but a reflection of a long-standing American self-representation that fosters short-term narcissistic gratification at the expense of long-term self-interest. If this critique can be made, then the national dialogue would shift from geopolitical strategy to the American self-definition that manifests itself in international policy. The argument is then positioned not as a conflict between "liberal" and "conservative," but between those who insist on America's entitlement versus people who are willing to assess the nation's strengths and weaknesses.

Such a debate will not be easy to win, but at least it is the right dialogue. If the issue of American grandiosity will come into national discourse, the country will be well ahead of where it is now.

REFERENCES

Alexander, M., & Bruning, J. (2008). *How to break a terrorist: The U.S. interrogators who used brains, not brutality, to take down the deadliest man in Iraq.* New York: Free Press.

Arendt, H. (1969). *On violence.* Orlando, FL: Houghton Mifflin.

Arrigo, J. (2003). *Torture is for amateurs: Report on the 2006 seminar for civilian psychologists and Army interrogators.* Retrieved from www.usfa.edu/isme/SMB08/Arrigo08.html.

Baceurch, A. (2008). The limits of power, New York: Holt.

Benedetto, A. (2003, March 16). Polls: Most back war, but want UN support. *USA Today.* Retrieved from http://www.usatoday.com/news/world/iraq/2003-03-16-poll-iraq_x.htm

Clark, R. (2003, September 7). *Ramsey Clark answers Bush's speech.* Retrieved from www.uncommonthought.com/mtblog/archives/2003/09.

Culberston, J. (1989). *The trade threat and U.S. trade policy.* New York: Twenty First Century.

Danchev, A. (2006). Like a dog! Humiliation and shame in the war on terror. *Alternatives: Global, Local, Political, 31*(3), 259–283.

Dean, J. (2005). *Worse than Watergate: The secret presidency of George W. Bush.* New York: Grand Central Publishing.

Dean, R. (2001). "They'll forgive you for anything except being weak": Gender and US escalation in Vietnam 1961–1965. In M. Young & R. Buzzanco (Eds.), *A companion to the Vietnam war* (pp. 367–382). New York: Blackwell Publishing.

Denbeaux, M. (2006). *A report on Guantánamo detainees: A profile of 517 detainees through analysis of Department of Defense data.* Retrieved from www.law.shu.edu/guantánamofinalreports/guantanemo_reportfinal_2_08_06.pdf

Emerson, S. (2008, April 9). *Testimony before the United States House of Representatives Permanent Select Committee on Intelligence.*

Fein, R., Lehner, B., & Vossekuil, B. (2006). *Intelligence science board study on educing information. Phase one report.* Washington, DC: NDIC Press.

Hixson, W. (2008). *The myth of American diplomacy: National identity and U.S. foreign policy.* Ann Arbor, MI: Sheridan Books.

Hoff, J. (2008). *A Faustian foreign policy from Woodrow Wilson to George W. Bush: Dreams of perfectibility.* New York: Cambridge University Press.

Hoffman, S. (2005). American exceptionalism: The new version. In M. Ignatieff (Ed.), *American exceptionalism and human rights* (pp. 225–240). Princeton, NJ: Princeton University Press.

Human Rights Center. (2008). *Guantánamo and its aftermath: U.S. detention and interrogation practices and their Impact on foreign detainees.* Retrieved from http://hrc.berkeley.edu/pdfs/Gtmo-Aftermath.pdf

Ignatieff, M. (2005). Introduction: American exceptionalism and human rights. In M. Ignatieff (Ed.), *American exceptionalism and human rights* (pp. 1–26). Princeton, NJ: Princeton University Press.

Kagan, R. (2003). *Of paradise and power: America and Europe in the new world order.* New York: Vintage.

Kernberg, O. (1975). *Borderline conditions and pathological narcissism.* New York: Jason Aronson.

Kernberg, O. (1976). *Object relations theory and clinical psychoanalysis.* New York: Jason Aronson.

Kernberg, O. (1984). *Severe personality disorders.* New Haven, CT: Yale University Press.

Kinzer, S. (2006). *Overthrow: America's century of regime change from Hawaii to Iraq.* New York: Times Books.

Kleinman, S. (2008, September 25). *Testimony before Senate Armed Services Committee.*

Kohut, H. (1966). Forms and transformations of narcissism. *Journal of the American Psychoanalytic Association, 14,* 243–272.

Kohut, H. (1971). *Analysis of the self.* New York: International Universities Press.

Kohut, H. (1984). *How does analysis cure?* Chicago: The University of Chicago Press.

Lewis, N. (2004, November 30). Red Cross finds detainee abuse at Guantánamo. *New York Times.* Retrieved from http://www.nytimes.com/2004/11/30/politics/30gitmo.html

Lockhart, C. (2003). *The roots of American exceptionalism: Institutions, culture, and policy.* New York: Palgrave McMillan.

McAskill, E. (2005, October 7). George Bush: God told me to end the tyranny in Iraq. *The Guardian.* Retrieved from http://www.guardian.co.uk/world/2005/oct/07/iraq.usa

Milbank, D., & VanderHei, J. (2003, May 1). No political fallout for Bush on weapons. *Washington Post.*

Mills, C. (1958). *The causes of World War Three.* New York: Simon and Schuster.

Military Balance. (2004, October 24). *Annual report.* London: International Institute for Strategic Studies.

Moore, M. (2007, June 19). Guantánamo Bay puzzles candidates. *USA Today.* Retrieved from http://www.usatoday.com/news/politics/2007-06-18-gitmo-candidates_N.htm

Nixon, R. (1970, April 30). *Address to the nation on the incursion into Cambodia.* Washington, DC.

Office of Inspector General. (2004). *Final report of the independent panel to review DoD detention operations.* Retrieved from www.defense.gov/news/Aug2004/d20040824finalreport.pdf

Osgood, R. (1965). *Limited war: The challenge to American strategy.* Chicago: The University of Chicago Press.

Reagan, R. (1979, November 13). *Announcement for the presidency.* Hilton Hotel, New York.

Reisman, M. (1999–2000). The United States and international institutions. *Survival, 41*(4), 62–80.

Rich, F. (2007). *The greatest story ever sold: The decline and fall of truth in Bush's America*. New York: Penguin.
Rove, K. (2005, June 22). Remarks of Karl Rove at the New York Conservative Party. *Washington Post*. Retrieved from www.washingtonpost.com/wp-dyn/2005/Ar2005062400097.html
Ruggie, J. (2005). American exceptionalism, exemptionalism, and global governance. In M. Ignatieff (Ed.), *American exceptionalism and human rights* (pp. 304–339). Princeton, NJ: Princeton University Press.
Sands, P. (2008). *Torture team: Rumsfeld's memo and the betrayal of American values*. New York: MacMillan.
Stam, J. (2003, December 22). Bush's religious language. *The Nation*. Retrieved from www.thenation.com/doc/20031222/stam
Stephanson, A. (1995). *Manifest destiny: American expansion and the empire of the right*. New York: Hill and Wang.
Stephens, M. (2003, May 1). From Long Kesh to Guantánamo Bay: Containing the Terrorist Body. *Popmatters.com*.
Stiglitz, J., & Bilmes, L. (2008). *The three trillion dollar war: The true cost of the Iraq conflict*. New York: W.W. Norton & Co.
Stiles, E. (1783, May 8). *The United States elevated to glory and honor: A sermon*. Connecticut State Assembly, Hartford, CT.
Suskind, R. (2004, October 17). Faith, certainty, and the presidency of George W. Bush. *New York Times Magazine*. Retrieved from http://www.nytimes.com/2004/10/17/magazine/17BUSH.html
Tocqueville, A. de. (1835). *Democracy in America*. New York: Penguin.
United Nations Commission on Human Rights. (2006, February 15). *The situation of detainees at Guantánamo*. Report E/CN.4/2006/120.
Watt, S. (1987). *The republic reborn: War and the making of liberal America, 1790–1820*. Baltimore: Johns Hopkins University Press.
Wills, G. (2006, November 16). A country ruled by faith. *New York Review of Books* Retrieved from www.nybooks.com/articles/article–id=19590.
Wolfowitz, P. (2000). Statesmanship in the new century. In R. Kagan & W. Kristol (Eds.), *Present dangers: Crisis and opportunity in American foreign and defense policy* (pp. 318–320). New York: Encounter Books.

Part 3

War and militarism deconstructed

Historian Eli Zaretsky (Chapter 9) charts the complex, creative, and reactive relation of psychoanalysis to war, set in the context of changing relations of the individual, to the state and to globalization. Considering three crucial experiences of international warmaking and the construction of what Zaretsky terms "the war imaginary"—the First World War, the London Blitz in the Second World War, and the onset and evolution of the post-9/11 war on terror—Zaretsky really utilizes the paradox that underlies the premise of this entire book. Psychoanalysis, as ultimately a theory of motivation, is part of the apparatus of patriotism, military recruitment, and civilian management (rhetoric and propaganda). The evolving nature of the neoliberal state, globalization, and new forms of warfare create new needs, new requirements for understanding, and individuality, and psychoanalysis alters in relation to these historical trends. Zaretsky traces shifts in psychoanalytic understandings of sexuality, of motivation, of human relatedness, and human vulnerability. State formation and warfare exist in a dialectical relation to psychoanalysis. The evolution of practices and theories within psychoanalysis, the move from a focus on aggression to a focus on relatedness and attachment, shape both reactionary and progressive reactions to state policy and political life.

Casus belli is the precise moment when trauma or psychoses intervene as a core search in the transference with the analyst, to investigate areas of a "nameless dread." Revisiting the symptoms of prominent researchers in the 1920s: Aby Warburg and Ludwig Wittgenstein, in their relationship with native rituals—Françoise Davoine's chapter (Chapter 10) elaborates how traumatic and psychotic symptoms wage an invisible war against psychic annihilation and cognitive distortions.

When nations prosecute foreign wars, they create an eroticized distance between soldiers and civilians, which occludes the reality of war. Sue Grand (Chapter 11) traces this alienated desire, as it appears in the analysis of a Vietnam veteran. Grand's work has always centered on the complexity, transferentially and countertransferentially, of metabolizing evil and heroism. This chapter has her signature careful self-reflection. It is the practice of an ethics of responsibility, here dyadic but ultimately collective. The veteran is a decorated war hero who committed an atrocity. He addresses a civilian, antiwar analyst. Together, they recreate the erotic rhetoric of war, and together they challenge that rhetoric.

Donald Moss (Chapter 12) gives us an intimate and challenging look at the transmission of war culture, a war imaginary, through three generations. Anxiety, excitement, and eros attend and are suspended as Moss talks about war and violence—actual, unconscious, and projected—with his father and years later with his son. We are in the presence of intergenerational transmission of ideals and needs, of use in warmaking and in the deconstruction of war. Moss walks us through the paradox of psychoanalytic ideas as our guide to the rivers of unconscious life, but understanding that this mode of understanding makes soldiers and pacifists.

Ruth Stein* (Chapter 13) presents research work into coercive persuasion (brainwashing or mind control), spanning brain processes such as trance induction, various practices that induce altered states of consciousness, but focusing in particular on a psychodynamic understanding of processes of personality change that serve for extrinsic purposes, such as domination, conversion, and exploitation. She illuminates the uncanny similarity and symmetrical difference of these mental-war processes from psychotherapeutic change processes.

Nancy Caro Hollander's chapter (Chapter 14) explores the impact of state terror on women, which is complex and somewhat discrepant in that the strategies as well as the traumatic impact of the terrorist state reinforce woman's traditional role. These strategies include the normative aspects of male violence against women, in some respects while they simultaneously alter it in others. Hollander explores two dimensions of Latin American state terror and its gendered meanings: the first related to the extreme, chronic trauma resulting from the horrific conditions of everyday life, which come close to the experience typical of combat; and the second related to the acute trauma of torture, which takes on dynamics similar to those of the battering and rape situation.

---

* Ruth Stein died very suddenly on January 2010. We have added a dedication to her.

Chapter 9

Psychoanalysis, vulnerability, and war

Eli Zaretsky

> What do people think when they come across photographs in a newspaper of young men and women killed in Iraq? I take it for granted that they feel pity and horror that someone so young is no more, but it would be interesting to know just how far they venture to imagine the lives of the young as they read the few lines of biographical information that accompany the picture. Here's Fred Something-or-Other, born in a small town out west, or in some city in the east, whose name and face recall someone we used to know in high school, looking at us, out of a photograph taken by the military, with the usual swagger of young men wearing a uniform. Many are making an effort to smile, some appear grim and determined, and only a few have the vulnerable, worried look of kids who think they might come back to their parents in a coffin.
>
> Charles Simic
> "The Nicest Boy in the World" (2008)

THE PROBLEM OF MOTIVATION IN DEMOCRATIC WARS

In 1932 Sigmund Freud inquired of Albert Einstein, "How long shall we have to wait before the rest of mankind become pacifists too? There is no telling ... But one thing we can say: whatever fosters the growth of civilization works at the same time against war." Although Freud's inquiry had been instigated by a League of Nations request, his interest in war was not new. As an adolescent, he affixed colored pins to a map hung on his bedroom wall to follow the battles of the Franco-Prussian war. In *The Interpretation of Dreams* (1900), he described a dream as the outcome of a series of battles. The first great paradigm shift in the history of psychoanalysis, the structural theory, emerged from its reflections on shell shock during World War I. The second, the turn toward the mother, found its fullest expression in the welfare states created during World War II. Psychoanalysis has always been involved with war.

In his letter to Einstein, Freud effectively called himself a pacifist. In this way, he sought to identify psychoanalysis with political movements and tendencies that opposed war. In fact, however, psychoanalysts could always be found on all sides of this issue. They have been not only doctors and healers but also advisors on psychological warfare and, soon enough, on torture. Prowar and antiwar, social democrats, communists and reactionaries, they have produced a huge analytic literature on war, taking up such questions as aggression, trauma, mourning, and reparation. This vast and heterogeneous literature has no overall unifying strand, but one thread does run through much of it. Historically, in its encounters with war, the experience of psychoanalysis has revealed an increasing sensitivity to human vulnerability. This sensitivity, in turn, has shed light on the problem of motivation in democratic wars.

This problem of how to motivate citizens to kill and be killed became pressing in the Freudian epoch, as a result of historical shifts in the nature of war. Until the 19th century, men fought either because they were aristocrats, bred for war, or because they were serfs or peasants, conscripted regardless of their will. The rise of popular democracy changed that by creating mass civilian armies. Older problems of motivation persisted, but a newer one was created. How could ostensibly free men, and later women, be persuaded to lay down their lives, often in their earliest youth? How could their parents and fellow citizens tolerate their deaths? How could free men and women be induced to kill and maim their fellow humans? What made democratic wars acceptable?

The 19th century supplied an answer to this question, namely, the nation. The nation, many argued, contained a unique and irreplaceable spirit or soul, which was the counterpart to the individual soul. To die for one's country was therefore to *sacrifice* one's life, not merely to lose it. Romantic nationalism was a complement to natural rights liberalism, one that rested on the belief that the soul of the individual and the soul of the nation could converge. While this belief certainly survives in oppressed nationalities and religions, and has considerable purchase still in democratic societies, it has not proved sufficient for modern, mass armies. In these societies, other solutions to the problem of motivating citizens for war were sought and found, including universal ideals, communal and group solidarity, the protection of loved ones and revenge for supposedly unprovoked attacks and humiliations.

Psychoanalysts were sometimes involved in formulating these and other rationales for war, sometimes not. But whatever their involvement, their special area of expertise was and remains motivation, especially unconscious motivation. In 1907 Freud wrote that from analytic case studies we learn "what is really going on in the world ... analyses are cultural historical documents of tremendous importance" (p. 251). The same is true of historical episodes or events. Through studying the encounters of psychoanalysis with war, I hope to illuminate not the analytic theory of war, but the world of affect, intention and meaning that psychoanalysts encountered during war, and in particular in their explorations of motivation. To this end, I will examine three episodes: the shell shock episode during World War I, the response of British analysts and of American political figures to the Nazi bombing of London in 1939, and recent psychoanalytic reflections on "the war on terror" by Judith Butler in her 2003 book, *Precarious Life*. The episodes do not by themselves fully comprehend the multiple sources of motivation for democratic war but they do shed a special light on the problem.

The light is not the one most readers will associate with analytic explanations of war, namely, the aggressive drive. The theme that emerges here is not aggression but vulnerability. Vulnerability is central to the discussion that follows in three respects. First, the three episodes show a deepening sense of the role that vulnerability plays both in weakening earlier motivations for war, especially the honor ethic of the warrior, and also in providing a new motivation, namely the need to protect those who are especially vulnerable, one's loved ones and children. Second, the episodes suggest that vulnerability—or rather defenses against awareness of vulnerability—underlies and explains at least some aggression in war. Finally, the history I am about to recount suggests a deepening sense of the depth at which the awareness of human vulnerability connects individuals to one another, while also broadening the circle of those who feel solidarity with others through shared feelings of vulnerability. In each case we will see that Freud had at least some reasons to connect the advance of civilization with the end of war.

WORLD WAR I AND THE COLLAPSE OF CLASSICAL LIBERALISM

The problem of mass, democratic motivation during World War I can be divided into two phases. In the first, states used nationalism and quasi-racial propaganda to stir up military sentiment. For example, the term *Hun* was ubiquitous in American propaganda posters, as was *Boche* in French. In the second, which followed the Bolshevik Revolution, appeal was made to universal ideals, both Communism and Wilsonian liberalism. Thus, Lenin argued that war was inevitable as long as capitalism survived, while Wilson called for a global revolution in liberal ideals in order to prevent future wars. Neither nationalism nor universalism proved truly adequate, however, and over time the war became more and more intolerable. The unprecedented scale of the disaster, the defensive stalemate symbolized by the trenches, the new landscape—psychic as well as geographic—of mines and no-man's-land, the fear of being buried alive, the deafening sound and vibration, the insidiousness of gas, the lack of visual clues at the front, the disappearance of the distinction between night and day, identification with the enemy, narrowing of consciousness: these were all symptoms that the war was increasingly being perceived as meaningless.

Shell shock, broadly considered, was one such symptom. Within a year of the outbreak of war, hundreds of thousands of cases appeared on both sides. Although originally explained as an impairment of the nervous system resulting from an explosion, psychiatrists increasingly turned toward psychological explanations, especially after the symptoms of shell shock were successfully removed by hypnosis. But even then the persistence of the older warrior ethic revealed itself in such diagnoses as malingering, "greed neurosis" and "pension-struggle neurosis." In 1918, however, Ernst Simmel, a young German student of Freud's, cast the matter in a different light:

> It is not only the bloody war which leaves such devastating traces ... it is also the difficult conflict in which the personality finds itself ... Whatever in a person's experience is too powerful or horrible for his conscious mind to grasp and work through filters down to the unconscious levels of his psyche. There it lies like a mine, waiting to explode. (pp. 5–6, 82–84)

Simmel's statement suggested another approach.

One of the main reasons the war had become intolerable was that the trench warfare generated a stalemate leading to long periods of passive waiting. Not surprisingly, by 1916 the psychiatrists' favored explanation for shell shock became the enforced "passivity" of the trenches. According to W. N. Maxwell, a chaplain and the author of *A Psychological Retrospect of the Great War*, a

> high degree of nervous tension is commonest among men who have ... to remain inactive while being shelled. For the man with ordinary self-control this soon becomes a matter of listening with strained attention for each approaching shell, and speculating how near it will explode ... An hour or two ... is more than most men can stand. (1923)

The British War Office also concluded that the primary cause of shell shock was "prolonged danger in a static position." Such explanations seemed vindicated when the incidence of shell shock plummeted after the German offensives of 1918 (Leed, 1981).

The younger psychoanalysts were all in the military, and the attempt to explain shell shock led to a major alteration in the theory and practice of psychoanalysis. The problem in shell shock was in the soldier's present experience, not the repressed or infantile past. Far from repressing their experiences, the victims of shell shock repeated them compulsively, in dreams, symptoms, and anxiety attacks. How could the repetition of a painful experience be explained? Freud had already addressed this problem in the clinical situation. In the transference, patients feared being in a passive, vulnerable position. They expressed their "resistance" to analysis by refusing to *remember* their experiences; instead they *repeated* them. Similarly, Freud had already explained a familiar childhood experience. His grandson was traumatized when his mother left him. He responded by inventing a game in which he alternately hid and produced a ball, *da/fort*, here/gone. "At the outset," Freud noted, his grandson "was in a *passive* situation—he was overpowered by the experience; but, by repeating it, unpleasurable though it was ... he took on an *active* part" (1905, p. 134).

By analogy, shell shock appeared as an attempt to master the experience of passively waiting in the trenches by compulsively repeating the experience of being bombed. As Freud noted, the upsurge in cases resulted from the recruitment of a conscript as opposed to a mercenary army. The draftee, unlike the professional soldier, was a divided soul; he was the victim of "a conflict ... between the soldier's old peaceful ego and his new warlike one ... which [the old ego] sees as threatening its life" (Paskauskas, 1993, p. 334fn). Developed to account for a wartime condition, Freud's considerations on shell shock led to a reformulation of psychoanalytic theory. In shell shock, Freud wrote, "the ego is defending itself from a danger which threatens it from without. ... In the transference neuroses of peace [by contrast] the enemy from which the ego is defending itself is actually the libido, whose demands seem to it to be menacing" (as cited in Eagle, 1987, pp. 81, 109). In both cases, the ego was defending itself against the experience of helplessness in the face of an accumulation of excitation that could not be managed by the normal workings of the pleasure principle (i.e., fantasy). The ego, accordingly, feared passivity; it had "a preference for the active role."

The fear of passivity, the preference for the active role, underlay the experiences of a conscript army, but this had a further consequence. Here, too, Freud drew on his clinical experience. Patients, Freud believed, did not come to analysis to get well; they came to serve neurotic ends, which would express themselves in the transference. For this reason Freud insisted on "abstinence." In refusing to palliate the patient's situation, the transference would be frustrated, intensified, and brought into sharper focus, but it was especially in the negative transference—anger, defensiveness, aggressivity—that the underlying neurosis emerged. The reason was that as the neurotic conflict came more and more into focus, the patient sought to satisfy simultaneously both the underlying wish and the resistance against that wish. The fear of passivity, then, the fear of being caught unprepared and vulnerable, would not reveal itself easily. On the contrary, anything that threatened to reveal the vulnerability of the soldier would result in more aggression, not less.

The encounter of psychoanalysis with shell shock, and the reformulation of psychology the encounter provoked, revealed the exhaustion of at least one earlier motivation for war, the elite ethic of masculine honor. In 1895 Oliver Wendell Holmes, a member of the Civil War generation, could still tell the Harvard graduating class that only in war could men pursue "the divine folly of honor." When World War I turned out to be a matter of mass, insensate violence rather than individual heroism, however, Holmes's platitudes wore thin. In 1929, when Ernest Hemingway noted that words like *honor* and *glory* had become obscene, his observation converged with a new, psychoanalytically inflected contempt for bluster and a search for honesty and directness in both personal life and artistic expression (Fussell, 2000).

Similarly, the encounter revealed the weakening of romantic-nationalist motivations for war. Thomas Mann, reading a 1921 newspaper review of *Beyond the Pleasure Principle* in which Freud's new account of the fragility and defensiveness of the ego was developed, wrote in his diary that the book signaled the "end of Romanticism [including] a weakening and dying of the sexual symbolism that is virtually identical with it" (1984). Mann, who was especially sensitive to the role that romantic self-sacrifice might play in bringing demonic-nationalist (as opposed to conservative) forces to power in Germany recognized the novelty in Freud's account of the tiny, fragile, contingent body who stood in a force field "of destructive torrents and explosions."*

As the war was ending, a new generation born in the 1880s and 1890s came to the fore. This generation contained strong antiwar currents. Incited by Dadaists in neutral Zurich, New York, and Munich; socialists and syndicalists in Russia, Germany, central Europe, and the United States; sexual reformers and conscientious objectors in London and Berlin; pacifists, suffragists, and flappers in London and New York; and surrealists and literary experimenters in Paris, they considered that the war experience demonstrated the bankruptcy of the older, liberal and Victorian order. Psychoanalysis was at the center of its consciousness. In its ranks, the new awareness of vulnerability appeared in at least three different guises.

First, the shell shock episode had offered a new picture of male vulnerability. Garfield Powell, a British army diarist incensed by antiwar talk during the Somme Offensive, exclaimed: "Shell shock! Do they know what it means? Men become like weak children, crying and waving their arms madly, clinging to the nearest man and praying not to be left alone" (Ecksteins, 1989, p. 173). Captain McKechnie in Ford Madox Ford's *Parade's End* pleaded: "Why isn't one a beastly girl and privileged to shriek?" In the following passage, the novelist Pat Barker reconstructs Dr. W. H. R. Rivers's reflections as he began to apply analytic methods to shell shock patients:

---

* The phrase is Walter Benjamin's.

In leading his patients to understand that breakdown was nothing to be ashamed of, that horror and fear were inevitable responses to the trauma of war, and were better acknowledged than suppressed, that feelings of tenderness for other men were natural and right, that tears were an acceptable and helpful part of grieving, [Rivers] was setting himself against the whole tenor of their upbringing. They had been trained to identify emotional repression as the essence of manliness. Men who broke down, or cried, or admitted to feeling fear, were sissies, weaklings, failures. Not *men*. And yet he himself was a product of the same system ... In advising his young patients to abandon the attempt at repression and to let themselves *feel* the pity and terror their war experience inevitably evoked, he was excavating the ground he stood on. (1992, p. 48)

In fact, Rivers treated the British poet Siegfried Sassoon for shell shock. Suicidal, wounded, and deranged, Sassoon recalled Rivers walking into his hospital room: "Without a word he sat down by the bed; and his smile was benediction enough for all I'd been through. 'Oh Rivers, I've had such a funny time since I saw you last!' I exclaimed. And I understood that this is what I had been waiting for" (Fussell, 2000, p. 102).

Second, the new awareness of male vulnerability was accompanied by a transformation in women's self-understanding, especially in regard to sexuality. When Vera Brittain began nursing at the front in 1915, she had "never looked upon the nude body of an adult male," but "from the constant handling of their lean, muscular bodies," she not only found herself at home with physical love, but also was led "to think of the male of the species not as some barbaric, destructive creature who could not control his most violent instincts but as a hurt, pathetic, vulnerable, patient, childlike victim of circumstances far beyond his control" (1993, pp. 165–166). Women's new awareness of men's vulnerability fostered interpersonal tenderness and responsiveness, especially in sexual relations. One result was a conflict between the postwar generation of Freudian-influenced flappers and their prewar suffragist mothers who had spoken regularly of the "sex war," called marriage "legalized prostitution," and campaigned against prostitution, alcohol, and nudity in films. The flapper, the *New York Times* noted, could "take a man's view as her mother never could" (Douglas, 1922, pp. 246–247). The older distrust of men persisted, as in Virginia Woolf's 1938 explanation of war as the result of the fact that men "desire other people's fields and goods perpetually ... make frontiers and flags, battleships and poison gas ... offer up their own lives and their children's lives" (Zwerdling, 1987, pp. 294–296). Nonetheless, a long-term tendency toward reconciliation between the sexes had set in, based on each sex's awareness of its own and the other sex's vulnerabilities.

Finally, by highlighting psychological vulnerability, the shell shock episode also helped foreground the significance of economic and material vulnerability, although this significance would only find full expression much later, in the social-democratic welfare states that emerged after World War II. In 1918 the Bela Kun Communist regime in Hungary offered government support for a clinic to treat the war neuroses, while inviting analysts to hold their first postwar congress in Budapest. Although short lived, the Communist government nationalized hospitals, sanitariums, and drug companies; sponsored free medical care for children; improved salaries and conditions for teachers; and introduced sex education in the public schools, removing crucifixes and forbidding school prayer (Eckelt, 1971). Meeting in the middle of a revolution, analysts resolved to prepare for "mass," that is, publicly financed, therapy. Freud explained: "the poor man should have just as much right to assistance for his mind as he now has to the life-saving help offered by surgery ... the neuroses threaten public health no less than tuberculosis" (1919, p. 167). Contra Peter Gay's interpretation, this was not a momentary lapse. Rather, as Elizabeth Danto has demonstrated in her biography of Freud, in postwar Berlin, Budapest, Prague, and Vienna, anyone "who was ill had the right to get help."* Analysts worked in nursery schools, child guidance clinics, teenage consultations, and social work agencies; militated against corporal punishment; urged teachers and parents "to reflect rather than be angry"; and distributed such works as August Aichhorn's *Wayward Youth*, Hans Zulliger's *From the Unconscious Life of Our School Youth*, and Paul Federn's *The Fatherless Society*.†

Overall, then, the experience of World War I demonstrated a difficulty that mass democratic governments might face in motivating their citizens to kill. Older motivations such as honor and self-sacrifice waned while liberalism seemed unable to generate new reasons to kill and be killed. Challenged by the rise of fascism and communism, the liberal tradition had gone into crisis, leaving a legacy of disillusion, skepticism, and pacifism in democratic societies. Antiwar sentiment became widespread, as exemplified in the "neutrality" acts and antiwar pledges of the late 1930s. Because of its repudiation of the older masculine ethic, its determined preference for reflection over action, and its sensitivity to the vulnerability, especially of the human mind, an elective affinity developed between psychoanalysis and disillusion. But analysis was also tied to positive currents: a softening of the relation between the sexes, and an increasing awareness of psychical, bodily, and material vulnerability. As a result, psychoanalysis would become integral to later attempts to institutionalize the recognition

---

* Hungarian analysis of the 1920s, Balint added, was "a very left-wing thing." Interview with Michael Balint, August 6, 1965, Oral History Collection, Columbia University Library.
† Ernst Federn (1990), son of Paul, writes: "Were we brought up psychoanalytically? Certainly corporal punishment was excluded" (p. 283).

of vulnerability in the regenerated alliance of democratic forces against fascism during World War II.

WORLD WAR II AND THE TRIUMPH OF SOCIAL DEMOCRACY

The weaknesses and caesurae in the efforts of democratic governments to win the support of their populations during the First World War was very much on the minds of such figures as Winston Churchill and Franklin Roosevelt in the late 1930s. The collapse of France in the face of the fascist onslaught, the result of internal conflicts over the Popular Front and, ultimately, over the republican form of government itself, served as an object lesson. Had England and the United States not been directly attacked the problem of motivation for democratic war would have proven far less tractable. As it happened, the 1939–1940 German bombing of London, one observer wrote, was experienced as "a natural disaster which fosters a single spirit of unity binding the whole people together," while the Japanese bombing of Pearl Harbor in 1941 rallied Americans in a similar way (Stansky & Abrahams, 1994). Nonetheless, the upsurge of national feeling that followed the attacks may not have been sufficient over the long run to motivate men and women to give up their lives and to destroy the lives of others.

At first, both Britain and the United States sought to further strengthen citizen motivation for mass, democratic war by stressing universal ideals of individual freedom. Aiming to define the war goals in this way, Franklin Roosevelt's Four Freedoms speech, delivered to Congress in January 1941, asserted two classical liberal freedoms, those of speech and religion, as well as two more recent ones, namely, freedom from fear and freedom from want, which resonated respectively with aspirations for disarmament and social-democratic reform. A few months later, these same ideals were espoused by Britain as well, in the jointly issued Atlantic Charter. The problem, however, was that universal ideals of this sort remained abstract. Far from inducing the great majority of citizens to kill and be killed, they motivated only a small proportion of the population—idealists and leftists. How, then, to motivate ordinary men and women to die for their country?

The United States found a distinctive solution: connecting universal, democratic ideals to the neighborhood, family, and personal life.* The idea was that while relatively few Americans might be expected to sacrifice their lives for a general ideal, many more would do so in order to protect their loved ones. Thus, it was necessary to forge a link between the willingness

---

* This paragraph and the next two closely follow Robert Westbrook, *Why We Fought: Forging American Obligations in World War II* (2004), from whom the quotes are taken.

to sacrifice for particularistic, family- and neighborhood-based ties and the American way of life. A vast propaganda apparatus emerged to forge this link. Part of the strategy involved portraying the enemies as collectivist societies that were supposedly indifferent to private and individual forms of love. Thus, the Japanese were depicted as constituting a single, great, hierarchically differentiated mass family, led by the emperor, in which individual loyalties were swamped. As one "expert" explained, due to the "combined influence of Buddhism, Confucianism, and the primitive Shinto," of "ancient compulsions for unlimited self-sacrifice [and] the exaltation of the community over the individual," the Japanese soldier had been rendered "indifferent to suffering—his own or others." Later variations of this supposed indifference to personal and familial ties were worked out for Nazi Germany and, after the war, for Soviet Communism. All were portrayed as authoritarian, collectivist, and, eventually, totalitarian. In contrast to these regimes, government propagandists appealed to Americans to join the war effort so that they could "discharge a set of essentially private moral obligations to individuals and interests similarly threatened—commitments to families, children, parents, friends and neighbors." Precisely this stress on "a rich and rewarding private sphere of experience" was held to define the "American way of life."

Norman Rockwell's *Four Freedoms* painting series epitomized the identification of universal ideals with those of the family and personal life. As Rockwell explained in his autobiography, he "wanted to make something bigger than a war poster ... make some statement about why the country was fighting the war." In place of the highfalutin rhetoric of Roosevelt's Four Freedoms speech he resolved to express its "ideas in simple everyday scenes. Freedom of Speech—a New England town meeting. Freedom from Want—a Thanksgiving dinner." Freedom of religion became an older couple praying at a local church. Freedom from fear became a husband and wife tucking their children in at night. Aiming to express democratic ideals "in terms everybody can understand," Rockwell linked them directly to a set of intimate or neighborly relations permeated by an ethic of care.

Rockwell's covers had long adorned *The Saturday Evening Post*, but its writers produced a further twist on the painter's vision: they linked the four freedoms not just to intimate ties but also to privatized mass consumption.* "Weren't you bragging just a little, Yamamoto?" taunted the editors. "Your people are giving their lives in useless sacrifice. Ours are fighting for a glorious future of mass employment, mass production and mass distribution and ownership." Countless ads for new consumer products elaborated the same theme, featuring images of children with such titles as "These are the

---

* *The Saturday Evening Post* produced 322 covers by the artist.

things we are fighting for."* Overall, the result was a modification of the Four Freedoms. As the economist Leland Gordon explained in 1943,

> The concept of democracy includes our economic as well as our political life. ... Although you may never have thought of it the concept of representative government in our economic life includes your freedom as a consumer to choose whatever you wish in the way of economic goods or services to satisfy your wants. To satisfy such fundamental freedoms as freedom of worship, freedom of speech, freedom of assembly, should be added economic freedom of choice. (pp. 7–8)

Psychoanalysis, with its intense investment in private relations, was important to the American military effort during World War II. For example, Roosevelt was counseled not to respond "hysterically" to hostile interpretations of government policy, but rather to "defeat them in the manner of a therapist whose non-responsive behavior [undermines] the patient's neurotic perceptions by withholding confirmation from them" while the Surgeon General's Office ordered that every family doctor in the military be taught the basic principles of psychoanalysis (Howells, 1975; Menninger, 1948; Parsons, 1954). After the war analysis invested mass consumption with a new ethic of "maturity," "responsibility," and "adulthood." Implying men's reorientation from the homosocial, adolescent world of "mates" or "buddies," to the nuclear family, the maturity ethic also applied to women, as in the 1956 movie *The Man in the Grey Flannel Suit*, in which Jennifer Jones accepts financial responsibility for her husband's war-child, thereby channeling America's financial responsibility for Italy under the Marshall Plan.

Nevertheless, it was England, which provided the fullest analytic exploration of the ties between the private family and the liberal freedoms for which the war was ostensibly fought. In the British psychoanalytic milieu, with its strong social democratic background, its passionate—Bloomsbury-inspired—interest in the new middle-class currents of personal life, and its refugees from Germany and central Europe, Freud's initial insight into the "preference for the active role" was turned into a full-fledged and complex developmental theory centered on the infant's relation to the mother. Melanie Klein was the central figure in the formulation of this theory, in which the classical Freudian concepts of ego, sexuality, and the individual gave way to new post-Freudian concepts of object, mother, and group.

The difference between Freud and Klein—the difference between the psychoanalysis of World War I and of World War II—was a difference in the conception of the overall problems facing modern men and women, of

---

* According to Westbrook (2004), even Betty Grable, the most celebrated American pinup of World War II—famous for her legs—was more a young mother than a sex object.

which war was one example. For Freud the key problem was strengthening the ego so as to give the individual freedom from impulses, from social pressures, and from impersonal representations of internalized authority. For Klein, in contrast, the problem was to build up an internal world of whole objects, that is, to forge and sustain personal connections. For both thinkers, the subject struggled to achieve a certain "goodness," but for Freud the struggle was Kantian and moral, whereas for Klein it was concrete and relational. For Freud, accordingly, the internal world was dominated by conflicts over authority; for Klein, it was dominated by responsibility to particular others to whom one had incurred obligations, not in virtue of being generically human, as in Kant, but because one had found oneself in specific relations and circumstances. For Freud, finally, the moral core of the person was formed in conflicts deriving from the "laws" that constitute our humanity, such as the incest taboo; for Klein, in contrast, the core conflicts reflected frustrations in obtaining basic needs, such as milk or attention, from immediate others and in the context of real or imagined rivals or enemies.

Expressing a new orientation toward personal life, the Kleinian ethic represented, in effect, a "feminine" alternative to Freud—an ethic of care instead of an ethic of justice. Largely, although not entirely, encompassing women, the Kleinian circle elaborated a new, mother–daughter, mother–son, and sister–sister discourse that shaped Britain's self-understanding during World War II. In June 1940, while the Battle of Britain raged, psychoanalyst Joan Riviere wrote to Melanie Klein:

> When the first official mention of invasion began, the possibility of our work all coming to an end seemed so near, I felt we should all have to keep it in our hearts. ... as the only way to save it for the future ... Of course, I was constantly thinking of the psychological causes of such terrible loss and destruction as may happen to mankind. So, I had the idea of your telling me (and then a group of us) everything you think about these causes ... First what you think about the causes of the German psychological situation, and secondly, of that of the rest of Europe and mainly the Allies, since the last war. To me the apathy and denial of the Allies, especially England, is not clear. (I never shared it). How is it connected with what I call the "Munich" complex, the son's incapacity to fight for mother and country? ... One great question is why it is so important to be brave and to be able to bear whatever happens? Everything in *reality* depends on this. (Riviere, 1940)

What is most striking in this letter is that Riviere cast Klein in the role of the mother. Thus, she asks Klein to instruct and, in so doing, to protect her children in the face of the emergency. Correspondingly, the most significant male role is not that of the father but that of the son. The key question is whether the son has the capacity to fight for his mother, his sisters, and their children, which is to say, for those who are vulnerable. The son should have learned from his own vulnerability in childhood, that is, from his relation to his mother—his "depressive position" to use Klein's language—to feel responsibility for others. But the English sons are "absent," caught up in what Klein termed "hidden, passive homosexual phantasies, plotting and scheming with the destructive father," and in manic efforts at "control." The "homosexual phantasies" to which Klein refers are men's passive relations to phallic, "hard," and dangerous figures, as the Nazis portrayed themselves. The same weakness that leads to what Klein sees as British men's unconscious complicity with fascism prevents them from recognizing their responsibilities to women and children. For her, accordingly, the relation to the mother means recognizing vulnerability and dependence. The relation to the mother is the key to ethical responsibility.

Klein's paradigm resonated with the matricentric overtones of the British war imaginary. The latter were especially apparent after the East End was bombed in 1942 and approximately 3.5 million people were evacuated to the countryside, provoking a sense of national emergency. Many of the evacuees were children, many of them poor, many separated from their mothers and fathers who remained behind to do war-related work (Perkin, 1988). Several British analysts, including D. W. Winnicott, John Bowlby, and Emmanuel Miller, sought to alert a wider public sphere about the psychic dangers of separation, creating a sense of Britain as a community organized around the need to protect vulnerable children. Likewise, among the most dramatic images of the Blitz, memorialized in Henry Moore's drawings, were of the individuals and families who occupied the London tubes, against official orders, during the bombing. These drawings symbolized the mingling of public and private in a city under siege, as well as the attempt to care for children in a semicommunal environment.

Representations of the unity of the English people sometimes had Christian as well as matricentric overtones. The most celebrated work of art produced during the war, Moore's *Madonna and Child*, was unveiled in 1943. The sculpture resulted from the initiative of Reverend Walter Hussey, who wanted to see the Church of England retake its leading role in the arts. Moore's shelter drawings seemed to Hussey "to possess a spiritual quality and a deep humanity as well as being monumental and suggestive of timelessness." At the dedication of the sculpture, Hussey told the congregation:

> The Holy Child is the centre of the work, and yet the subject speaks of the Incarnation—the fact that the Christ was born of a human mother—and so the Blessed Virgin is conceived as any small child would in essence think of his mother, not as small and frail, but as

the one large, secure, solid background to life. (Stansky & Abrahams, 1994, p. 65)

Nonetheless, the mother-centered imagery heralded a social revolution. George Orwell's essay, "Socialism and the English Genius," written in London at the height of the bombardment, argued for the welfare state by calling Britain "a family with the wrong members in control" (Hennesy, 1993, p. 37). The 1942 Beveridge Report, with its focus on mothers and children, was more explicitly mother-centered. After the bombing of the East End, the queen announced her personal support for socialized medicine, remarking, "The people have suffered so much" (Perkin, 1988). When the National Health Service was finally created in 1948, Britain became the first Western country to provide free health care to the entire population, the first based not on the insurance principle, in which entitlement follows contribution, but on the principle of social citizenship, and the first capitalist welfare-state to pay for psychological counseling, thanks to the lobbying of the British analysts.*

During World War II, then, both Britain and the United States were successful in solving the problem of democratic motivation. While in both countries, the cultivation of the will to fight and the acceptance of death benefited from an attack from the outside, in neither was this sufficient. In both a new connection between investment in one's family and neighbors and the larger national and, at times, even transnational cause laid a more lasting basis for the willingness to die. In England, the sense of family was more communal and matricentric, providing a special resonance with the Kleinian themes. By contrast, the American solution, including its psychoanalysis, was oriented toward a more privatized and consumption-based way of life.

In forging their solutions, both countries laid a new stress on the sensitivity to human vulnerability that had emerged in World War I. In Britain, this led to a recasting of psychoanalysis so that it began to converge with child development studies and pediatrics. Donald Winnicott, whose theory of the "good-enough mother" validated working-class familial practices against middle-class "advice," exemplified the change. In the United States, wartime publicists had emphasized the uniquely private character of the family but postwar mass culture featured vulnerability. The 1950 film *The Men* centered on Marlon Brando's struggle to accept his war-generated paraplegia. Discharged from an all-male Veterans Administration hospital against his will but for his own good, he is shown in the final scene dragging his useless body up the walk of his single family, suburban home. His wife asks if he needs help. "Please," her husband replies (Graebner, 1991, p. 15).

---

* According to Rudolph Klein: "At the time of its creation it was a unique example of the collectivist provision of health care in a market society" (Hennesy, 1993, p. 132).

In both Britain and America the awareness of vulnerability also led outward to create new forms of human solidarity. In Britain, the Christian symbology of the mother and child led to identification with core Western—that is, "human"—values. When the pianist Myra Hess played Bach's "Jesu, Joy of Man's Desiring" at a lunchtime concert at the National Gallery, emptied of paintings, with bombs bursting overhead, Kenneth Clark remembered, that "in common with half the audience, I was in tears; this is what we had all been waiting for—an assertion of eternal values."*
In the United States, the ideal of the private family allowed the country to present itself as the core of the "free world." In Hannah Arendt's 1948 *Origins of Totalitarianism*, for example, the destruction of private life distinguished totalitarianism from earlier forms of tyranny while in Stanley Elkins' 1959 *Slavery*, the absence of private space—a house, a garden— made American slavery much more virulent and destructive than that of Brazil or Haiti.

In both countries, finally, psychoanalysis found a place within the continuing evolution of core liberal values, as spelled out in Roosevelt's Four Freedoms speech and in the Atlantic Charter. In both countries, psychoanalysis was integrated into the new social-democratic welfare states, which incorporated the new understanding of vulnerability, risk, and insecurity. In the 1960s, however, the New Left, and the women's and gay movements, targeted the welfare state, including psychoanalysis. Although the analytic profession survived a new, more vibrant terrain of analytic cultural and philosophical thought opened up—what the French call *le champs freudien*—often feminist in inspiration and no longer restricted to the clinic.

THE WAR ON TERROR AND THE DAWN OF GLOBAL JUSTICE

On September 11, 2001, 19 radical Islamists hijacked four U.S. airplanes and used three of them to attack the World Trade Center and the Pentagon. The immediate response of people throughout the world was one of sympathy and identification with the United States, exemplified by the famous *Le Monde* headline, "We are all Americans." The Bush administration, however, used the opportunity to pursue an illegal and misguided war in Iraq, which turned the world against the United States. Neither the sympathies for the United States nor America's desire for international support disappeared, however. Beginning in 2006, Barack Obama's primary campaign for the U.S. presidency began to reopen lines of communication

---

* In contrast to World War I, when German music had been frowned upon and even banned, the allied symbol for victory was the opening bars of Beethoven's Fifth Symphony (V in Morse code).

that had been aborted. Here, then, is a classic case of (a) a trauma; (b) a self-destructive reaction to the trauma—presumably a repetition in Freud's sense; and (c) a self-correction or reparation. Can we situate this episode in relation to our first two episodes?

To answer this question we should note one important difference from the preceding cases. By 2001, the United States had adopted a mercenary or so-called volunteer army, and so the problem of motivation centered not so much on the soldiers, many of whom signed up for material benefits, as on the American people, whose diffuse support for "the war on terror" the government needed. In gaining this support, the government relied in part on what can only be described as socially organized ignorance concerning America's role in the world. For decades the United States had propped up authoritarian governments in the Middle East, which pursued military power at the expense of social and economic development, resulting in a mass of alienated, angry, and jobless young people, some of whom had turned to anti-U.S. terrorism before the 9/11 attacks. Had this been known, the attacks might not have been such a surprise. In fact, however, the American people did not have a clear view of their government's actions and hence were shocked. As in the case of Pearl Harbor, they were prone to support the war on terror unthinkingly. But here too an additional, deeper motivation kicked in, namely, the desire to ward off realization of their own nakedness and vulnerability, which the attack revealed. In understanding this "extra" factor, psychoanalysis sheds special light.

Several analytically inspired essays by the philosopher Judith Butler explore this terrain. Written in almost immediate "response to conditions of heightened vulnerability and aggression" that followed 9/11, Butler's essays describe a period when "US boundaries were breached [and] an unbearable vulnerability was exposed." Why, Butler asked in 2003, did vulnerability lead to military action and aggression, rather than to the higher level of global communication that seemed possible in the immediate aftermath of the attack?

Although deeply analytic, Butler's answer departs from classical psychoanalysis in one crucial respect. Classical analysis suppressed interpersonal and social relations to isolate what might be called the individual's relation to himself or herself. Butler rejects this distinction. Instead, she views ethical responsibility toward others as the core of the personality. Certainly, this resonates with the object-relational tradition of psychoanalysis. But whereas Kleinian analysts explored responsibility in the context of the early mother–infant relationship, Butler's analysis of 9/11 situates responsibility in relation to dying and death. Here she draws upon Emmanuel Levinas's idea of "the approach to the face," which Levinas saw as "the most basic mode of responsibility." According to Levinas, the face "is the other before death, looking through and exposing death ... the face is the other who asks me not to let him die alone."

The idea that we are faced with an ethical command to be present at the death of our fellow human beings changes our understanding of war. A clue to how can be found in the close etymological link between *grievance*, as in the cause of war, and *grief*, as in war's consequence.* Both words come from the Old French, *grever*, which meant causing sadness, suffering sadness, but also complaining about its causes. Similarly, our modern English word *grieve* has both active and passive connotations, meaning both to lament and to file a complaint. These linguistic links are psychologically revealing. On the one hand, men and women support war because of *grievances*: the assassination of an archduke, the bombing of Pearl Harbor, the attacks on the Twin Towers; indeed Butler concedes that the wanton murder of 3,000 people in the Twin Towers gave the American people a grievance. On the other hand, she adds to the idea of a grievance the element of *grieving*, especially our responsibility to prevent others from dying alone and, by extension, to mourn them after they have died. Examining the near aftermath of 9/11, Butler attempts to locate both impulses: the grief that accompanied the grievance and the aggression that supplanted the grief, amid the chaos that followed the attacks.

In fact, the imperative to grieve and not to let others die alone had considerable force in New York in the fateful days that followed the attacks. Presumably, the hijackers chose the World Trade Center and the Pentagon as icons of U.S. greed, arrogance, and impiety. The initial response, however, was to dereify those iconic objects by dissolving them into the individuals that comprised them. One could see such attempts at dereification in several forms: the pictures of the "missing" found on public walls in New York for many weeks afterward; the *New York Times* project of publishing capsule biographies sometimes with pictures of each and every victim; and the attempt to find every shard or bone fragment by which an individual could be identified, even though this prolonged the cleanup process and deferred the achievement of what was called "closure." In all these cases dereification, by focusing on individuals, sought to wrest meaning from an act that was initially perceived as incomprehensible.

As in World War II, when a mass event was personalized, and so in the immediate aftermath of 9/11, the ordinary and the everyday were infused with personal meaning. Typical titles of the *New York Times* obituaries were "Outdancing her Husband," "She also Tended Bar," and "He Wanted More Children." The shopping emporia and the office suite no longer appeared only as degraded sites of consumerism and money-grubbing, but as life-worlds, suffused with individual aspirations and personal dreams. Particularly striking was the revelation that the many brokerage firms housed in the World Trade Center were staffed by upwardly mobile

---

* I take this suggestion from Paul Hoggett, "Politics and the Relative Autonomy of Affect" (unpublished paper).

working class men and women from Brooklyn, young immigrants from Latin America and East Asia, and Middle Easterners, both Israeli and Arab. Equally important was the face-to-face recognition that marked the normally anonymous streets of New York, as strangers recognized, with a new level of depth, that they inhabited a common world.

These phenomena suggest that the immediate response to the traumatic event included a range of different possibilities, not just shock and a desire for justice. Almost immediately, however, the element of grief—the desire to mourn, the need to connect to others, the responsibility not to let people die alone—was shut off, as the Bush administration turned to nation building and martyrology. Ten days after the attack, Bush addressed Congress saying, "Our grief has turned to anger and anger to resolution. Whether we bring our enemies to justice or bring justice to our enemies, justice will be done." American foreign policy, supposedly the product of generations of reflection, was swiftly transformed. Special interests vaulted into the vacuum, positioning themselves to profit from the coming wave of "disaster capitalism." The characteristic symptom of trauma appeared, namely, repetition, the constant replaying of images of the towers in flames, the towers collapsing, the compulsive visits to ground zero. Reason fled. Jacques Derrida, in New York several weeks after the attack, remarked: "Not only is it impossible not to speak on this subject, but you feel or are made to feel that it is actually *forbidden*, that you do not have the right, to begin speaking of anything ... without making an always somewhat blind reference to this date" (Borradori, 2003, p. 87).*

Certainly, any response to violence would have mixed grief and grievance, the desire for justice and the need to mourn. But the American response went one sidedly in the direction of what Butler calls the "foreclosure of alterity." Historians will eventually have to explain why. What is clear even now, however, is that two different currents of feeling—grief and grievance—remained mixed up in the response, even as the second seemed exclusively to prevail. It is likely, too, that this is part of a general pattern, that is, that every grievance that leads to war also includes an element of grief, which can mitigate the drive to war and perhaps avert further violence.

To render this insight useful, a larger—political—framework is needed. According to Butler (2009, Chapter 1), liberalism—broadly conceived as the framework of modern political claims making—presupposes an ontology centered on self-ownership and self-sufficiency. As a result, "we have to present ourselves as bonded beings—distinct, recognizable, delineated subjects before the law." In lieu of this presupposition Butler asks, "what form political reflection and deliberation [might] take if we take injurability and aggression as [our] points of departure for political life," that is, if

---

* "I gave in regularly to this injunction," Derrida adds.

we began from "an understanding of how easily human life is annulled," rather than from the idea of a delineated, rights-bearing and autonomous subject per se. Doing so, she suggests, implies an "insurrection at the level of ontology."

If we start by acknowledging not just the full force of our grievance but also the full weight of our grief, according to Butler, we will feel our commonality not only with the 3,000 who died on 9/11 but also with the hundreds of thousands, perhaps more, who have died in the Middle East as a result of our policies. Solidarity with these deaths has nothing to do with *excusing* the attacks of 9/11. Rather, it has to do with rebuilding the world after the attacks. Everyone in the United States remembers where they were and what they were doing when they learned that the planes hit the towers. Everyone begins the story of that day by invoking a first-person narrative point of view. That narrative is part of the healing process. As Butler writes, "a narrative form emerges to compensate for the enormous narcissistic wound opened up by the public display of our physical vulnerability." But although necessary, this narrative form is not sufficient. Equally necessary is a process of decentering to which mourning is key. The goal, Butler argues, is "the ability to narrate ourselves not from the first person alone, but from, say, the position of the third, or to receive an account delivered in the second." It is in achieving this goal that grief and mourning play a critical part.

Why is mourning so important to decentering? Butler answers this question by considering how grief is processed and distributed at the center of community. Thus, she problematizes the community's economy of grieving, which permits its members to mourn the loss of some subjects, but forecloses the mourning of others. In principle, the experience of mourning based on a recognition of our common, bodily vulnerability, can deepen and intensify our understanding of community. Yet "the differential allocation of grievability" operates to produce and maintain exclusionary conceptions of "what counts as a livable life and a grievable death." Many, excluded from the circle of grievability, die alone. Whereas American deaths are "consecrated in public obituaries that constitute so many acts of nation-building," the names, images, and narratives of those whom the United States kills are suppressed. Our capacity to mourn more broadly, beyond the nation, is foreclosed by our failure to conceive of Muslim and Arab lives *as lives*. Guantánamo becomes the purgatorial locus of "unlivable lives." The result, Butler suggests, is "a national melancholia," the result of "disavowed mourning," a melancholia that the Obama presidency has presumably the responsibility of relieving.

Butler furthermore argues that the unique place that mourning occupies in the building of community rests on the way it entails a loss of control, or surrender, that allows us to recognize and accept the other—in the 9/11 case, the Muslim world. Grief, she writes, entails "moments in which one undergoes something outside one's control and finds that one is beside oneself, not at one with oneself." One thereby does not only undergo, one *submits* to a transformation that "de-constitutes choice at some level." Mourning then disrupts the bounded sense of self and the overvalorization of choice that is at the base of the liberal tradition; it catalyzes the insurrection at the level of ontology that Butler believes is necessary. In mourning one "accepts that by the loss one undergoes one will be changed, possibly forever." We lose not just the person but also our former sense of identity; "When we lose someone we do not always know what it is *in* that person that has been lost." Acceptance of this loss, she argues, is crucial to the new terms of ethical responsibility entailed by globalization:

> Many people think that grief is privatizing, that it returns us to a solitary situation and is, in that sense, depoliticizing. But I think it furnishes a sense of political community of a complex order, and it does this first of all by bringing to the fore the relational ties that have implications for theorizing fundamental dependency and ethical responsibility.

Grief, she concludes, "contains the possibility of apprehending a mode of dispossession that is fundamental to who I am." It exposes "my unknowingness, the unconscious imprint of my primary sociality."

Butler's essays by no means constitute a finished political philosophy, but they do demonstrate the continuing illumination provided by psychoanalysis. As with Freud in 1919, the revolt against passivity is at the center of Butler's explanation of why the American people so wholeheartedly followed triumphalist and aggressive policies that were obviously mired in weakness and insecurity. As with Klein and her followers in 1939, Butler argues that a mature response to aggression draws on our experience of infantile dependence on others, an experience that turns us into ethically responsible individuals. Incomplete though Butler's account is, it also provides a clue to the long-term response to 9/11, scarcely imaginable in 2003 when Butler wrote, namely, a candidate for president in 2008 who at times came close to apologizing to the global community for the one-sided American response, however differently he behaved once he became president. In that sense, time provided an answer to Butler's question: "Is there [not] something to be gained from grieving, from tarrying with grief, from remaining exposed to its unbearability and not endeavouring to seek a resolution for grief through violence?"

At the same time, Butler's essays leave us with a larger question. If we do take "an understanding of how easily human life is annulled" as our point of departure for political life, this might lead in two different directions. Either our deepening understanding of vulnerability *displaces* the idea of a liberal polity based on individual rights and social justice in favor of a politics of recognition and mutual support. Or it *imbues* the idea of a liberal polity with an understanding of the inherent vulnerability of the subject, and thereby encourages a tense but dialogic relation between the pole of separateness and individual rights and the pole of mutual need, dependence and recognition. The account given here of the mutually imbricated evolution of liberalism and psychoanalysis suggests that the second alternative is preferable. Most likely, we will be in a better position to avert, limit, and ultimately end war if we consider our fellow human beings not only as vulnerable bodies, but also, and at the same time, as rights-bearing individuals and members of polities.

CONCLUSION: TOWARD A POLITICS OF VULNERABILITY?

In this chapter, I have posed the question of how democratic societies motivate their citizens to fight and support wars, especially given that these wars are often organized by elites for self-interested ends and only justified post hoc through the manipulation of a frequently corrupt and shallow media. Unlike those who consider the acceptance of war easy or natural, I have suggested that motivating a democratic citizenry has posed a continuing problem. In facing that problem, democratic societies have become increasingly aware of human vulnerability in its multiple forms: infancy; psychological vulnerability, bodily and economic vulnerability; old age, dying, and mourning among them. Over time, the awareness of vulnerability has converged with social-democratic efforts to cushion the risks of life, both natural and social. Throughout that unfolding history, psychoanalysis has played a central role.

Their involvement with war provoked analysts to deepen an insight into vulnerability that first arose in the clinic. With hindsight we can identify three stages in this deepening. In the first stage, that of World War I, Freud and his followers showed that the modern conscript soldier's traumatic experiences often provoked or intensified an irrational and defensive wish to avoid vulnerability. In the second, that of World War II, Kleinians sought to transform that wish into the obligation to protect vulnerable loved ones, an obligation that led to support for the community and the nation. Finally, in the third stage, that of the war on terror, Butler showed how anxiety concerning vulnerability contributed to an outward-directed aggression.

In all three moments, we can also see that increased awareness of vulnerability has been linked to an expanding sense of human connection. In World War I, racial and national casus belli increasingly failed. In World War II, personal relations, centered on the family, connected the soldiers to national and sometimes even to transnational ideals. Reflecting on the war on terror, finally, Butler glimpses the possibility of a global community, in which Muslim lives receive coequal attention with American lives. In this sequence, we see progressive efforts to extend local loyalties to embrace that ever-larger—extrafamilial—unity we call humanity, and which Freud called civilization, meaning the work of eros or binding.

If the three moments show a broadening of human connection, what conception of the mind underlay this shift? Whereas the Enlightenment viewed the subject as the locus of reason in the sense of universal and necessary truths, Freud situated reason amid the confusions of what he called "the riddle of the neurosis," a force that was "inexpedient, and running counter to the flow of life." In World War I, this led to the insight that the apparent rationality of the ego was shot through with the resistances, repetitions, and denials of vulnerability exemplified by shell shock. During World War II, analysts' empathy for the inner worlds of bombed civilians, devastated soldiers, and bereaved children precipitated the collapse of the metapsychology (i.e., id, ego, and superego) and the turn toward the Kleinian themes of attachment, connection, and rupture (Homans, 1989; Suttie, 1945).

Butler, finally, operating in the broader *champs freudien* and drawing on the cultural revolutions of the 1960s and 1970s, redefined what Klein called internal object relations to include the ethical self-understandings that could become commitments (commands or Mitzvoth for Levinas) to the human species as a whole, and not merely to immediate others.

In these respects, the trajectory leading from World War I to World War II and then to the war on terror represents an advance. Yet we may wonder whether it risks losing what is finally specific to psychoanalysis, namely, the investigation of what Butler calls "the bounded subject." In the shift to object relations, psychic structure—ego, id, and superego—threatened to disappear in favor of a focus on internal object relations (i.e., the psychical representation of ethical relations to others). But a robust conception of psychic structure is necessary to understand reason as situated in an internal environment characterized by competing impulses, self-criticisms, fears, and demands. Butler's work, too, harbors an analogous risk. Can we really base a conception of the subject on dependence, vulnerability, and pre-Oedipal condition of merger? Or should we also hold fast to Freud's idea that "Where id was, there shall be ego?" In the latter case, the task would be not to abandon the bounded subject, but to stress its constrained, connected, vulnerable, surrounded, and endangered character.

Whatever the answers to these questions, we are only at the beginnings of the encounter between psychoanalysis and war. The history so far shows us that in that encounter psychoanalysis did not stand alone. In World War I, it had deep and complex relations with communism, social democracy, feminism, and modernism. In World War II, it was inseparable from the social-democratic welfare states. If Butler's initiative is to bear fruit, psychoanalysis will have to forge connections with new, as yet undefined but comparably powerful political currents.

REFERENCES

Barker, P. (1992). *Regeneration*. London: Dutton.
Borradori, G. (2003). *Philosophy in a time of terror: Dialogues with Jurgen Habermas and Jacques Derrida*. Chicago: University of Chicago Press.
Brittain, V. (1993). *Testament of youth: An autobiographical study of the years 1900–1925*. New York: Macmillan.
Butler, J. (2009). *Precarious life*. London: Verso.
Douglas, A. (1992, 1996). *Terrible honesty*. New York: Farrar, Straus & Giroux.
Eagle, M. (1987). *Recent developments in psychoanalysis: A critical evaluation*. Cambridge, MA: Harvard University Press.
Eckelt, F. (1971). The internal policies of the Hungarian Soviet Republic. In I. Völgyes (Ed.), *Hungary in revolution 1918–1919* (pp. 61–88). Lincoln, NE: University of Nebraska Press.
Ecksteins, M. (1989). *Rites of spring: The Great War and the birth of the modern age*. Boston: Houghton Mifflin.
Federn, E. (1990). *Witnessing psychoanalysis: From Vienna back to Vienna via Buchenwald and the USA*. London: Karnac.
Freud, S. (1933). Why war? In J. Strachey (Ed. & Trans.), *The standard edition of the complete psychological works of Sigmund Freud* (Vol. 17, pp. 197–215). London: Hogarth Press.
Freud, S. (1920). Beyond the pleasure principle (SE 18, pp. 7–64). In J. Strachey (Ed. & Trans.), *The standard edition of the complete psychological works of Sigmund Freud*. London: Hogarth Press.
Freud, S. (1919). Lines of advance in psycho-analysis psycho-therapy. In J. Strachey (Ed. & Trans.), *The standard edition of the complete psychological works of Sigmund Freud* (Vol. 17, p. 167). London: Hogarth Press.
Fussell, P. (2000). *The Great War and modern memory*. New York: Oxford University Press.
Gordon, L. (1943). *Consumers in wartime: A guide to family economy in the emergency*. New York: Harper & Brothers.
Graebner, W. (1991). *The age of doubt: American thought and culture in the 1940s*. Boston: Twayne.
Hennesy, P. (1993). *Never again*. New York: Pantheon.
Homans, P. (1989). *The ability to mourn: Disillusionment and the social origins of psychoanalysis*. Chicago: University of Chicago Press.
Howells, J. G. (1975). *World history of psychiatry*. New York: Brunner-Mazel.
Leed, E. J. (1981). *No man's land: Combat and identity in World War One*. Cambridge, MA: Cambridge University Press.
Mann, T. (1984). *Diaries 1918–1939* (H. Kesten, Ed.; R. Winston & C. Winston, Trans.). London: Robin Clark Ltd.
Maxwell, W. M. (1923). *A psychological retrospect of the Great War*. London: Macmillan.
Menninger, W. C. (1948). *Psychiatry in a troubled world: Yesterday's war and today's challenge*. New York: Macmillan.
Nunberg, H., & Federn, P. (Eds.). (1963). *Minutes of the Vienna Psychoanalytic Society*. New York: International Universities Press.

Parsons, T. (1954). Propaganda and social control. In *Essays in sociological theory* (pp. 89–103). Glencoe, IL: Free Press.
Paskaukas, R. A. (Ed.). (1993). *The complete correspondence of Sigmund Freud and Ernest Jones, 1908–1939*. Cambridge, MA: Harvard University Press.
Perkin, H. (1988). *The rise of the professional society: England since 1880*. New York: Routledge.
Riviere, J. (1940). Letter Joan Rivier to Melanie Klein, June 3, 1940. pp/KLE/C95. British Psychoanalytic Society.
Rockwell, N. (1988). *My adventures as an illustrator*. New York: Abrams.
Simic, C. (2008). The nicest boy in the world. *The New York Review of Books, 55*(15) (October 6).
Simmel, E. (1918). *Kriegs-neuroses und "psychisches trauma."* Munich: Otto Nemnich.
Stansky, P., & Abrahams, W. (1994). *London's burning: Life, death, and art in the Second World War*. London: Constable.
Suttie, I. (1945). *The origins of love and hate*. London: Kegan Paul.
Westbrook, R. (2004). *Why we fought: Forging American obligations in World War II*. Washington, DC: Smithsonian Books.
Zwerdling, A. (1987). *Virginia Woolf and the real world*. Berkeley, CA: University of California Press.

Chapter 10

Casus belli

Françoise Davoine

BLOOD ON OUR HANDS

Adrienne Harris and Steven Botticelli, the editors of this volume, ask us to address a very provocative statement: "Psychoanalytic, mental health, and psychiatric institutions are responsible for warmaking, aids to propaganda and torture, as well as war resistance and postwar healing."

The same tools that are used for healing can be used to torture. As Jonathan Shay (1994) concludes his powerful analysis of combat trauma and the undoing of character, "Unfortunately, the same scientific knowledge that can define a science of human rights can be used to perfect the science of tyranny. Anyone who understands the causes of PTSD [posttraumatic stress disorder] and character damage, can train torturers" (p. 209). In the same vein, according to Simon Epstein's (2008) recent book, many pacifists in France who had been honestly defending human rights during the 1930s, were, after the defeat in June 1940, ready to find themselves colluding with the collaborationist regime of Pétain and Laval. As they would tell themselves, it was for the sake of peace that they contributed to the iniquitous "Jewish Laws," as well as to deportations.

Quite a number of them had endured World War I as veterans or civilians. Perhaps they had been confused by the wishful thinking: the "Nevermore!"; the *plus jamais ça*, endlessly repeated during "the long weekend" between the two wars, as Bion (1982) called it. However, even in the comfort of peacetime, Shay (1994) warns us: "We have blood on our hands" (p. 197) if we cannot face and try to heal—"yes, narrative heals!"—the traumas of veterans, civilians, and of their descendants.

ENTERING THE FIELD OF CATASTROPHES

As an analyst who works with trauma and psychosis, I find myself in agreement with Shay. My focus on matters that have been erased from history is linked to my early childhood spent in the Alps during World War II in a war zone, where civilians at that period were just as expendable as they are now. So, I tend to trust disturbed patients as people who have experienced the frail limits of reliable speech and are best qualified to teach us about matters uninvited to be written in history. I consider them as coresearchers aiming at that inscription.

But this is easier said than done. Oftentimes, as Harris and Botticelli notice, psychoanalytical tools betray their claimed purpose to free the subject of speech, and instead support propaganda for war and are enlisted in creating prowar slogans, for instance in 1933, the aryanization of the German Institute for Psychoanalysis in Berlin, as we will mention later on.

Still, without choice, psychoanalysis seems to deal with war. At times, sessions look more like a combat zone than like a laboratory, especially when dissociated material comes to life in the relationship between patient and analyst. We called this necessary pass a casus belli in the analysis, which will be described. At such times we fight against annihilation, and experience the same kind of failures, errors, cowardice, and cruelty, as well as the luck and joyful coincidences, that occur on the front line.

Like a military march, the rhythm of our work follows a binary beat, with ups and downs, hope and despair, progress and regress, victory and sudden collapse. The chronology also is warlike, starting with the illusion that it will not last long, and then for no reason it lasts for ages, with flat periods of no action, leaving us to watch the frontier while nothing happens.

Finally, the unavoidable crisis occurs, bringing forth the ghosts from the past. The analyst may suddenly embody a strange kind of ruthless otherness. In my experience, this unpleasant episode occurs just as we start to feel that the treatment is going well, as if psychological torture had to wait until this relaxed moment to catch us unawares in the transference.

At that moment, the Erinyes, the Greek goddesses of vengeance, also known to the Romans as the Furies, with their hairdos full of snakes, enjoy a manhunt across the generations, under the cover of a scientific warrant, or a classical psychoanalytic approach. They declare: "No way! This disturbed patient is genetically or structurally, mentally sick. He is done for!" So analysts, beware those Ladies of the Structure or of the Gene; they want to cudgel our brains into what the historian Dominick La Capra (2001) calls "a desperate melancholic feedback loop" (p. 21).

My purpose in this chapter is to show how to get the hell out of that war and escape the temptation to, with the best of intentions, misuse our powers as therapists.

Long ago, Greek tragedies used cathartic techniques to interrupt fate's repetition across generations. They used drama to enact that fight, *agôn*, so that citizens of Athens could express and expel the unsayable and unimaginable issues of trauma. Nowadays such cathartic practices are still in use, for instance, among aboriginal people, who speak of exploring "the world of the dreaming." But strangely enough, developed countries seem quite underdeveloped in this respect. The skills we have gained in technology, consumption, and positivistic science have been lost regarding the science of healing traumas.

Shocks and drugs are no better than the prehistorical good old "bump on the head." By saying so, I am probably being unfair to our prehistoric colleagues, who actually made sophisticated use of language in their cave paintings. Although we are unable to understand them, they probably had a healing function.

To our credit, we who have become so humane, antiracist, civilized, and democratic are baffled by the inhuman dimension and intensity brought on by trauma or by psychosis. So, even if we strive to acknowledge our common humanity with our psychotic patients, we also have to validate that they deal directly with hell. To reach them there, we have, so to speak, to be willing to enter hell, keeping one foot outside while putting the other into it as says Gaetano Benedetti (1980), a psychoanalyst who used to treat psychotic patients in Basel, Switzerland.

Ira Steinman (2009), a psychoanalyst based in San Francisco, is not at all repelled by the inhuman side of history. He repeatedly asks his purportedly untreatable patients the question: "But what happened to you?" After a long while, they finally bring their tortures into the sessions, then slowly recover. This is true also for war survivors who stop talking when they meet general condemnation of the apocalypse, which for them is still ongoing, in the endless present. Sparing their analysts the horrors they have experienced, they are self-censoring, silencing what they have to say.

My own method is to face, enter, and interact with the field of catastrophes, when time stops, and when limits between men and things are blurred, as in the case of petrifaction. I tell anecdotes, bits of tales, lines of songs or poems, passages of books from my own background, which have always been my saviors in difficult times, "my favorite things" as Julie Andrews sings in *The Sound of Music*. Telling stories appeared to me as the only means to weave together the human and the inhuman, to make connections between cultures or between people who seem to share nothing with each other and have nothing to tell each other.

For instance, once I quoted Remarque's novel *All Quiet on the Western Front* (1928/1995), which spoke from the German trenches, to forlorn diaries written in the French ones by discarded grandfathers, on the limits of the no-man's-land where their delusional great-grandchildren cried out unspoken truths. Fragments of those diaries make an appearance in my office, shedding light on what had until then been the incomprehensible symptoms of their descendants. Those stories also bear witness to an unexpected bond between men in the mud, the beauty of the skies, the colors of the coming spring, and a kind of love between fighters and survivors, which the ancient Greeks called *philia*.

Storytelling is the most ancient and universal way to communicate the paroxysm of energy stemming from trauma, encoded within what Damasio (2003) calls "body receptors." Difficult as they are to explain, those encoded memories of trauma nevertheless uncannily make themselves known. Through myths and fairy tales, we learn how to make sense of images saturated with this inexplicable energy emerging from the Real, which Lacan defines as the impossible to be named or imagined. Oftentimes, in these stories, man is at the bottom of the ladder of the species, and has, for the sake of his body and his psyche, to learn from a "non-human environment" (Searles, 1960) how to relate to the wilderness and its monsters, a wilderness with which psychosis is so familiar.

So I am going to tell you stories, in an attempt to show how war traumas and psychotic symptoms represent a search for unprecedented links, not so much to tame the Real as to invent from scratch new social links that the best specialists could not have detected. I will try to show how the most disturbed patients can offer fresh understandings that may help renew worn-out paradigms.

I will tell those stories, following their wandering paths, in exactly the way I would tell them to people I work with. And in so doing I will bear witness to Ludwig Wittgenstein's (1930–1933/1993) statement, "One could almost say that man is a ceremonial animal" (p. 129). This statement speaks to the power that psychoanalysis holds with respect to warfare: to create a way of healing the traumas of every sort of war, those raging at home, as well as civil wars, religious wars, or new-fashioned wars. By bearing witness, analyst and patient create the possibility for exiled subjects to come back, as demonstrated by Felman and Laub (1992), connected through the social link of transference.

PSYCHOANALYSIS AND TORTURE, BERLIN 1937

One remembers the hijacking of the Psychoanalytical Institute of Berlin by Mathias Göring, cousin of Marshall Göring, which became the German Institute of Psychological and Psychotherapeutic Research, ruled by Nazi ideology (Brecht, Friedlich, Hermann, Kaminer, & Juelich, 1985). Jewish psychoanalysts had then to search for exile in order to save their lives. We know that now other totalitarian ideologies over the globe continue to use psychiatry and even psychoanalysis to confine and brainwash political opponents. At the German Institute, Werner Kemper was charged with responsibility for treating the psychogenic reactions of the soldiers in the Luftwaffe. Fleeing to Brazil after the war, he became (under Ernest Jones's protection) one of the founders of the Brazilian Psychoanalytic Institute and the supervisor of Amilcar Lobo (Besserman Vianna, 1997), who collaborated with perpetrators during the period of the military dictatorship in that country (1964–1985).

In France during World War II, prominent psychiatrists like Georges Heuyer and Daniel Lagache (Grim, 1998) were asked to screen "deficient, morally endangered children." In order to produce an official classification, they had to choose between a cultural and a biological diagnosis. The latter was of course preferred, as a biological diagnosis was in keeping with a "scientific" approach that obeyed "objective" criteria. This scientific rigor was especially appreciated by Laval, premier in the collaborationist regime of Maréchal Pétain. In order for France to play the role of "a brilliant second" to Nazi victory, he had personally initiated the creation of a technical council for deficient childhood and in moral danger, presided by George Heuyer, founder of child neuropsychiatry, who launched this global classification.

In 1944, Daniel Lagache issued a report that specified criteria for distinguishing children as having "potential, semi-potential or no potential for rehabilitation." A threatening system of detection was organized at the very moment of the "final solution" (Rossignol, 1998). Fortunately, this diagnostic scale was designed too late to be implemented, so that the children determined to have "no potential" escaped extermination, which was the typical fate of children so designated. As a matter of fact, after the war, some of the same psychiatrists who had been involved in this project shifted their support to another "science" called historical materialism. Psychoanalysis, to them, became a "bourgeois," despised, "subjective" practice and "Makarenko," Soviet pedagogy became a model (Vexliard, 1951). Within this framework, to be truly humanistic psychotherapy had to be collective, combining psychoanalytic recipes with political slogans and shock treatments.

The common seductive trait of both Nazi- and Communist-inflected psychology was its "scientific" appeal, supported, as noted by Hanna Arendt (1956, p. 306), "by a carefully elaborated system of scientific proofs." Nevertheless, they each seem to represent a remake of familiar understandings of the relationship between mind and brain, and of mechanistic versus transferential ways of healing, that had been prevalent at least since the Middle Ages. At that time, as today, a purely biological approach prevailed. Madness was considered to stem from brain deficits, and was to be treated with shocks and drugs. This approach coexisted and also conflicted with a literary approach, which used romance of chivalry, courtly love, carnival rituals, and the theater of the fools as a cathartic tool.

The war between these two points of view on madness is made even clearer when considering their differing perspective on war trauma, as shown in Ben Shephard's *A War of Nerves: Soldiers and Psychiatrists in the 20th Century* (2000). A brilliant example of this war is also illustrated by Pat Barker's trilogy (1992, 1994, 1995). Drawing on the notebooks of the famous anthropologist and neurologist William Rivers, the novels show

how he fashioned a psychoanalytical approach to treat soldiers in World War I, eschewing the commonly used electric treatment called faradization. Among those he treated was the poet Siegfried Sassoon.

TODAY

The scientific religion of "neuronal man" (Changeux, 1983) considered as a pure brain, without even a body, prevails nowadays. To parody Sullivan's "one genus postulate": "we are all simply much more human than otherwise," one might say that today we are all much more depressed, bipolar, schizophrenic than otherwise. Yet the worship of the old mechanistic model, still in the name of science, pushes us to regress toward biological treatments, as a contemporary version of *The Extraction of the Stone of Madness* painted at the turn of 16th century by Hieronymus Bosch.

Fortunately, another point of view rooted in history is still available. Cathy Caruth (1996) considers traumatic symptoms as "symptoms of history ... claiming for Unclaimed Experiences" (p. 5). At first glance, the ghosts of history, which Bruce Reis (2007) calls "the keeper of secrets, and they are the manifestations that something has gone wrong" (p. 621) may seem terrifying. Otherwise why all that fuss about staying silent? But usually, as soon as they are allowed to speak their truth in therapy sessions, like Hamlet's father, they carry us away into a whimsical dialogue, oftentimes in a Bakhtinian and "liminal" way, as Adrienne Harris (2007) observes.

This edgy vertex is not often considered a serious treatment for psychological problems. Just yesterday, as I was happy and joking at dinnertime, I turned on the television to watch the news. A cartoon warned that millions of us were in danger of succumbing to depression and did not know how sick they were, how ignorant of the meds anxious to cure them.

My joy turned immediately into a professional concern. Was the listless guy in the cartoon haunted by some ghosts of historical catastrophes, even one that took place many generations before? Was he stuck into a frozen present with no future? Perhaps he already looked at himself as a disease: a "bipolar," a "schizophrenic," a "paranoid." How was I to help him get out of his "voluntary servitude" (La Boétie, 1548/1942) to the craziest experiments that have been undertaken throughout history for the treatment of the mad?

Honestly, I would advise this little guy rather to follow the path opened by Erwin Schrödinger (1956/1990), one of the inventors of quantum physics. In his book *Mind and Matter*, he advises psychologists to reconsider "the initial gambit," which permitted the creation of traditional science among the ancient Greeks, by exiling the subject from its field, thereby inevitably leading to objectification. As a result, "the world of science has become so horribly objective as to leave no room for the mind and its immediate sensations"

(p. 129). Even if the latter are studied and quantified, "nowhere, you may be sure, however far physiology advances, will you ever meet the personality, the dire pain, the bewildered worry within these souls, though their reality is to you so certain as though you suffered them yourself" (p. 134). In the chapter "The Principle of Objectivation," Schrödinger points out that the observed object is modified by the act of being observed and extends that statement from the field of subatomic particles to psychology. I take his argument to support my contention that we eschew objectification in our field of psychology in favor of the view that the psyche pertains to relationships.

Objectification is indeed currently being nurtured by efforts to annihilate the subject of speech, in a soft or harsh way. It can also be resisted, as for instance by men in battle (e.g., Gray, 1959/1998) who fight for their lives and for the life of their minds, using talking cures among themselves on the battlefield and eventually with a therapist after war's end.

The problem is that the psychotherapy of trauma takes a long time. But traditional scientific research proceeds slowly as well. Scientists repeat their experiments until they are ecstatic when their positive results are published, only to sing a different tune when they have to start all over again. The psychoanalysis of traumatic and psychotic symptoms move with the same rhythm, oscillating between elation and despair, enigma and breakthrough, swiftness and petrifaction, especially when they step on the toe or the tail of some snake-haired Furies hurling themselves into the sessions.

TELLING STORIES

Having evoked the crazy atmosphere in which such therapy may take place at some precise moments, I am going to tell stories about snakes. I took them at first as a riddle, until they came to offer me a useful tool in engaging patients who, in the transference, were at some moments my coresearchers into this experience, and at others my challengers, aiming to transform the Furies wrath into epics and history, so that they become, as in the end of Sophocles' trilogy, the "Eumenides," the benevolent ones.

The first story took place in Benin in 2002, the second 19 years earlier in Los Angeles. As they are quite enigmatic, I looked for clues to unraveling their meaning from two prominent researchers, who were also disturbed veterans.

My first coinvestigator is the historian of Renaissance art, Aby Warburg, who became crazy during World War I and emerged from his madness by performing a lecture on "the Ritual of the Snake" in 1923. The second is Ludwig Wittgenstein, a soldier and a POW during the same war, whom Jacques Lacan (1991) labeled with a "ferocious psychosis" (p. 69). Although he was greatly indebted to his work, Lacan never quite grasped Wittgenstein's breakthroughs regarding transference in the case of trauma, a specific game of language that may function, as Adrienne Harris and Steven Botticelli might say, to promote "resistance and postwar healing."

THE GOD PYTHON, BENIN, OCTOBER 2002

Five years ago, a conference was organized in Benin, Africa, by our friend Paulin Hountondji (2004), dean of philosophy at the University of Cotonou. The title of the conference was "The Encounter of Rationalities." After the conference, Paulin sent us to visit the ancient capital Ouidah. On the beach a monument built by UNESCO, "The Door of No Return," stands to commemorate that at this very place millions of people were sent to slavery in the New World over a period of at least three and a half centuries.

Near the cathedral sits a temple for the god Python. The house of the snakes is round, like in Greek Antiquity the mysterious *tholos* in Epidaure, supposed to have sheltered the sacred snakes in the temple of Asklepios, the god of medicine. In their healing function, the snakes used to wander at night among the patients, who were reclining in a place called *abaton*. Their dreams were interpreted the following day, and were believed to offer precious indications for their somatic and psychotherapeutic treatment by the priests who were also medicine men. Engraved stones discovered quite recently bear witness to this ancient *Traumdeutung*.

When I least expected it, a young guardian of the temple asked me if I wished to take one of the snakes in my hands. Being a member of a healing profession in a healing place devoted to the gods of healing, I answered yes without thinking. Then I immediately thought: "My god!" as he was serious about my handling the snake. So there I was with a huge python—perhaps not so big after all—around my neck. Contrary to my expectation, the animal was fresh, dry, soft, and full of strength. Its head was at alert, as I was myself, exploring with his tongue our new common situation. I felt peaceful, perhaps at having overcome my fear and repulsion at the thought of making contact with this strange animal, suspended on the boundary between two worlds, two cultures, two histories, beyond already-made phallo-freudo-lacanoid interpretations. The impression I was left with from this experience resonated immediately with another strange snake story that had been told to us in California 20 years earlier.

A SNAKE IN BEVERLY HILLS, JULY 1983

We are at a gathering with a few analysts, at Beatriz Foster's place in Los Angeles. She had been a medical director at the Austen Riggs Center in Stockbridge, an open hospital for the psychodynamic treatment of psychosis. At that time, she was working in private practice in Los Angeles. We are both in the kitchen where she is preparing our dinner. She asked me abruptly "Françoise, why do you do this job?" Without thinking, I answered with a gesture indicating a toddler's size: "Because I have done it since I am that high." "Me too," she said, with the same simplicity with which she was tossing the lettuce in the salad bowl.

We were meeting for the first time and I knew only that she was a well-respected Argentine psychoanalyst who treated psychotic cases. It would be more than 10 years before I realized that my toddler years had been spent in the Alps among people in the Resistance during World War II, during "the Battle of the Alps," which raged in that region until May 1945. By then I had gotten interested in those chaotic times and spaces situated at the "borders," the times and spaces explored by my patients.

After dinner, Beatriz told us a story from her practice, a case that had puzzled her some months before. During a session, a patient suddenly pulled out of her bag a snake, which she threw on the floor between them. Head erect, hissing so to speak with its threatening tongue, the beast had moved all around, looking at Beatriz with an unempathic stare. The analyst had been terrorized and unable to articulate whatsoever. Untroubled, the patient took the animal in hand and put it back in her bag.

"It's amazing what people carry in their bag," Beatriz said tersely.

"And then?" we asked in chorus.

"Then I decided to get myself used to snakes. So I went to a farm in the Mojave Desert and learned how to handle them. I had somehow to humanize the monster that was inside this patient."

To be honest, I remember Beatriz's interpretation less well than the objection of one of us, who until then had stayed silent.

"I don't agree. The point is not to have the snake become human, but for us, humans, to become the snake."

Quite stunned, we turned toward Art Blue, an Athapascan Indian who had come from Canada. Their native tongue, he said, is akin to that of the Tarahumaras in Mexico, which were so dear to Antonin Artaud. Art Blue was at that time a psychotherapist in Manitoba, after having served as a fighter pilot in the Korean War.

Beatriz's doubtful face indicated her impression that she was in the presence of some new age Californian. But Art Blue continued unabated:

"I speak from my experience of the Hopi's snake dance. I had to dance with a live rattlesnake between my teeth."

"And then?"

"And then, either you become the snake or you die."

I will come back later to that second snake story, which had presented us with this riddle. For it asks a question: Was Beatriz's case not to be treated like a psychiatric case, and was it not related to humans either? What did that mean? Or rather, how were we to handle the nonhuman without reducing it to our stupid categories?

A possible answer, as I said previously, comes from two traumatized veterans, one a soldier, the other a civilian. The latter is Aby Warburg, who nearly died as a child, just after the war of 1870 between France and Prussia, and who became completely crazy during World War I through the revival of his earlier trauma. The former is his contemporary, Ludwig Wittgenstein, who investigated the enigma of the rupture of speech and concluded that "language is not a cage" after going through the hell of the war.

They both resumed their research, almost at the same period, in the 1920s, when monsters, broken faces, broken people, and ghosts were wandering all around during that uncanny "surrealistic" period. Both Aby and Ludwig shared a similar background. They both came from very wealthy families, in Germany and Austria. Both were outstanding researchers and became very disturbed after the war. Both were ambivalent about their Jewish origins.

ABY WARBURG'S MADNESS, 1914–1923

The scene takes place in the early 1920s in Kreuzlingen, on the banks of Lake Constance in Switzerland, at the Bellevue clinic, which belonged to the Binswanger family. The director, Ludwig Binswanger is a friend and disciple of Freud's. The place is cozy and welcoming. Many celebrities are treated there, for instance, the dancer Nijinski; the painter Ernst Ludwig Kirchner; and the feminist Bertha Pappenheim, alias Anna O., who had been one of the first patients of Freud and Breuer. Founder of the discipline of iconology, Aby Warburg was a famous art historian of the Renaissance period. His career was considered eccentric in a family of Hamburg bankers and started with a noteworthy event.

In 1879, at age 13, he relinquished his birthright as the head of the bank to his younger brothers, on the condition they give him as many books as he wished for the rest of his life. Warburg's brothers kept their word, even during his madness, so that his famous library contained 85,000 books at his death in 1929. At the dawn of the Third Reich in 1933, the books were smuggled to London Library so that they would not be burned at the hands of the Nazis. The collection has been established as a working library—the Warburg Institute, in London. Still organized according to its founder's original classification, the library attracts researchers whom Warburg used to call, at the end of the 1920s, his patients. At that time he had returned in Hamburg after a long hospitalization.

By the time of World War I, Warburg had become utterly mad and finally confined after the war in Kreuzlingen. One lovely afternoon, he exploded in a tantrum against his psychiatrist, Ludwig Binswanger. Having just caught sight of a recently published book about the artistic creation of mental patients on his desk, he shouted that he did not want to be treated like a case. He hollered that this book had been displayed there intentionally for him to make him believe that he was only an object for psychiatry: "Everybody conspired against him, and wanted to make a case out of his state of mind" (Warburg, 2003, p. 28).

Indeed, his intuition about Binswanger's assessment of him was accurate, for in 1921, Binswanger had conferred with Freud on the case and concluded that for a psychotic patient such as Warburg the prognosis was desperate and that psychoanalysis was not indicated in his case. They were doubtful that he would ever recover his intellectual potential. Ironically enough, Kraepelin, the famous psychiatrist, consulted by Warburg's brothers for help, offered a more hopeful assessment and a less dire prognosis. Perhaps, as was true of Bleuler during the same period, he was more open to engaging psychotic patients than psychoanalysis was at that time.

Now what was the content of Warburg's madness? During the Great War, his library was a battlefield, with Warburg the general-in-chief, receiving thousands of reports selected from the newspapers, which he ordered his wife and his children to bring him urgently. He came to believe that World War I had been lost because of him. Some of his crazy rantings would prove prophetic. In the early 1920s, he went around shouting that Jews soon would be deported, tortured, and exterminated under the orders of Binswanger the butcher. The "nice" Bellevue clinic was hell, where the inmates were given human flesh to eat. He saw himself as "a ghost coming back from the dead" while, at the same time, Hitler was busy writing *Mein Kampf* in prison.

At the same time Aby violently struggled against his madness. There is a parallel here with the future Nobel Prize winner John Nash, who while hospitalized at McLean Hospital, refused to take his prescribed medications, fearing they would impair the delicate "seismograph"—as says Aby Warburg—of his soul and intellectual faculties (Nasar, 1998). Like Nash, Aby took seriously the crazy ideas emanating from the same cells that had led to so many powerful discoveries.

To consider the truth in craziness: How did Warburg foresee the objectifying process that would lead inexorably to the "scientific" reification and actual eradication of the mad among millions of people, and presage the genocide of the Jews?

ABY WARBURG'S DELUSIONAL RESEARCH

Warburg's delusion may be considered a revival of the trauma he experienced in the Franco-Prussian War of 1870, during which he and his mother almost died of typhus. Moreover, he was 13 years old—the age of the pledge sealed with his brothers—when he witnessed, in the aftermath of the Franco-Prussian War, a new kind of mass anti-Semitism, especially when the Hamburg anarchist Wilhem Marr launched the first Anti-Semitic League in 1879 (Chernow, 1993). Was the symbolic alliance between children meant to resist an ominous political perversion prompted to surge in times of financial crisis?

At the end of that war, France had to pay compensations for its defeat and did so in a surprisingly short time, in 5 years with gold money. German banks were destabilized, the Warburg bank nearly bankrupt, and Jewish bankers held responsible for the financial crisis. At the same time, as Hannah Arendt (1948/1978) described, new rights for Jews fostered, paradoxically, a new kind of massive anti-Semitic hatred still unimaginable. With a delay of 25 years, Warburg's delusion showed the way to explore the trend leading to mass murders.

Trauma has a way of reappearing years after it has taken place. At the eve of World War I, Warburg launched his research, deemed psychotic by others but actually bearing on areas of death, taped during his youth on the delicate apparatus that he called "the seismograph of my soul." As is always the case, his delusion was not only a sophisticated tool of investigation, but also a theory for healing. Let us state some of his issues.

During the First World War, he was indeed showing, pointing out what could not be said, nor heard. This specific knowledge—denied by the classical language game with signifiers, allowing repression—was conveyed through sensorial images, which he called "surviving (*nachleben*) images." His research tracked them through different cultures and different ages of humanity. By using the impressions of his "seismograph," he was keenly registering the uncanny with the corner of his eye, at the margins of the pictures, where he described young maids who seemed to enter the frame like *nymphas*. He called *pathos formel* some shapes saturated with a huge energy, such as, for instance, the *snakes* of the statue of Laocoon. During his delusion, the fluttering moths or butterflies appeared to him as the souls of the dead.

Then comes the day when Warburg's family is tired of paying so much money to the chic Binswanger's institution with no improvement in Warburg's condition. Fed up himself, Aby shouts that he wants to get the hell out of here. Finally Ludwig Binswanger agrees to discharge him, on one condition: Warburg has to prove his sanity by giving a talk on his research to the staff and the patients. Does he believe his patient can do it? The challenge seems impossible. However, to everybody's surprise, Warburg succeeds in picking up the gauntlet and gives on the 13th of April 1923 a one-hour lecture on one of his favorite topics, titled "The Ritual of the Snake." True to his word, the psychiatrist discharged him one year later. Warburg then resumed his scientific research and his teaching in an academic position that his brothers had created for him in Hamburg. He relished the library they had built for his books, which he called "Mnemosyne." He died of a heart attack in Hamburg five years later.

Binswanger bragged of this success to advertise his family's clinic. But the truth lies elsewhere. The actual agent of Warburg's recovery was the faithfulness of his disciple, Fritz Saxl (Biswanger & Warburg, 2005), who later became the director of the Warburg Institute in London. Saxl never lost faith in Warburg's intellectual capacities and brought him books regularly during his hospitalization to keep him in touch with his research. He was Warburg's real therapist—etymologically *therapôn*, which means in Homeric Greek "a ritual double, a second in combat," as Patroclus was for Achilles (Nagy, 1979) and as Sancho Panza was for Don Quixote. Don Quixote and Warburg were both mad with books and for books, and not the easy kind. Their *therapontes*, seconds in combat, each showed them the same unconditional philia and the same faith in their endeavors, whatever the outcome.

APRIL 1923: THE LECTURE ON THE RITUAL OF THE SNAKE

As Aby Warburg feverishly prepared his lecture, he did not know if he would be able to make a successful presentation. The talk was half improvised. In his lecture he tried to find a transitional space between madness and sanity, war and peace, and between several cultures. His topic was the dance of the snake among the Hopis in New Mexico. He attempted to link snakes as they appear in Native American rituals to the snakes that appear in Renaissance art and, earlier, to snakes in pagan rituals during antiquity. He argued that their use in Indian rituals in Arizona or New Mexico should be considered a "living archeology."

This is where we return to the snake dance that Art Blue had spoken of at Beatriz Foster's place in Los Angeles.

Warburg was not interested in providing an "objective" description of this ritual but rather wanted to demonstrate the power it held for him, in the very process of analyzing it for the audience. "To speak about this dance was at the same time dancing with words and with surviving images, during that hour" (Warburg, 2003, p. 19). The confrontation with terror was at the center of Warburg's lecture. His subject was the surviving culture of Pueblo Indians, threatened by extinction, remarkably like his vision of the wiping out of the Jews.

In a lecture on the snake dance ritual of a different Indian tribe, the Hopis, he gave the following description: For half an hour, Hopi dancers hold rattlesnakes in their mouths and hands, representing a terror otherwise unspeakable. Afterward, they throw the snakes back into the wild space of the desert, where they are said to take on the shape of lightning and become messengers for rain and conduits for a power coming from the ancestors, who creep between our world and the world underneath, the world of the dead. The ritual enacts a shift between the threat of an imminent death—the venomous strike of the snake, as fast as lightning—and the renaissance of language from an unspeakable anguish.

Adrienne Harris (2007) calls such a passage "the bridge world," which connects in a new social link, "a particular province of unlinked speech" (p. 665) and the creation of an otherness. It brings back the subject from exile, as well on the analytic scene, as in ritual forms of theater promoted by Antonin Artaud (1936/1964) in order to heal plagued societies.

The dance performs the origin of speech as it emerges out of dread, at the intersection of the concrete and the abstract, at the birth of metaphor, since the deadly creature actually fills the mouth and the hands of the dancers. As Warburg (2003) puts it:

> The Indian opposes his will to understand incomprehensible phenomena, by transforming himself and becoming the cause ... of an unexplained sequence, which he produces as comprehensibly and visibly as possible. ... By embodying symbolism in the real, he takes the snake at full hands as the cause and the substitute of the lightning, he puts it in his mouth so that a real union takes place between the snake and the man. (p. 126)

So Art Blue was right. He became the snake in a specific field, where surviving images of dread, haunting the primary process (or bêta elements, as Bion might call them) have to be filtered through a language game, which Warburg calls "a danced causality." When time stops in catastrophic areas, the normal dimension of causality is defeated. On the border of what Bion calls the "nameless dread," and Lacan the Real, it is possible to enact a special kind of causality in the spatial dimension of a ritual space.

In that lecture, Warburg came very close to the ideas in Wittgenstein's almost contemporaneous *Remarks on Frazer's Golden Bough* (1930–1933/1993), where the philosopher demonstrated our kinship with "primitive" practices. His lecture perfectly illustrates Wittgenstein's theorem, which I quoted in my introduction: "One could almost say that man is a ceremonial animal."

We could also very well say that such rituals present "family resemblances" with the psychoanalytical discourse regarding psychosis and trauma, when we are dealing with a dread that cannot be named. Nobody can bear to listen to it. It, the Real, is not linked to any word or image, for words proceed from the Other, and no otherness is available for it. Warburg calls it the field of investigation of "a nameless science." Only his crazy investigation, supported by his disciple Saxl who never lost hope or trust in Warburg's research, allowed "his own renaissance."

With his performance, Warburg unwittingly followed Wittgenstein's (1930–1933/1993) advice to proceed through "an association of practices, similar to the association of thoughts, and related to it" (p. 143). By doing so, he won the war against reification, at a time when science was being pressed into service by totalitarian ideologies to justify themselves.

At this juncture I am reminded of a patient called Gilda, who was quoted in *History Beyond Trauma* (Davoine & Gaudillière, 2004). Like Warburg she "had explored all ages of mankind," imagining herself as the Great Goddess, of whatever culture, sometimes as the Cretan goddess holding snakes with both hands, like the chryselephantine statue from 1600 BC exhibited in the Museum of Fine Arts in Boston. She used to say: "We, the mad, wage a ferocious war against nullification."

According to Wittgenstein (1958), this war is, like his philosophy, "a battle against the bewitchment of our intelligence by means of language" (p. 109). It took a long time for him to acknowledge and analyze the process leading from objectification to the birth of a subject, in the shadow of totalitarian systems.

1918: LUDWIG WITTGENSTEIN, A VETERAN OF WORLD WAR I

The opening sentence of Wittgenstein's *Tractatus Philosophicus* (1918/1999) reads: "The world is all that is the case." It was written during World War I, in "the agony of hell," partially while he was under fire on the eastern front. The "case" indicates that *casus*, *der Fall* in German, comes from the Latin *cadere*, "to fall": The case is what has fallen.

But on whom? On what kind of other? After the war, Wittgenstein was looking, at times, like a psychiatric case, suicidal, violent, and at least eccentric. He had a hard time readjusting to his former life and to Viennese society. In those circumstances, how to consider the relationship between the silenced case and some kind of otherness? In 1918, Wittgenstein had no answer to that question. Especially when this other may be a ghost of the war, a thing or a beast, like Beatriz Foster's snake that falls in between her patient and herself. The case indeed falls upon us as analysts, when real catastrophes open the field for petrifaction and freezing silence.

Turning to Greek roots, the same "fall" is present in the verb *sumpiptein*, from which our word *symptom* is derived: from the preposition *sun*, "with," and *piptein*, "to fall." Could we say that traumatic or psychotic symptoms make the analyst *fall with* his patient? But where? Into the wild space beyond language, haunted by strange creatures. Such symptoms occur when all laws have collapsed, when the given word is betrayed. In the Middle Ages, the "mad man" was represented by "the wild man," who flees in "the space of the marvel": forests or mountains, where he meets with fairies, monsters, untamed beings, and learn from them. At the same time, his departure in the wild space is an act of freedom.

Cathy Caruth (1996) links the German world *unfall*, standing for a traumatic accident, "to an opacity which generates the surprising force of a knowledge...an unconsciousness of leaving that bears the impact of history" (pp. 22–23). Quoting Freud in *Moses and Monotheism*, written when he fled his home for England under the threat of the Nazis, she emphasizes "another aspect of the act of leaving, in what Freud calls 'freedom (p. 23).'"

For Wittgenstein, it took 10 years between the traumatic event and his departure to England to be able to elaborate a knowledge that constitutes what is commonly called his second philosophy.

His first philosophy stated that "Whereof one cannot speak, thereof one must stay silent." By this formula establishing the limits of speech, Wittgenstein closes the same *Tractatus*. He publishes it in 1918, after coming back from one year of captivity in Monte Cassino, Italy. His return to peacetime is unbearable. He continues to wear his old uniform, which he has worn out during five years. He decides to abandon philosophy and to become a schoolmaster in the Austrian mountains, where he brutally treats the children under his charge. In the meanwhile, he is very suicidal.

To his elder sister Hermine, who scolds him because he is wasting his genius so dear to his master in Cambridge, Bertrand Russell, he answers quoting Descartes: "You remind me of somebody who looks through a shut window and cannot explain the strange movement of a passerby. He can neither tell which kind of hurricane is raging outside, nor that this person has probably a hard time staying on his feet" (cited in Monk, 1991, p. 170). By the way, Descartes also wondered at the same age, in the same mood, about the same doubts, during and after fighting in the Thirty Years War (1618–1648). "The Discourse of the Method" is the outcome of the nightmares of a young soldier.

During the next 10 years, Wittgenstein entered what he calls the "bloody rough way," which led him to elaborate on the question of the limits of speech and reification. When he went back to Cambridge at the end of the 1920s, the final sentence of the *Tractatus* became, so to speak, "his second philosophy": Whereof one cannot speak, thereof one cannot stay silent, but shows what cannot be said. But again, to whom?

That is the point of transference that Lacan (1958/1977) could not address. Immobilized by his structural approach, he refused to go "beyond Freud" in the handling of transference for trauma and psychosis, whereas Freud in fact had pulled back from his research on trauma and therefore psychosis.

In a letter to Fliess—later removed from the official record of their correspondence—Freud admitted his father had sexually abused some of his (Freud's) siblings. We might speculate that Freud's turning away from some of his earlier findings on the impact of trauma was part of an effort to deny this terrible reality. "Unfortunately, my father was one of those perverse who is responsible for my brother's hysteria, and of some of my younger siblings. The frequency of this relationship leads me to think" (2006, p. 292).

Nevertheless, contemporaries of Lacan, such as Sullivan, Frieda Fromm Reichman, and Bion, had already gone beyond this limit to explore the transference with psychosis using the tools forged through their own experiences of trauma. To go beyond Freud and Lacan is to ask: To whom is shown what cannot be spoken of? Well, to some other, although not necessarily human. It may be, as the fabulist La Fontaine (1668/1694) wrote, "a god, a table or a basin," or even a rattlesnake. But how to handle the transference in that case?

FIRST, DO NOT HARM

How do we act, as analysts, when the absence is mere disappearance, as in the case of the *desaparecidos* in Latin America, or genocide like in Rwanda, or in ethnic cleansing as in Bosnia where it was led by a psychospecialist? When no other is available, except a ruthless agency crushing all forms of alterity, how not to harm in the very midst of cruelty, when compassion may turn into voyeurism and indignation into an empty speech meaning passive collaboration?

Wittgenstein (1934/1965) states that what cannot be said, cannot either escape being shown and pointed out. He thereby opens the dimension of a possible transference, when all otherness seems impossible, with trifling beings, such as "a piece of furniture or an empty spot" (p. 50). This sounds odd, but would be quite familiar to a bereaved person staring at the chair of her missing loved one.

By that time, in the 1920s, Wittgenstein had also lost three of his older brothers to suicide. His attunement to that petrifying experience was not merely theoretical. I choose here to read literally his statement: "there are several philosophical methods, like different therapies" (1958, p. 133), and consider his tools "like a therapy," for trauma and psychosis.

In *The Blue and Brown Book* (1934/1965) Wittgenstein condemned our addiction to generalities: "Instead of 'craving for generality'... I could have said 'the contemptuous attitude towards the particular case'" (p. 18). Examples are given from his own background, stressing a plural body with humans or things, as a relationship more reliable than an empty speech: "An innumerable variety of cases can be thought of in which we should say that someone has pains in another person's body, or say, in a place of furniture, or in any empty spot!" (p. 50).

But then, "how is speech possible when no other is present?" The answer is given by the process of stating an address by talking *to* and not *about*: "I don't really speak *about* what I see, but *to* it" (Wittgenstein, 1934/1965, p. 175). Therefore, the silence following the impossibility of speech may be very eloquent, as is the case in the following story where the philosopher analyses the resonances, on his side, of this strange silence.

At the beginning of his "Notes on the Private Experience and the Sense Data," Wittgenstein (1935–1936/1993) shares with us a clinical story of fright beyond words. The philosopher encounters somebody with "a far away look and a dreamy voice, which seem to be (his) only means for conveying (his) real inner feeling" (p. 202). This person cannot tell him what is the matter. The philosopher offers him the only way to get them both out of this impasse: transference. He tells his spaced visitor about his impression of their encounter: "The philosophical problem is: what is it that puzzles *me* in this matter?" (p. 203).

The philosopher is not a neutral witness to an enigmatic far away look and a dreamy voice, but he stands in for this addressee, the one to whom this look is directed, offering hospitality to this strange encounter. In such a case, speech can only proceed from the analyst's or the philosopher's impressions. The enunciation of his puzzlement on the part of the philosopher (or the analyst), constitutes a possible pole of otherness to the unspeakable "which plays the hell to us" (Wittgenstein, 1935–1936/1993, p. 204). On the next page, we find in a sequence of three words: "feeling, thought, transference," the Freudian word: *Übertragung*.

So now we are able to define this special kind of transference. Its unavoidable first step is to occupy the place of this impossible other to whom the despair of hell is shown. For a short period of time, the analyst cannot help becoming the monster, say the snake. The second step is to quickly get out of there with the patient by addressing this surviving image, kill it, or make it speak, so that it enters a game of language that can be shared from now on.

Those images and words, at the nexus of the abstract and the concrete, are called in Latin *fata*, meaning literally speeches proceeding from the sacred. *Fata* is etymologically related to the word for fairies, *fées* in French, who speak from the wild space at the limit of our world. Should we add that in Provence, the popular word *fada* comes from the same etymology and means "nuts"?

So, in this kind of transference a space and time is built up where "thoughts without a thinker," as Bion (1967, p. 165) says, can be expressed and therefore thought. "Taming wild thoughts" (again, Bion's expression, 1977) becomes possible. Wittgenstein (1930–1933/1993) insists we are "much more savage" (p. 131) than those we call primitive, for we are often unable to use "the mythology which is stored in our language" (p. 131). Not only words, but also sensorial images, which include the language of our impressions.

Therefore, "not to harm" is not to take refuge into generalizations and not to be blind to what the philosopher calls "changes of aspect." For instance "imponderable evidences … of the genuineness of an expression

... such as subtleties of glance, of gesture, of tone" (Wittgenstein, 1958, p. 195), to which children, even babies, are so sensitive.

Not to harm is to trust the most imponderable evidences, if we are to bear witness to silent or crazy languages. Those impressions are usually kept secret and hidden from colleagues. When speaking about patients, we are trained to make believe that we are neutral, that the transference is "waterproof" on the side of the analyst, that nothing of his own life story is ever evoked during the sessions.

This pseudoneutrality fuels war, for the hypervigilance of patients is unforgiving, as they have been trained from a very young age to stay on alert for the slightest change in the other person, having been lied to by the people they most should have been able to trust. To allow them to deny their perceptions of us is to engage in mystification, another betrayal by "fictitious" discourses, in the words of Hannah Arendt.

At such moments, when the task is to create the basis of a new loyalty when all laws have collapsed, the number one danger is then to stay silent, as if in front of a snake, fascinated, and mute, reproducing the dumbfounded silence of a passerby in this strange world on the verge of language, blind to the fact that we have become part of it.

To some extent, psychoanalysis can easily slide toward propaganda and torture, when it dismisses, as structurally psychotic, people who have been mute witnesses of major upheavals in the story of their country or their family. At those points, when time stops and may even flow backward, Erwin Schrödinger (1956/1990) advises us not only to escape the "principle of objectivation," but also to disobey the mechanism of causality.

Aby Warburg fought and survived his crisis "in spite of the good medical care" that a century before almost condemned Auguste Comte to a mental death had he not escaped the good intentions of the famous Esquirol Clinic in Paris. Instead of the mechanistic causality of medical discourse and even of psychoanalytical discourse that would have declared them hopeless cases, we have understood the principle "First, do no harm" as a singular experience of transference that fights objectivation at whatever scale, especially in our field, where the shame on veterans is conveyed throughout the reification of their descendants who teach us how not to harm the delicate apparatus of their seismograph.

REFERENCES

Arendt, H. (1948/1978). *The origins of totalitarianism.* New York: Harcourt, Brace and Co. (Original work published 1948).
Artaud, A. (1964). *Le théâtre et son double.* Paris: Gallimard. (Original work published 1936).
Barker, P. (1992). *Regeneration.* New York: Penguin.

Barker, P. (1994). *The eye in the door*. New York: Penguin.
Barker, P. (1995). *The ghost road*. New York: Penguin.
Benedetti, G. (1980). *Alienazione e Personazione nella Psicoterapia della Malattia Mentale*. Torino: Giulio Einaudi.
Besserman Vianna, H. (1997). *Politique de la Psychanalyse face à la Dictature et à la Torture*. Paris: L'Harmattan.
Binswanger, L., & Warburg, A. (2005). *La guérison infinie. Histoire clinique d'Aby Warburg* (M. Renouard & M. Rueff, Trans.). Paris: Payot & Rivages.
Bion, W. R. (1967). *Second thoughts*. London: Karnac Books.
Bion, W. R. (1977). *Taming wild thoughts*. London: Karnac Books.
Bion, W. R. (1982). *The long weekend*. London: Karnac Books.
Brecht, K., Friedich, V., Hermann, L., Kaminer, I., & Juelich, D. (Ed.). (1985). *Hier geht das Leben auf eine sehr merkwürdige Weise weiter Zur Geschishte der Psychoanalyse in Deutschland*. Hamburg: Kellner.
Caruth, C. (1996). *Unclaimed experience: Trauma, narrative, and history*. Baltimore: Johns Hopkins University Press.
Changeux, J. P. (1983). *L'homme neuronal*. Paris: Fayard.
Chernow, R. (1994). *The Warburgs*. New York: Vintage Book.
Damasio, A. (2003). *Looking for Spinoza*. New York: Harcourt.
Davoine, F., & Gaudillière, J.-M. (2004). *History beyond trauma*. New York: Other Press.
Epstein, S. (2008). *Un paradoxes français: Antiracistes dans la collaboration, anti-Semites dans la résistance*. Paris: Albin Michel.
Felman, S., & Laub, D. (1992). *Testimony*. New York: Routledge.
Freud, S. (2006). *Lettres à Wilhem Fliess, 1887–1904*. Paris: PUF.
Gray, J. (1998). *The warriors: Reflections on men in battle*. Lincoln, NE: University of Nebraska Press. (Original work published 1959).
Grim, R. (1998). Inadaptaton et Handicap de la Clinique à l'Histoire. *Le Carnet Psy, 39,* 19–20.
Harris, A. (2007). Analytic work in the bridge world: Commentary on paper by Françoise Davoine. *Psychoanalytic Dialogues, 17*(5), 659–669.
Hountondji, P. (2004). La Rationality une ou plurielle? *Diogenes, 51,* 3–4.
La Boétie, E. de (1942). *Anti-Dictator: The Discourse sur la Servitude Voluntaries of Etienne de la Boétie* (H. Kurtz, Trans.). New York: Columbia University Press. (Original work published 1548).
La Capra, D. (2001). *Writing history, writing trauma*. Baltimore: Johns Hopkins University Press.
La Fontaine, J. de. (1668/1694). *Fables. Le statuaire et la statue de Jupiter*. Paris: Hachette.
Lacan, J. (1977). *Ecrits*. New York: Norton. (Original work published 1958).
Lacan, J. (1991). *L'envers de la psychanalyse*. Paris: Seuil.
Monk, R. (1991). *The duty of genius*. Penguin Books.
Nagy, G. (1979). *The best of the Achaeans*. Baltimore: Johns Hopkins University Press.
Nasar, S. (1998). *A beautiful mind*. New York: Simon & Schuster.
Reis, B. (2007). Witness to history: Introduction to symposium on transhistorical catastrophe. *Psychoanalytic Dialogues, 17*(5), 621–626.
Remarque, E. M. (1995). *All quiet on the western front* (A. W. Wheen, Trans.). New York: Fawcett. (Original work published 1928).

Rossignol, C. (1998). Quelques éléments pour l'histoire du "Conseil Technique de l'Enfance Déficiente et en Danger Moral" 1943. Approche sociolinguistique et historique. *Le temps de l'historie, 1*.
Schrödinger, E. (1990). The principle of objectivation. In *Mind and matter* (pp. 126–137). Cambridge University Press. (Original work published 1956).
Searles, H. (1960). *The nonhuman environment.* Madison, NJ: International Universities Press.
Shay, J. (1994). *Achilles in Viet Nam.* New York: Touchstone.
Shepherd, B. (2000). *A war of nerves: Soldiers and psychiatrists in the 20th century.* Cambridge, MA: Harvard University Press.
Steinman, I. (2009). *Treating the untreatable.* London: Karnac Books.
Vexliard A. (1951). L'éducation morale dans la pédagogie de Makarenko. *Pédagogie, Neuropsychiatrie, Sociologie 3* (May–June), 251–268.
Warburg, A. (2003). *Le rituel du serpent.* Paris: Macula.
Wittgenstein, L. (1999). *Tractatus philosophicus.* London: Dover. (Original work published 1918).
Wittgenstein, L. (1958). *Philosophical investigations.* Oxford: Blackwell.
Wittgenstein, L. (1965). *The blue and brown book.* New York: Harper and Torch Books. (Original work published 1934).
Wittgenstein, L. (1993). *Philosophical occasions* (J. C. Klagge & A. Nordmann, Eds.). Indianapolis, IN: Hackett. (Original works: *Remarks on Frazer's "Golden Bough."* (pp. 115–155), 1931–1933. and *Notes for Lectures on "Private Experience" and "Sense Data"* (pp. 200–288), 1935–1936).

Chapter 11

Combat speaks
Grief and tragic memory*

Sue Grand

Peter has an unquestioned identity as a war hero. Now, in his 60s, he stands naked before his mirror. He gazes at his battle scars. Wrinkles, gray hair, withered muscles disappear. Each morning, in the mirror, a golden warrior awakens to look upon himself. And then, a good officer moves out among men. A year after arriving in Vietnam, he ordered the firebombing of a village. His orders set mothers and children and infants on fire. While he watched. While he, himself, shot the elderly and the unarmed. There is some sadness about the burning of children. But his soldiering is always told with a quiet, prideful gravitas to a civilian who knows nothing, who was protesting the war when he was fighting it. Conditions were chaos. He was fighting for American freedom. It was guerrilla war; the enemy was everywhere. There was nothing for it, but a scorched earth.

In the months after the firebombing, his troop blew up tunnels; they were hit by snipers, they were killed by land mines. He risked his life to save the lives of his men. Until he, himself, was blown up. He was hospitalized for months, and received both medals and scars. He lost his right foot and part of his right leg. He had a prosthesis, phantom pain, and a crutch. Athletic discipline restored his ability to walk. For Peter, movement is will. Will is his heroic, now as it was then. Since his return from Vietnam, he has been a guiltless man of modest desires. He pursued what he wanted, and what he wanted he got: business success, home, wife, children, grandchildren. Now, he wants a retirement full with love and health and ease and prosperity. "Golden" years that move easily backward and forward through time. But this wish is disrupted by the coldness of his wife. What he wants is to make her desire him again.

In Peter, there is no conflict over history, no registry of regret. And there is no real perception of self or of other. As Peter exalts the heroic myth of combat, as he erases the bestiality of that combat, he can only glimpse the I–Thou relation (Buber, 1923) in an increasing state of rupture. Each

---

* This chapter was previously published as "Maternal Surveillance: Disrupting the Rhetoric of War," in *Psychoanalysis, Culture and Society*, 12, 305–322. Reprinted with permission.

morning, the mirror resuscitates a youthful Adonis, an officer who did not die, who will never die, and the gap widens between himself and his real, aging wife. Peter's life story has no elasticity, no evolving new perspective; it admits no grief or moral or political complication. It does not respond to time or distance, to changing gender roles, or new historical/cultural epochs; nor to aging, or fatherhood, or to any other kind of intimate experience. His body, like his vision, is perfected. Soldiering is a hypermasculine phantasm linked to hyperbolic splits—to "gooks" and men, and the purifying necessity of violence. He has never looked at his own grandchildren and seen those other children. He has never seen the young boy within himself. There are no depths to his mirror. Looking is a condition of blindness.

REFRACTING CULTURAL BLINDNESS

Eventually, Peter and I would restore his tragic vision and share his tragic memory. But we, too, began in a condition of blindness. I was a pacifist-civilian, and he was a patriotic Vietnam veteran. I don't believe in his heroic phantasms, and he resented antiwar protestors. We were alienated from each other, but we were also secured by our mutual isolation. Neither of us had ever really known what had happened in that war. When I glance at Peter, I'm only seeing what I remember from the 1960s. We resisted the war and we didn't know the soldier who fought in that war. While he has been gazing at his heroic phantasm, I was gazing at my own hero: the draft dodger, who fled to Canada, or went to prison, or to the Peace Corps, or became a conscientious objector. Forty years later, Peter and I are meeting in my office. Civilian and soldier, pacifist and combatant: Our identities and our positions are fixed. Neither of us has ever been informed by the other.

When nations prosecute foreign wars, our war rhetoric always conjures this mutual silence and blindness. If civilians could somehow *see* into live combat, they might be more hesitant about war. To prosecute war, our governments must loosen our empathic and ethical constraints. They must allow us to objectify soldiers; they must control our vista of war. Toward that end, ideologies regulate what citizens can, and cannot, see. As Sontag (2003) notes, war photography has been used to ignite arousal and to suppress reflectivity. As a result of this manipulation, our gaze is directed away from the pain of others. But it is also directed away from the real pain of ourselves. Peter could not ask about the suffering of his victims. I could not ask about what drove a boy to kill innocent victims. For decades, neither of us had access to divergent, complicating images. We only saw what our own polemics had prescribed or imagined.

When citizens (soldiers *and* civilians) have access to divergent images of combat, humane responsiveness tends to appear. Soldiers and civilians start having complicated conversations. To exhort us toward war, this conversation must be foreclosed. Humane responsiveness interferes with mass violence. To diminish empathy and tragic guilt, our ideologies direct our gaze toward what *the enemy* is doing to *us*. We are frequently prevented from seeing what *we* are doing to our *own* children, and we cannot see what *our* soldiers are doing to *their* children. We all become complicit in what I have elsewhere described as "malignant dissociative contagion" (Grand, 2000).

Warfare has long known what psychoanalysts have theorized: that access to simultaneous and divergent images would enhance the capacity for linking (Bion, 1965), reflectivity and mentalization (Fonagy & Target, 1998; Winnicott, 1963/1965), and an awareness of our own destructive agency (Benjamin, 2006; Klein, 1948/1975). When multiple perspectives appear, malignant dissociative contagion (Grand, 2000) breaks up. We become able to relinquish our faulty heroic mythologies (see Grand, 2010), and we seek an I–Thou relationship. To sustain malignant dissociative contagion, our communal sensorium must become an insulated field of totalizing knowledge.* To achieve this, war rhetoric "transforms the whole society into a field of perception" (Foucault, 1979, p. 214). As Feldman suggests, this requires strict visual controls, which act as a closed "circuit of vision and violence [which] is itself circumscribed by zones of blindness and inattention" (2000, p. 49).

In the reportage of war, we can find this wholesale control of our vision and knowledge. After the American invasion of Iraq, the American media was forbidden to photograph the flag-draped coffins of returning soldiers.† Injured troops are similarly kept outside the camera's view, "'flights carrying the wounded arrive in the United States only at night' ... both Walter Reed and the National Naval Medical Center in Bethesda barred the press from 'seeing or photographing incoming patients'" (Mark Benjamin, as quoted in Rich, 2007, p. 14). In these selective and absent images, the war wound is replaced by imaginary whole bodies. At the home front, our eyes seem to inform us, but we are not informed.

When citizens are caught in the cultural predicament, we are unable to direct or redirect our vision. And we are unable to know we are not regulating our sight. We imagine that we see through our own eyes. But our critical capacity is bewildered.‡ As Feldman (2000) describes it, this is a condition in which we seem to know "who and what can be watched" (p. 52), but we cannot seem to know what sees. Our ideologies have produced a "blindness of the seeing eye" (Freud, 1893, p. 117).§ This holds the soldier

---

* This creates what Mahler (1968) described as an omnipotent autistic orbit.
† This policy has been changed by President Obama.
‡ According to Reis (2006), Klein views this as an unconscious process that involves, "a phantasized sequence of projection and introjection so complicated that its actual workings appear unfathomable" (p. 183).
§ This condition has resonance with the psychotic processes described by Williams (1998).

hostage. It dictates simplistic percepts of the soldier-as-war-hero or of the soldier-as-war-criminal. Throughout much of our society, civilians are excited by a ritualized portrait of combat. Civilian desire is aroused by heroic performance. Soldiers cannot subtract themselves from that desire. Instead, they perform for that desire. Civilians and soldiers have an alienated relationship, which is infused with distance, objectification, and idealization. Entire nations can be emptied of remorse and ambivalence, and filled with noble causes. Magic omnipotent denial (Klein, 1946/1952) sustains us in dangerous states of arousal. We act as a collectivity of unthinking bodies, who cannot think about other bodies, wounded. Or we repudiate our soldiers for their war crimes. Whether or not we believe in the "just cause," we tend to objectify our combatants and claim our own innocence. To break out of this system, civilians must engage in a radical action: We must construct an I–Thou relationship with our own soldiers. This was the challenge I encountered with Peter.

PETER: THE WAR ZONE BREAKS INTO PSYCHOANALYSIS

In response to my inquiries, Peter was compliant, disinterested, unruffled. His aging, the atrocity, his limp, his war: These were unrelated to his problem. He wanted to talk about his wife. She blames him for his sons' problems. She thinks he is too "hard"; he thinks she is too "soft." His two sons are in their 30s. They are divorced, with children. Unsuccessful in marriage, adrift in their careers, financially unstable, they keep returning home. She gives them money, takes them in. To Peter, she is too indulgent; she makes them passive; they've got to "stand on their own two feet." I thought that this metaphor spoke volumes. He wasn't interested in it. What he knew was that his wife wouldn't have sex with him. He didn't know why. Or what he should do. He attributes her erratic behavior to menopause. About his family, he seemed innocent, but cooperative and earnest; concerned about his wife and sons, but blank about the conditions for their mistrust. For Peter, there was no reflectivity or interiority (see Slochower, 2006), or interpersonal comprehension. Ogden's (1986) "autobiographical subjectivity" did not exist. Time and memory skipped atrocity; they skipped decades; they rearranged themselves in the mirror. He lived on a dual track in a continuous present: He was 20 years old, an able-bodied good officer; he was a mature solid citizen. The same man, in two different eras: muscular, self-disciplined, every limb still intact.

He had no thoughts about his childhood and could tell me little about it. He had a good life, a good marriage, a good family, until now. His wife said that he should go to therapy. He wasn't sure what she meant, but he would try if it would fix his marriage. He would come in to his sessions, sit forward, and say, "So, OK, let's get going." I was quiet, and then he would say it was my show; he didn't know how this worked, but he wanted to "take care of this business." "Go ahead" he would say, "shoot." Shoot? An unnerving invitation from this veteran. I would hesitate, he would get restless: "I hear you really know your business. I'm paying a lot for this. So go ahead. Shoot." Not rude, really, but commanding, ready. This was not what I anticipate in an analytic conversation. Do your job. Fire your weapon. Earn your money, Doctor. As an opening gambit it said everything, and yet he would have claimed it meant nothing. If I waited, nothing more was forthcoming. And so, I would wonder aloud if he was lonely now that he was estranged from his wife.

Lonely? He would try to answer me, but he never knew what I was asking. I tried to listen, but I didn't know what I was listening for. Often, there seemed to be nothing to hear. I was silent about the firebombing, although it haunted my imagination. Every time he said "shoot" I saw the elderly and the unarmed. The more I saw them, the more they fell outside my excavations. I am a pacifist. But I didn't think I should impose my own political/moral values on his marital agenda. I didn't know how to sit with what he had done. I didn't know how to help him or even what "help" might mean. Or what ethical compromise it would take for me to assist him.

From the first, I held myself apart from and above him. I knew nothing about his war. I didn't want to know anything about his war. But I was certain that I could never do what he had done. There was no shared humanity between us. There was not even a common body. Just as Peter needed an exit from maternal accusation, I needed an exit from any shared destructive capacity. For both of us, that escape had to be embodied. Hormonal. To Peter, his wife was just being menopausal. I saw the firebombing as testosterone gone wild. Estrogen was the bedrock of my nonviolence. I have never believed that biology is destiny or that sex equals gender. But our estrangement felt cellular, structural, sexed, biochemical. Sometimes, I felt that we were not even of the same species. He sat quietly in his chair. But when I looked at him, I saw an animal poised for predation. His body would mutate; he was not even a "him" but a beast "red in tooth and claw."

I didn't think I could tolerate seeing him even once a week. I could not identify Peter as a human subject. Or want him as a human object. Oddly enough, he wanted to come twice. He didn't know about this psychology business, but he wanted to solve his problem and quick. I knew I could not proceed with him in such a state of repugnance. Somehow, I had to defer the atrocity and contain the atrocity, while I found some empathic pathway to his interiority. I pursued the only intervention that seemed possible: a detailed inquiry into his relations with his wife and children. It was a dull but reasonable conversation. Peter is bewildered: He was a good father and husband, and suddenly he isn't. His wife and sons are now united in a cold front against him. When he comes into the room, they seem angry and fall silent. He says, "It's like I was some kind of beast."

"As if I was some kind of beast." Ordinarily, of course, I would ask him for his associations. But this is not an ordinary conversation. I don't pick up the referent to the beast. Instead, I engage in psychoeducation about familial communication, roles, and interactions. He is a serious, if concrete, student. He is not interested in his own psychodynamics or in the patterns learned in his own childhood. He has a very narrow lexicon for his own affects; he has no dreams, no explicit transference material for me to attend to. But he practices new interpersonal strategies. I am resigned to a superficial psychotherapy. His family life will get patched up, and then, he'll move on.

Then suddenly the frame goes into breakdown. I see Peter at my home office in New Jersey. While things were plodding and dull with him, another case had been heating up. A woman patient is about to separate from her husband. Paranoid, intermittently psychotic, addictive, raging, and dependant, the husband has been in and out of treatment for many years. They have three children. Recently he tried to get a gun permit under a false identity. She has decided to get out. She fears his incipient violence. The police have recommended that she get a restraining order. This man knows where I live; he has left me threatening messages. My local police recommend a security system; they advise me to carry an emergency alert in case I am accosted on the driveway. I lock the door to the waiting room, and a new security camera is installed.

A few days after this installation, I am waiting to see Peter. My last patient has left, and I have about 30 minutes before Peter is due to arrive. He is rarely early. I go out to get the newspaper. And there he is, my patient's husband, pacing on my driveway. He rushes over, loud, cursing, begging. I am trying to be firm, to get him to leave. Peter picks today to come early. He walks right into this altercation before I even realize he is there. He confronts the other man, and barks at him, once, to get out. My patient's husband hesitates, blusters, while Peter seems bigger, straighter, broader, more fluid, and more powerful than he seems in my office. There again is the stern voice telling him to get out and not to come back. The husband shuffles away, muttering. Peter asks if I am okay. I ask him to meet me at the backdoor to the office. I go in the front door and buzz him in through the new security system. I have a few minutes to collect myself. When Peter comes in, it's clear he's got the story. The altercation, the new security system: It's a crazy patient or a crazy relative of a crazy patient. "You look pale," he says, "do you want time to yourself? Do you want to call anybody? Can I do anything?" He wants to know what I'm doing to secure my own safety. Have I spoken to the police? When is my husband coming home? He's protective and pissed as all hell at the jerk on the driveway. I am touched by Peter for the first time. But I'm appalled at the disintegration of the frame. In 5 minutes on my driveway, his protected, private space has been eliminated. I feel unprofessional, out of control. How can I go back to my therapeutic position? Who is the patient here, in this moment? I was his doctor, now he's my bodyguard? I feel like Dr. Melfi who fantasized about Tony Soprano's protection.

I try to shift the conversation away from me, and ask him what he is feeling. As usual, he can't recognize his affects, especially his own vulnerability, but simply refers to danger and action: "The guy's a nut. He could've taken a swing at you." A pause. Then: "You might have been all right but it's a good thing I was here." There is no language for his subjectivity. But as I am trying to recoup my therapeutic position, I am aware of a new sensation. I have enjoyed his soldierly protection. I have seen him as he sees himself in his mirror. I am riveted by his assurance in danger. I want to imagine my heroic future as he remembers his heroic past: muscular, fearless, incontestable in danger. Once, he had evoked my revulsion. I had been unable to imagine him as an erotic object. Now, I want to be him, and to have him.

As a heterosexual woman of a certain age, I am inhibited in my aggression. After the altercation on the driveway, I despise that inhibition. Suddenly, that hate is mutating my desires. For 35 years, Peter has recycled himself as a war hero. He has an opportunity for this performance when he sees me accosted. Now the story of killing "gook" children can become the rescue of "our" women and children. So that his body can change in its erotic significance. I can want to *be* him, and to *have* him, without knowing that the *him* I want to *be* and *have* is the same *him* who commissioned an atrocity. For the first time in my memory, I can admire a military uniform. I see a "good soldier" and imagine that soldier unsoiled by war.

In the moment when I join Peter in his mirror, he sets me loose from my feminine constraints. Over the next few sessions, he refers back to the incident on the driveway. I know he's worried about my soft female body. He figures, correctly, that my husband was off at work when the "jerk" accosted me. I am moved by Peter's concern and his kindness. But I feel imprisoned by this feminine impotence and dependence. Peter, my husband, the police: Why do I need their protection? Why aren't my muscles good enough? Just then, Peter says, "You know, you were really pretty tough out there. I was watching from the street for a minute. You were doing okay." Really? He could not have given me a better gift. I do not subject it to analysis, because it might turn out to be a wish or a fantasy or a projection, an appeal to my narcissism, or a strategy of false reassurance. For himself and for me. Maybe this is what soldiers tell each other after they have just shit themselves in combat. I don't know, and I don't want to know. Right now, I need my toughness to be real.

Together, we make that toughness real. I am becoming hard-bodied, because a hard-bodied old soldier sees and admires my prowess. In the same session, he praised my incisive "muscle" about familial communication. Increasingly, he uses military lingo: After our meetings he says he implements his "marching orders." I stopped being an armchair, timid, bookish know-nothing; I stopped being a vulnerable girl who needed a man to protect her. I was someone who could "cut through" interpersonal disorder. I got the job done; with my own precision weaponry and a minimum of "collateral damage"; without arousing any latent shell shock in this decent old veteran. He made me feel like a good officer. And he got to be a grunt in my army. He told me war stories, and I, the pacifist, found myself laughing. My patient's husband hadn't been back. But he was still calling and screaming. It was a difficult few weeks of management. Peter was there with me in the "war zone." He mixed black humor and concern with phallic solidarity and recognition. I knew he was still taking care of me by getting me laughing; I tried to ask him about it, but he just cracked another joke.

In the rest of life, I am antiwar. But in the moment on the driveway, Peter saw into me. I became Peter's because he knew, and confirmed, what I desired. I yielded the division between his body and mine. I was enraptured, and I was gone. Gone "for a soldier." Because, in his eyes, I found my own force. I found my own aggression, and I felt victorious. And in my gaze, he, too, would become enraptured. Gone "for a soldier." Because treatment would make him feel that he could be loved by a man, and love a man, without ever having his sexual identity unravel.

All of this was potentiated by the electronic eye at my door. Once, the waiting room had been open to all. Now, each time Peter arrived, an ocular apparatus functioned as a persecutory system: it discriminated between "bad men" and "good men," and then it let the "good man" come in. Peter's destructiveness was extruded onto a real, alternate personification: my other patient's husband. Badness was outside, goodness was inside, and our aggressive bonding was legitimized. Sometimes, the absurdity of this security arrangement would strike me: The man I was admitting to my waiting room had killed countless victims. The man I was keeping out was nasty and crazy, but he hadn't killed anyone. My patient's husband was a real danger. But he also seemed like the personification of a traumatized vet: paranoid, addictive, hypervigilant, armed, ill-adjusted to peace. My patient's husband had never been in combat. But he signified that other Peter who never came home from his war. Each session, the security system kept that Peter out. So that the one who came in was neither damaged nor damaging: He was sane and potent and alive and intact.

Like me, this Peter was a heterosexual "of a certain age." In going to war, he had a fixed vision of a man. If I was inhibited in aggression, he was inhibited in love: He had to be the man whom he could not want to have. When male armies go into combat, love between men is deep and physical and tender. They can touch because they are killing; they can touch because they are dying. But that touch must be asexual, narrowly constituted by violence. About homosexuality, the U.S. military is still stuck with "don't ask, don't tell." But now, in the analytic dyad, Peter anoints me as a male officer, who is also a heterosexual woman desirous of a man. He has me looped in a loop of erotic complication, so that he can desire a man, and be desired by a man, and imagine that the man is female. All of this emerged in unconscious derivatives. When he talks about wanting sex with his wife, I inquire exactly how he wants to have sex. When I asked this earlier in his treatment, he would say, "What do you mean? Like before." Before meant him arousing her, then the missionary position, him on top, her below. Now, when I ask this question again, he is embarrassed. He's been having dreams of someone sucking his penis. He's on his back. Silence. Then, "I'm not gay you know. Jesus. This therapy shit." I ask him what he is dreaming, and he tells me that sometimes it's a man, sometimes it's a woman sucking his penis. "Jesus. I'm not gay am I? How could I be gay?"

He is cringing with shame. So I say that I don't think this is about being straight or gay. I say, "Maybe passivity has something exciting to it." He's straining to get this. "Oh you mean I'm not gay!" I don't say anything. Then he says, "Oh, I think I get it. If someone else is on top, it's gotta be a man?" Yes, I say, you want to lay back and it frightens you at the same time. It's lonely and tiring the way you are living. Wouldn't it be nice if someone else got to know you, took care of *you*? But in your mind, they would have to be sort of a man and sort of a woman. Tough enough so you could lay back. "Oh, yeah" he says, "I do sort of lay back sometimes in here. I get to talk and you listen." There is a silence, and he gets worried: Do I think he wants me to suck his dick? "Jesus, you Freudians, everything is about sex." More cringing and then a lightbulb goes off. For the first time, he grasps at a metaphor as if it is a life raft. Not a moment too soon: What a relief! "Oh, you mean maybe I want you to take care of me, sort of." His emergence from concrete thinking saves him from gender disorganization and sexual humiliation. He is still heterosexual. And he's not soliciting fellatio from his therapist. He just wants some rest. To stop giving the orders, trying to save his unit. But he's afraid of lying down—some jerk might fuck with him or fuck him up. I say that the desire that would fulfill him is associated with violence.

After this, his associations become more fluid. I suggest that his erotic dreams refer, in part, to lying on my couch. After some anxious joking, he decides to lie down. Grief emerges about all of his lost men: his war buddies, his sons, his father. He is less shy about wanting male love. At home, he behaves better with his wife and sons. Like his leg, the analysis seems intact, even though the frame was blown up. Indeed I feel somewhat magical: I am a warrior and a pacifist; racing ahead of contradiction, so that contradictions are seamless. I have tutored him in vulnerability and compassion, while our unconscious field continued to be captivated by war. At that time in the analysis, my dreams were idealized transcripts of firepower, in which the firebombing did not exist. By day, I was protesting the war in Iraq. But at night, combat became exaltation without horror. I was a ready conscript to the heroic imaginary. I was a teenage boy signing up to "be a man"; I was a woman who always wanted to be a man, and I was a romantic girl captivated by uniforms. In the area of our shared seduction, Peter and I only saw our perfected bodies. Regardless of whether I was being him or having him, regardless of whether he was being me or having me, we were "systematically sanitized ... everyone has his limbs, his hands and feet and digits, not to mention expressions of courage and cheer" (Fussell, 1989, p. 268). For each of us desire had become more elastic; but it had an affinity with killing, with that "awe, fascination with power, and feelings of violence and boundlessness that transgression arouses" (Stein, 1998b, p. 256).

PETER: MATERNAL SURVEILLANCE AND THE RETURN OF THE REPRESSED

Two months passed since the altercation on the driveway. My patient's husband was no longer menacing. In his treatment, Peter moved between martial excitement, an unconscious wish for dependence, and a narrow band of grief that recognizes "us" but not "them." For both of us, real external threat was fading. The electronic eye began to shift its meaning. Previously, the eye had been confirmatory of our mutual phallic goodness, legitimizing our solidarity in a "just" war. But now, the eye seemed to grow breasts. And without a real external threat to absorb badness, badness started threatening to "come in."

Peter was preoccupied with the security system, with its one-way field of vision. On the way I could look into, and at him, without him being able to see me seeing him. It was winter by now, and the days were getting shorter: what if dusk obscured his image in the eye and made me mistake Peter for *him*? What if I thought he was the "nut" and I wouldn't let him in? What if I called the police? I said, "On the couch, and in the eye of the camera, I can see you without you seeing me see you. You can't be sure what I am seeing, and I might somehow see *you* as the 'beast.'" For the first time, I have uttered that word that links us to the atrocity. He cannot formulate any kind of response. And then, he reports a series of dreams.

He is at home, in his bedroom. The room with the mirror. A woman is just there: gloved, demure, coifed. Eyes obscured by the black mesh veil of a pillbox hat. A 1960s, white, ladylike presence, reminiscent of Jackie Kennedy's elegance. At first he barely notices that she is in the room. But her presence is both fixating and obscure. She sits, and she is silent. He doesn't know her. He cannot see her eyes behind the veil. The dream recurs, and his dread seems to increase with every repetition. I ask him for his associations.

He can only say that she reminds him of Jackie Kennedy at JFK's funeral, although he doesn't know what this means. The pacifist within me has been dormant, but now it is reawakened by her entrance—Jackie, the aristocrat of motherhood whose own children were half-orphaned by violence. Whose husband was sanctified by his assassination. Inserted into history as another "profile in courage." In a condition of eternal youth. Like Peter, in his mirror. In 1960s America, Jackie's widowhood was grace; her dignity and fortitude was used to confirm the man's heroic honor. But now, in Peter's unconscious, that widowhood is becoming an admonishment. Grieving, she watches Peter. She is saying nothing, doing nothing, but she casts the "evil eye" on his badness. Peter's dread is shapeless and wordless and insistent. The perpetrator in him seems to become a child, a toddler, an infant. I asked him what Jackie sees. His body became fetal and turned toward the wall. Then, he said, she knows what we did. Now, he begins to tell me about his war. In this telling I am implicated: she knows what *we* did. We began to communalize combat (Shay, 1997): *his* atrocity was becoming *our* atrocity, and *his* war was becoming *my* war.

MATERNAL SURVEILLANCE: WHEN GRIEF RESTORES ABSENT IMAGES

When the war zone has been obscured by our warrior mythology, soldiers often find that they cannot tell the truth to civilians. Civilians cannot ask for the truth of combat. When we are both sealed into this arrangement, our "mirror" can only be broken by a third perspective. In Peter's analysis, this third perspective arrived when we had reached another moment of stasis. It appeared as another *set of eyes*: the eye of an all-seeing mother. This dream figure was not unique to Peter's *personal* unconscious. It can be traced to the structure of war rhetoric, as it resides in our "social unconscious" (Layton, 2006).

In the rhetoric of war, we always find an idealized, and sacrificial mother. Her loins have yielded the hero; she has relinquished him to destiny; she waits; she mourns his remains. Her eyes are always cast upon the war zone. In our warrior mythology, this maternal figure functions as a perfected projective object. She has the all-seeing eye of loss and grief. This figure refers to our fantasies of our primordial, infant mother. Once, her infant's body was entirely known to her, when that infant was still in a state of unknowing. Her eyes have always registered that which we could not see. What she sees cannot be disputed. During war, the warrior's mother seems to maintain this sacred claim to truth. She has sent her child off to war; she watches; and she *sees* the death of her hero son. In this construction, the hero's mother is nominated as a maternal hero. As Harris (1983) describes this figure:

> She waits always in the position of *mother*, to whom return is made, as the mirror reflects the bravery and valor of military men … (she is) the "angel of consolation" … in whose name war is made and danger is suffered, the one who holds the place for virtue and love, while men do the dirty work of violence and death. (p. 94)

In the polemics of foreign wars, her mourning is meant to witness *our* soldier's wound, and legitimize *our* aggression. Her grief and authority confirms the "just war." But she can only confirm the just war if her grief has superhuman powers. Those superhuman powers reside in her *maternal surveillance:* in her *omniscient look at the war zone.* In our social unconscious, this maternal surveillance has exceptional force. In war rhetoric, this force confirms the division between us and them. But maternal surveillance can slip outside the constraints of our ideologies. Real mothers are actually weeping in the war zone. Their pain can never coincide with our ideologies "in a moment of perfect, remainder-less, grasp" (Norris, 1987). One mother's grief eventually refers to other mothers. All of whom have lost their own children. Together, they seem to grieve for all children. Civilians and soldiers. Enemies and allies. The line blurs between *us* and *them.* Mourning becomes an exponential expanse, a universal restoration of war's absent images. If war rhetoric has obscured the war zone and silenced our soldiers, maternal grief seems to restore our tragic vision.

And so her gaze found us, in the United States on February 2007. The war in Iraq was entering its fourth year. We had never seen returning coffins or the bodies of the wounded. Political debate raged: Do we send more troops or pull out? A *New York Times* headline read: "Analysis of Iraq's Future is Bleak, but Both Sides in War Debate Find Support." The article is replete with data, political analysis, strategies, noble causes. But on the page, the text is visually displaced. It is a border that works itself around the edges of a large, black-and-white photo. The photo displays a street in Baghdad. There is a car in the background, shattered from a bombing. Debris litters the street. Behind that car, another car, mangled. In the recesses of the photo, men huddle, their backs turned to us. In a posture that is suggestive of despair. In the foreground, a woman faces us. Draped in black, head covered by a hijab. Her face is exposed. Her hand is covering her mouth. She seems to have witnessed the unspeakable. On her hand there is a simple gold band. Her face is round, she is not young; she gazes out at us in sadness. Next to her is a small child, perhaps 5 years old. White sweater, thick black hair, eyes round with fear and supplication. The woman seems to hold the child's hand. This child seems to refer to other children, no longer living. To orphans, whose mothers are not living. Woman and child are unidentified. To someone in Baghdad, they are, of course, particular.

To us, in the United States, this mother suggests a nameless series of mothers. Whose eyes have seen all, and know all, and address us with their grief. Below the photo, a caption reads, "A common scene in Baghdad: a car bomb in a Shiite commercial district. Violence against Shiites and among them is a theme in a new prognosis" (Mazetti, 2007). The text traces reasoned political discourse. It traces violence between Sunnis and Shiites. No photos attest to the detritus of American bombings. Our government only permits us to see the effect of insurgent bombings. But the maternal gaze points us toward every war wound. Looking out at us from the combat zone, she reinstates those absent images that our government is withholding from our eyes. Her grief reminds us that, "every document of civilization is also a document of barbarism" (Benjamin, as cited in Said, 2004, p. 23). So that we must see into our own destructiveness, and relinquish our pretense to the "chivalrous passage of arms" (Freud, 1915, p. 277).

In war, the omniscient maternal eye becomes a potent antidote to violence. This gaze is conjured in art forms, in media; it is activated in women's protest movements. It is conjured, as well, in the psychoanalysis of a war hero like Peter. Who never converted the war wound into narrative. When a soldier commissions atrocity without any guilt or remorse, when heroic mythology never converses with loss and grief: This is when the maternal phantasm arrives. This personification seems to know that, as Sartre (1964) put it, "through shame we confer on the Other an indubitable presence" (p. 251).

CONJURING MATERNAL ADMONISHMENT

When Peter started dreaming about Jackie, our phallic idealization fell apart. In Peter's material, I started to move between being the omniscient mother and being exposed to the omniscient mother. Sometimes, he imagined that I have her x-ray vision. He made more confessions. Sometimes Jackie seemed like a transferential referent to my feminine nonviolence. I certainly wanted to claim her as myself. But it became clear that her moral authority was not my own. The woman in the veil appears and reappears in a multiplicity of forms, ceasing to be Jackie, proliferating, morphing into almost any anonymous woman in a veil. Muslim. Catholic. Jewish. Pious, demure, sexless. Draped, anonymous, nameless, concealed. What she derives from me is my body configuration, my position of seeing without being seen. But as she moves further from Jackie, from my own sociocultural context, it is clear that this phantasm borrows from me, but that she is also not me. She is something outside of, and beyond, this treatment. She is the return of the repressed, a signifier for an unknown series of mothers and children; a personification of persecutory guilt, a precursor of Benjamin's (2006) "moral third."

This figure borrows my eyes. But she also sees in me what she sees in Peter: righteousness inflated to evacuate terror. Intoxicated aggression. Us versus them. When Peter is not imagining that I have the omniscient eye, he is dreaming of me as another soldier. The phrase repeats, "she knows what we did." Our crimes seem broadcast from one veiled woman to another. This is a transference representation that reflects perception, not projection. All through this treatment, I have been sacrificing her children. Insofar as I remembered the atrocity, I excluded Peter from the realm of the human. I never wanted to know the subjective reality of his war. Insofar as I was enticed by Peter's heroics, I forgot that innocents were burning. I forgot about traumatized boy soldiers. I was excited by combat. Once, I thought Peter was narcissistic, or schizoid, or suffering from a well-concealed version of PTSD. Now I knew him as a warrior-sacrifice, offered up at the altar of our civilian desires. If I had been in his war, perhaps I would have shot the elderly and the unarmed.

Peter and I were human, we were whole, and then we had become someone's enemy. Demonized. Threatened. Me, on my driveway. Peter, in boot camp and then in Vietnam. There, we lost the moral compass of depressive intersubjectivity. Like Joseph Conrad's Marlow, we found ourselves traveling "in the night of the first ages, without a sign and no memories ... in need of some deliberate belief." (Bhabha, 1985, p. 146). A belief that acquired the mythic properties of the heroic. Human frailty was replaced by violence. There was no war wound. There was no resonance with vulnerability, in the self, in the other. Our manic defenses (see Altman, 2005; Peltz, 2005) locked us into a persecutory system. Together, we were blind to the ordinary body, until a veiled woman returned that body to our vision.

Particular and nameless, she was both real and a phantasm. She invoked both the intersubjective, and the fantastic registers (see Benjamin, 1995). She was the trace of shared cultural and historical referents. And she was the trace of primordial body memory. She was a mysterious subject, who acted upon as like the return of the repressed. Her veil denoted a gaze which had seen all, but which could not be seen. As no one, she was a universal signifier for the abject. But she was also the "mythical" or "transformational" object (Bollas, 1986; see also Mitrani, 1994) whom we nominated as sacred. A beacon of light, a figure of dark enchantment, an "Angel" and an "omniscient Mother" (Benjamin, 1995): We imbued her with "symbiotic or telepathic knowing" (Bollas, 1986, p. 95). Her gaze was the font of life, from which we could be cast out.

In this figuration, an absent I enhanced the power of the omniscient maternal eye. Her veil seemed to lift the veil from our eyes. Our shared response to this trope was an "adoration [which] covers dread with awe and mystery" (Horney, as quoted in Benjamin, 1995, p. 81). To be recognized, to come into view, our violent "not me" selves (Sullivan, 1953) required this dark version of the transformational object. She was an object whose vision would not be deferred to the rhetoric of war.

This omniscient mother seemed immune to those "language rules and denials" (Arendt, 1955), which prevail in "atrocity-producing" (Lifton, 2005) situations. Placing "the presentation of fragile human life above the instrumentalities of technocratic power," her vista of war was "a rejection of amoral statecraft and an affirmation of the dignity of the human person" (Ruddick, 1989, p. 81). She restored wounded children to our vision.

PROTEST THEATRE: WHEN WAR MEETS MATERNAL SURVEILLANCE

For Peter, this was an intimate, relational transformation. But this transformation also occurs on the political stage when women protest. War and state terror are wanton in their erasure of mothers and children. But resistance movements can turn grief into maternal omniscience. In protest theatre, woman bodies become both reality and phantasm; they link the intersubjective, and the fantastic registers. Inserted into our vision, in the public domain, they disrupt the vista of war which is promulgated by polemics. During these acts of resistance, the visual field is appropriated by a transformational mother who "propels a regenerative cycle" through "mourning play" (Bassin, 1994). Women achieve this by "de-privatizing their mourning" (see Elshtain, 1994, p. 81) in public spectacles that conjure the woman in the veil. To transgress before the mother's immaculate authority; to have this moment engraved on the public transcript (Scott, 1992); to have that transcript dispute our highest authorities: here, destruction can meet the threshold of its own impossibility.

It was in this way that Cindy Sheehan kept vigil at Bush's Texas ranch. For weeks, we watched her watching him. In a nonviolent, public spectacle. And that Bush could not speak or move against her. Or even have her removed. Because her son, Casey, was killed in the war in Iraq. Cindy Sheehan was no one: She was a nameless mother, and a nameless soldier's mother* and, then, in her vigil, she became every soldier's mother. As Ruggiero (2006) puts it,

> Peace advocate, movement leader, passionate—Sheehan is all of these things. But after spending last Monday with her, I realize that above and beyond all else, Cindy is a Mother. Not just a Mother, but Mom Laureate, Subcomandante Momus, Nobel Peace Mom, Dr. Mom, Jr. Mahatma Momdi, National Mom, World Mom, Milky Way Mom. Which is to say, Casey's Mom. (p.)

---

* Sheehan states that her war resistance was mobilized, in part, by Bush's ignorance of her son's name when he proffered his condolence for her loss; see Ensler (2006).

In Ruggiero's description, Sheehan's protest becomes both real and a phantasm. It linked the intersubjective and the fantastic registers (Benjamin, 1995). She was an ordinary, particular, grieving mother, speaking on behalf of other ordinary mothers. She named her son, and she named herself. But she also draped herself in the veil of maternal deindividuation. As a particular mother, and as every mother, she gained mythic force. She had birthed the hero; she held sacred status in our warrior mythology (see Grand, 2010). And she refused to confirm the "just war."

As a particular and private mother, her loss had previously been enlisted by the president, in a ritual intended to confirm Bush's compassion (Sheehan, 2006). But in her vigil as mother, her authority exposed his lack of all compassion. Watching Bush, Sheehan restored our absent images of the war, while our president's agency was confounded. Belligerent, wanton, freely trafficking in violence, he found himself paralyzed. His advisors couldn't find anything to do to her or with her. Because the "omniscient maternal eye" had found them. If they made a public gesture against her, they risked that "dread that through her he might die or be undone" (Horney, 1932, p. 81). Cindy Sheehan was untouchable. She dignified doubt. She invited the testimony of soldiers,* and war resistance took off.

Sheehan's action drew on what Steiner (1984) refers to as our "antique imaginings" of a primordial mother. Who knew her infant's body, when her infant was in a state of unknowing. Who once contained the fetus within her own uterus. Whose umbilicus refers to our helpless dependence. Her gaze reduces us to children. It scolds us; it evokes a dread of retribution. To mobilize these "antique imaginings," women must conjure themselves as the umbilical body. They achieve this effect by donning markers of deindividuation. By mobilizing their lack of subjectivity, and then inscribing this lack on named particularity.

This performance is a frequent trope in war resistance, and in resistance to state terror (see Elshtain, 1994; Graziano, 1992; Hollander, 1997; Maathai, 2006). Its potency is derived from the very structure of violence, with its gendered "social unconscious" (see Layton, 2006). Because war (and state terror) are characterized by splitting, they tend to masculinize the aggressive subject, while diminishing a weak, feminine object. But they also anoint a mysterious, unknown, and beatific, primordial mother. Who stays home and awaits the returning hero. Who stays home to birth the next hero. This figure is obscured and domesticated, lest it emerge to dominate the infantile self states that are expressing their dominance through violence (see also Benjamin, 1988; Harris, 1983). As a primitive narcissistic object, this primal figure can confer a halo on the aggressor. But she is also the judge, the original arbiter of good and evil.

---

* See Zinn (in Sheehan, 2006).

PETER: DREAD, MATERNAL SURVEILLANCE, AND REMORSE

But this mother does not just want to be a signifier for exposure and retribution. She is always a signifier for loss and grief; for a multiplicity of bodies in the war zone. Her children are the victims of the phallic warrior. Her child is the phallic warrior, himself. What she wants is the sanctity of her children. At first, she finds her force in our shame and dread. But when her audience is ready, her gaze invokes the tragic. Now, a space opens between excitement and destruction, and a real, human other finally steps in.

And so it would be, in Peter's analysis. Once I thought I could never do what Peter had done. But the woman in the veil started breaking into our mirrors. And filling us with dread. He takes pills for insomnia and anxiety. I read histories and Vietnam diaries, and watch documentaries and war movies. My eyes fill with wounded soldiers. His fill with innocents, dying. I realize that these soldiers could be me. He realizes that Vietnamese children are his. In his sessions, Peter's past and present are becoming inseparable in grief. She knows about him, he says, and she's been protecting the kids from him. I don't know if he is referring to his wife. Or to the phantasm in the veil. Or to real mothers who once cowered before him. "The burning children?" I ask. He sobs, nods, yes. Then he says, she always knew. "Your wife?" I ask, and again, yes. Finally I am made to understand his arrival, now in my office. There has been the war in Iraq, the atrocity at Haditha. Iraq reignited Vietnam; Haditha made his wife see him as "the beast." I ask if his sons know. He thinks so, he is sobbing, "How can they touch me?"

Still the veiled woman appears. She is no particular mother, and she is every ferocious mother. She is entirely unlike the mother whom he grew up with. Who mimicked Jackie, and was flirtatious and glib and elegant and social climbing, deferring his care to sequential housekeepers, while his father moved between his business and his bar and his newspaper in the den. Peter has never known a protective mother or a tender father. If they had cared for him and sheltered him, he wouldn't have been drunk in college. He wouldn't have flunked out. He would never have been drafted, lost his leg, lost his men, lost his mind and his morals. At last, he really tells me about Vietnam. The way he went mad before the firebombing. His dream series ended. Dread fell away. Together, we evolved a more tragic worldview. With it, there was a sense of grief and departure. As if,

> I saw him for the last time. He was kneeling by a tomb of white marble and the shadow of a veiled woman rose out of the grave beneath and waited by his side. The unearthly quiet of his face had changed to an unearthly sorrow … the darkness closed round the pilgrim at the marble tomb—closed around the veiled woman from the grave … I saw and heard no more. (Collins, 1985, p. 281)

REFERENCES

Altman, N. (2005). Towards the depressive position. *Psychoanalytic Dialogues*, 15(3), 321–347.
Arendt, H. (1955). *The origins of totalitarianism*. New York: Meridian Press.
Bassin, D. (1994). Maternal subjectivity in the culture of nostalgia. In D. Bassin, M. Honey, & M. Mahnek-Kaplan (Eds.), *Representations of motherhood* (pp. 162–174). New Haven, CT: Yale University Press.
Benjamin, J. (1988). *The bonds of love*. New York: Pantheon Books.
Benjamin, J. (1995). *Like subjects love objects: Essays on recognition and sexual difference*. New Haven, CT: Yale University Press.
Benjamin, J. (2006, June). *Our appointment in Thebes: The analyst's fear of doing harm and the need for acknowledgement*. Paper presented at the IARPP Conference, Boston.
Bhabha, H. K. (1985). Signs taken for wonders: Questions of ambivalence and authority under a tree outside Delhi, May 1817. *Critical Inquiry*, 12(1), 114–165.
Bion, W. R. (1965). *Transformation*. London: Tavistock.
Bollas, C. (1986). *The transformational object in the British School of Psychoanalysis: The Independent tradition* (G. Kohon, Ed.). London: Hogarth Press.
Buber, M. (1923). *I and Thou* (R. G. Smith, Trans). London: Routledge & Kegan Paul.
Collins, W. (1985). *The woman in white*. London: Penguin Books. (Original work published in 1859).
Elshtain, J. B. (1994). The mothers of the disappeared: Passion and protest. In D. Bassin, M. Honey, & M. Mahnek-Kaplan (Eds.), *Representations of motherhood* (pp. 75–92). New Haven, CT: Yale University Press.
Ensler, E. (2006). *Insecure at last: Losing it in our security-obsessed world*. New York: Villard Press.
Feldman, A. (2000). Violence & vision: The prosthetics & aesthetics of terror. In V. Das, A. Kleinman, M. Ramphele, & P. Reynolds (Eds.), *Violence and subjectivity* (pp. 46–78). Berkeley, CA: University of California Press.
Fonagy, P., & Target M. (1998). Mentalization and the changing aims of child psychoanalysis. *Psychoanalytic Dialogues*, 8(1), 87–114.
Foucault, M. (1979). *Discipline and punish: The birth of the prison* (A. Sheridan, Trans.). New York: Vintage Books.
Freud, S. (1893). Studies on hysteria. In J. Strachey (Ed. & Trans.), *The standard edition of the complete psychological works of Sigmund Freud* (Vol. 2, pp. 1–313). London: Hogarth Press.
Freud, S. (1915). Thoughts for the times on war and death. In J. Strachey (Ed. & Trans.), *The standard edition of the complete psychological works of Sigmund Freud* (Vol. 9, pp. 275–300). London: Hogarth Press.
Fussell, P. (1989). *Wartime: Understanding behavior in the second world war*. New York: Oxford University Press.
Grand, S. (2000). *The reproduction of evil: A clinical and cultural perspective*. Hillsdale, NJ: The Analytic Press.
Grand, S. (2010). *The hero in the mirror: From fear to fortitude*. New York: Routledge.

Harris, A. (1983). Bringing Artemis to life: A plea for militance and aggression in feminist peace politics. In A. Harris & Y. King (Eds.), *Rocking the ship of state: Toward a feminist peace politics* (pp. 93–115). Boulder, CO: Westview Press.

Klein, M. (1952). Notes on some schizoid mechanisms. In M. Klein, P. Heimann, S. Isaacs, & J. Riviere (Eds.), *Developments in psychoanalysis* (pp. 292–320). London: Hogarth Press. (Original work published, 1946).

Klein, M. (1975). On the theory of anxiety and guilt. In *Envy and gratitude and other works, 1946-1963* (pp. 25–42). New York: Delacorte. (Original work published 1948).

Layton, L. (2006). Attacks on linking: The UCS pull to dissolve individuals from their social context. In *Psychoanalysis, clan, and politics: Encounters in the clinical setting* (pp. 101–118). New York: Routledge.

Lifton, R. J. (2005). *Home from the war: Learning from the Vietnam veterans.* New York: Other Press.

Mahler, M. S. (1968). *On human symbiosis and the vicissitudes of individuation.* New York: International Universities Press.

Mazetti, M. (2007, February 3). Analysis of Iraq's future is bleak, but both sides in war debate find support. *New York Times*, p. A6.

Mitrani, J. (1994). On adhesive pseudo-object relations, part I. *Contemporary Psychoanalysis, 30*, 348–367.

Norris, C. (1987). *Derrida.* Cambridge, MA: Harvard University Press.

Ogden, T. (1986). *The matrix of the mind.* Northvale, NJ: Jason Aronson.

Peltz, R. (2005). The manic society. *Psychoanalytic Dialogues, 15*(3), 347–367.

Reis, B. (2006). Even better that the real thing. *Contemporary Psychoanalysis, 42*(2), 177–197.

Rich, F. (2007, March 11). Why Libby's pardon is a slam dunk. *New York Times*, p. A14.

Ruddick, S. (1989). *Maternal thinking.* New York: Beacon Press.

Ruggiero, G. (2006). Editor's note. In Sheehan, C. *Dear President Bush* (p. ix). San Francisco: City Lights.

Said, E. W. (2004). *Humanism and democratic criticism.* New York: Columbia University Press.

Sartre, P. (1964). *Being and nothingness.* (H. E. Barnes, Trans.). New York: Citadel Press.

Scott, J.C. (1992). Domination, acting and fantasy. In C. Nordstrum & J. Martin (Eds.), *Domination, resistance and terror.* Berkeley, CA: University of California Press.

Shay, J. *Achilles in Vietnam: Combat trauma and the undoing of character.* New York: Scribner.

Sheehan, C. (2006). *Dear President Bush.* San Francisco, CA: City Lights.

Slochower, J. (2006). *Psychoanalytic collisions.* Hillsdale, NJ: The Analytic Press.

Sontag, S. (2003). *Regarding the pain of others.* New York: Farrar, Straus & Giroux.

Stein, R. (1998b). The poignant, the excessive and the enigmatic in sexuality. *International Journal of Psycho-Analysis, 79*, 253–268.

Steiner, G. (1984). *Antigones.* New York: Oxford Press.

Sullivan, H.S. (1953). *The interpersonal theory of psychiatry.* New York: Norton.

Williams, P. (1998). Psychotic developments in a sexually abused borderline patient. *Psychoanalytic Dialogues, 4*, 459–493.

Winnicott, D. W. (1965). The development of the capacity for concern. In *The maturational processes and the facilitating environment* (pp. 73–82). New York: International University Press. (Original work published 1963).

Chapter 12

War stories

Donald Moss

> She mimicked her mother's sharpness, but she was rather ashamed afterwards, though as to whether of the sharpness or of the mimickry was not quite clear.
>
> Henry James
> *What Maisie Knew* (1897)

I sit down. The shades are drawn in the windows across the street. Everyone, silent and somber, is at work on a text: everyone, every American adult. The country has taken a pause. All of us have been given the same assignment. Our writing, compiled and preserved, will constitute a monument to a citizenry thinking about war and harm. We will be remembered for this. We will have interrupted history.

This fiction lasts a second.

A shade opens. A woman appears on her terrace. The heating contractor knocks on the door. There will be no large compilation, no major monument.

I'm doing this alone, then. I want to stop more than I want to continue. I pause. I don't have to do this, I think. Yes I do. It's an obligation. It's about being a Jew, this obligation. You must keep telling about what was done, about what remained, about what came next. You do this, you keep telling, and it hooks you in. You find your place. Telling holds your place for you. You listen, you tell, you listen, you tell.

I once went to Terezin. Down the road from the concentration camp were the ovens. I walked there alone. You walk into an innocuous brick building and there they are: two ovens. The ovens were exactly like ovens, only mammoth. The doors of the ovens were open. You stand in front of the open doors, looking in. For me, "ovens" had been a metaphor for the Holocaust. But here, the metaphor collapsed. But not all the way. Though these ovens looked exactly like ovens, what most impressed was my incapacity to experience them as the actual ovens they apparently were; they remained likenesses. I couldn't actually achieve concreteness in front of them. There was a gap I couldn't navigate. I couldn't get real people into these particular ovens. I stared at them. No matter what I tried, they kept being "ovens," as though they were models, somehow, of other ovens into which people had been put, as though this were an exhibit, a rendering, of the ovens at Terezin. I stayed there a long time. I was surprised at this incapacity. I kept working to diminish the gap between what I was seeing and what was there. But I didn't really know what kind of work to do. After an hour or so, I think, I saw a small, torn piece of paper on the floor next to one of the ovens. On the paper, written in Hebrew, was a prayer: "Shma Yisrael ..." ("Here, oh Israel, the Lord our God, the Lord is One.") The paper had been left there by a Jew who had been there before me. Whether yesterday or 40 years earlier didn't matter. What mattered was that the torn paper opened up a time channel for me. I held the paper in my hand. This paper was concrete. The prayer was concrete. Neither was a likeness. Holding the paper and reading the prayer placed me in a historical line: me here now, and the person who wrote this before me, and all the others before both of us. The paper put me in contact with all the others. The contact was not a likeness to contact. This contact was concrete and overwhelming. What had been a feeling of insufficiency turned immediately into excess. The intensity was too much. I had to leave. I walked out and stayed away for a few minutes. When I came back, the gap reappeared. Although the ovens were still "ovens," I now felt like I had achieved contact with the people for whom they were designed. The gap had narrowed.

That text in Terezin not only did no harm; instead, it transformed harm—helpless disidentification—into connection. For now, here, that prayer written on torn paper is my exemplary text, the one that opens up a time channel and narrows the disidentificatory gap. The exemplary text binds reader to writer and each to a lineage of shared predecessors. How to write my own?

I have no personal experience of war. I have almost no clinical experience with people who do. Friends, colleagues, patients—we are all buffered. Although we may think about and imagine what it must be like, we know that we know almost nothing. What I mean by *know* here is, I think, directly tied to what I mean by *experience*. We know almost nothing because we have experienced almost nothing.

I was going to write about contemporary films about the Iraq war: *Stop Loss*, *Jarhead*, *In the Valley of Elah*. It seemed similar to writing about newspaper reports from a year or two ago: an inert project.

I was going to write about some books, comparing firsthand reports of the Iraq War with similar reports from Vietnam and from the Second World War. I knew there would be differences—the moral high ground vanishing over the decades—and I also knew I could make little of them.

I was going to write about a recently published surgical textbook filled with grim, nearly unbearable photos of traumatized people in Iraq and the recommended treatment procedures. I had read reviews of this book. The reviews said that it succeeded where most war writing did not—that it conveyed, more directly, what it is like, or what, from my point of view, I might imagine it must be like. The book left me cold: surgical treatments of generic gore. Nothing for me to say.

I was going to write about 9/11, about feeling frightened, crying, enraged, vengeful, about what I had already written, but I've said what I can.

How to write here, lacking a sense that there is something I know, something ready. The task, then, is to look, see what I can find. Whatever my method, the aim will be to keep the lever arm short: no confessions, conclusions, or conceptualizations.

Start with an inventory.

This war, old wars, my father at war, Vietnam, the next war, new wars, just wars, torture, restraint, reason, Freud, civilization, my kids, war games, video games, murder rates, the death penalty, doctors in war, doctors against war, evading the draft, my father at war, my father killing people and telling me about it, war and telling it, what happens when you tell it, when you show it, when you think it, when you do it, when you wonder about it, what do you wonder about, what's the harm, where's the harm, why's the harm, war as a fact like time is a fact, all the books and pictures, the films, all the showing, terrible, clever, sad, brilliant, enraged, heroic, resistant, courageous, all the ones who have wanted us to see, and all the killers and the killed, the dramas, the stories, the memories, centuries of them, no memory without war there's always a war, war like an illness, an infection, a death cell, a terror, gathering stories to tell no harm in that, is there?, how to know, really, what harm is, the need to remember, the harm of remembering, the harm of forgetting, which is it, being a psychoanalyst, remembering, forgetting, telling, not telling, staying silent, moving on, visiting, staying away, what to do, why, with whom, someone comes into the house and kills everyone and leaves and that's it, and my father did that, well there you are, and who do you tell, and why, and you hear about this, and who do you tell, and why?

There's a finding: my father—his war stories. I'll start with him.

THE ENTRANCE WOUND

"And I'll only say this one once. We didn't have time to fuck with prisoners. Five minutes before they were prisoners, they were killing us. My captain used to say things like take these prisoners back to the prisoner dump, which would be eight miles or ten miles back and be back in five minutes. So we would be back in five minutes.

"Anyway, I really can't talk about some of that stuff. It's 45 years. Perhaps I can. There were two, one instance still haunts me, two instances I can talk about, one maybe. We were foot troops and we would clean up any pockets. And we're running along trying to stay below the windows and guys behind us are trying to cover the windows here on the other side of the street. I don't know if I can talk … Jesus, 50 years, I'd think I'd be over it. We went underneath the window and we heard noises in a window. Now all the Germans knew. I mean they knew what we were doing. So, I pulled two grenades out and I tossed them both in the window, over my head. And then you wait for the blast. And I think the door was locked but we blew the door open and ran in and evidently there was a nursery and there had to be I don't know how many—maybe six, maybe three—babies in bassinets blew all over the room. At the time, I didn't think. I remember maybe I might have said oh shit and I got out of there …

"We had pot-bellied stoves. I was evidently warming up and this German officer pulls himself up and gives a Nazi salute. And he says I'm a German officer and I demand Geneva rights. Rominick is a Polack. There's all these dead people scattered around. Rominick kneed him twice with his 45 automatic and the guy is lying there. And Rominick announced to the people standing there, including me, which made me deathly afraid of Rominick from then on, first man touches him, I'll kill him. Nobody bothered him.

"One of the things that makes it hard, one of the first things was the delight I felt, I remember feeling in shooting things, in killing things, now not those babies, that was, Jesus Christ. But, I would volunteer to snipe, which is a, even now is kind of a sneaky way of doing anything. And you played with it a little bit, if you were careful, got the windage, how much wind there was, from what direction, how far your target was away, you could lay a bullet in, if you've got time. And I used to like that. I'd catch a guy, he'd be going to the can to take a shit. I mean, a guy might be thinking of his *Fräulein* going home, so far removed from the war, and I'd kill that poor fucker, or at least I'd hit him, and enjoy it, Jesus, Donald, I enjoyed it."

THE PASSAGE THROUGH, THE INFILTRATION

The strangest feature of this story—"And I'll say this only once"—is that my father, in fact, told this story to me 50, 60, 70 times. And each time, just like this one—the second to last—he would hesitate, resist, be unable to continue, be overcome. And then, always as though for both the first and last time, he would find a way to say it, to tell me about those babies, and those prisoners and the captain and the guy going to the can to take a shit. So each time, then, the real force of the story lay in the fact that my father had once again and for the first time, found the strength, the narrative muscle, to get the telling out. Each time he told it, the telling turned him into a hero, as if it weren't the deeds that were difficult and it certainly wasn't the hearing that was difficult, but only, really, the telling. As though I were witnessing something like a self-surgery he were performing, an extraction of a bullet, say, or an amputation. And I, each time, never an exception, felt exactly in accord with his premise. It was, for me too, always both the first and last time. I was always the only witness. I never said, never quite thought even, that he was doing this again, that we were doing this again. And yet, also, I always knew, always thought, that yes, here we are again, doing this again.

The telling began, I think, before I was 10 years old. It ended the year he died. He told it to me once more after this. I was 54 then. More than 44 years of this story, of this doing together what no one but us really knew. It was not exactly a secret. The information was not meant to be held in private. I never thought it mattered much who knew. I used to tell people. I was neither ashamed nor proud of what he had done. But, like him, I knew that, yes, I'd tell someone and enjoy it, Jesus. I enjoyed it. He never told me he enjoyed telling me, but I know he did. I never told him about my enjoyment either, in the hearing or in my own tellings. Whatever he had done, years before, served as a means for what he/we were doing now, what I'm doing now—now. And of this I never spoke, except for now, and, in the face of it all, yes, even now, I can say, like him, and with him, I'm enjoying it.

It wasn't, finally, information that was being transmitted; it wasn't trauma either. It was excitement—his and mine. Those babies, prisoners, the guy on the can, the captain—they were never quite the point. My father and I put the stories to our use. That—the stories' use—was the point. And that use had something to do with a shared sense that in each telling those blown-up babies, murdered prisoners, wasted shitter, and tortured officer were the price we willingly were paying. He—and I—as the tellings went on, would, we each knew, would have it all happen again, knowing we could make this out of it. That was our secret. Given the chance, he would do it again, and I would want him to.

One terrible part of this secret is that both of us would deny its truth had either of us tried to expose it. Therefore, now, when I write that it was a secret, I wonder whether I am writing an exculpatory fiction and whether it may have been I alone for whom both the deeds and the tellings were a means to a treasured end—the treasure consisting of the discovery, the repetition, and the deep enjoyment of this inverted fairy tale—in which the worst that can happen does happen and that only because it happened can you enjoy the intimacy that all of civilization aims to forbid: the pleasures and excitements of telling it. I know he was in on it. I know it. I also have my doubts. What kind of knowledge is this? It's madness to be certain of a mind not your own. But I know it.

So this, then, is, I guess, one of the harms engendered by war stories. Maybe all war stories are told by fathers to sons. You can't know that. When you encounter the admonition to "First, do no harm" you are meant to know, as a matter of course, what harm is. But here, in these stories my father told me, in my housing them now, in my telling them, I cannot distinguish pleasure from harm. I know with certainty that he told me too much, that I, like any child, was, in some sense, harmed by hearing what

I heard, by knowing what I knew. But only in some sense. I also feel that the stories were a gift providing pleasures unrivalled to this day. And I also know that the fact of those pleasures is, in large measure, what is meant by the stories doing harm to me. I've heard people speak this way about shooting up cocaine. They speak of an intensity of pleasure that they know the rest of the world will never provide. They speak, then, of the harm done to them by pleasure. This is what I am trying to do now: speak of the harm done to me by pleasure. But the contradiction knocks me off my feet. I don't feel on firm ground as I write this. I know there is harm. I know there is pleasure. But I'm writing from a zone in which both are absent. I don't think it would be possible to write from the zone in which either—meaning both—were present.

I remember hearing of some Navajo soldiers who, under no circumstances, would ever speak of war to either their wives or children. Instead of war stories there was silence. I wonder about the impact of that silence. Is it a way of telling a war story? I remember my admiration of these Navajo and of the forces that allowed them to retain what they knew, to remain silent. Admiration and reason saluted their self-control, their commitment to do no further harm to those who depended on them, those who would have had to listen if they had spoken. That what home is, what I was—the place where the others (at least one other, I think) have to listen to you. But admiration and reason actually did not, and do not, penetrate, do not disrupt, the flow and impact of these stories. The stories take on an admirable glow themselves, precisely because they are so shameful, their telling such a sign of helplessness and neglect, such a refusal, such an overflow, of all the constraints and limits established by the mere demands of being decent. The babies are blown apart and so is the mind of the child who hears of it. The story is a grenade, thrown this time with secret foreknowledge. OK, you play the soldier, I'll play the babies. Let's do it again and again. Actually, as I write this, I realize that the roles are not as clear as that. Yes, let's do it again and again. But, let's really be in on it together, each of us the soldier, each of us the babies, each of us telling each of us, without it mattering really who appears to be telling, who listening. Let's go beneath differences like that. Let's tap into force, into the joy, power and energy pulsing through and destroying all those arbitrary differences. Here it is, then, the impact of those stories—they eliminate difference and make it impossible to distinguish pleasure from harm. For a moment, then, the title of this book may seem like it could just as well have been "First, do no pleasure." This is hard won: the sense that the difference between the two is not fundamental, that it is the product of some activity. Knowing that, or thinking you know it, is itself both a marker of harm done and pleasure taken.

The moment you use a phrase like "eliminate difference" you have abandoned the project. The project is not to conceptualize an effect. The project is to tell an effect, show an effect, affect an effect. It's to do what my father did, to do it as his son, to find a listener, and to expose a pulse, an artery, just this once, that carries and transports these stories, these tellings. The tellings are immortal. The father's tellings are injected—not like venom but like syrup. No, the tellings are not injected. They do not penetrate. They are opportunistic, like water. The tellings flow. They slide in through an opening that can't be seen. They slide in at a point of attachment, a kind of umbilical attachment, binding father to son. *Binding* is not right. *Umbilical* is not right. Binding is a side effect. Perhaps the stories bind but before that the point of attachment is cool and open—a point through which things flow, back and forth: father to son, son to father. The son can make the father sick if what flows from son to father presents father with a mismatch or an excess. And for the son the flow is, at the beginning, unremarkable—as the lungs attach all of us to the air and therefore to the planet so this point, this opening, attaches son to father. The opening changes character when these stories flow through. The opening takes on muscle and force—here, then, the point of attachment begins to bind. I write this and of course think of bodies, male bodies, hard, muscular, forceful, binding. But this, too, this package of thoughts, of associations, seems a side effect, a flight more than a realization. First comes the point of attachment. It precedes. Whatever you try to say about it, it precedes that too. The point of attachment, the place where the stories come in, then, is behind you, behind what you can be conscious of. You can only be conscious of the tellings once they have arrived inside of you.

The stories leave from in front. I can direct them, aim them. What I can't do is tame them, turn them into words, into ideas, concepts, tell you about them. The stories are made up of primal words. They are disguised as words. The harm they do includes providing undeniable evidence that the primal is real and the word part is being put to use. They do harm by infiltrating all your efforts at thought, love, order, reason with the fragrance of uncertainty. You too, the stories affirm, you too. Fragrance is not right. Umbilical is not right. There is nothing like a cord, a conduit. You cannot catch the stories in passage. The stories are, in the hearing, the air you are breathing. You breathe the stories through a mouth or nose you cannot see and do not know you have. The stories are a heat or pain. They are a temperature.

THE PARTIAL EXIT, THE PASSAGE OUT: THE RETURN OF THE SNIPER

Tomorrow is Inauguration Day. Today our dog was diagnosed with lymphoma. The diagnosis was sudden and devastating. The dog now has a year or so to live. Late in the evening I was sitting with our 12-year-old son, trying to console him about the dog. I found nothing to say. He wouldn't let me touch him.

We were both silent for a few minutes—he, I think, focused on the dog, me focused on him.

Suddenly, my son asked me what I thought the odds were of Obama being assassinated. He wanted the odds on an assassination tomorrow, not those on the entire Obama presidency. He clearly felt I knew enough to answer his question. "No idea" would not suffice. I said, uncertain why, that the odds might be about a million to one. My son said he thought they were less—closer to one thousand to one, maybe less than that.

All you need, he said, is to sit two miles away with a high-powered sniper rifle and have a clear line of sight. Do that, he said, and Obama's dead.

ENJOYING IT—THE HARM TRANSFIGURED

"One of the things that makes it hard, one of the first things was the delight I felt, I remember feeling in shooting things, in killing things, now not those babies, that was, Jesus Christ. But, I would volunteer to snipe, which is a, even now is kind of a sneaky way of doing anything. And you played with it a little bit, if you were careful, got the windage, how much wind there was, from what direction, how far your target was away, you could lay a bullet in, if you've got time. And I used to like that. I'd catch a guy, he'd be going to the can to take a shit. I mean, a guy might be thinking of his *Fräulein* going home, so far removed from the war, and I'd kill that poor fucker, or at least I'd hit him, and enjoy it, Jesus, Donald, I enjoyed it."

So here it is, I thought. Here's my father's war story, being told to me, with variations, by my 12-year-old. My 12-year-old imagines the sniper and identifies with the victim. He's blown away by the technology—his version of "if you played with it a little bit." It somehow helps him out to get in touch with this, helps him deal with his dog's terminal illness. The story moves my son from the position of helpless object of an inexplicable disease to the position of an excited object of a determined man's capacity.

My father's story is about being a sniper, about how he liked it, and I wondered how that story is filtering through me to my 12-year-old. I know I've told him about sniper rifles like the one he imagines. In fact, I like hearing about these rifles, and until last night, I never thought about the line formed by that liking, the line binding me to my father, and my son to my father's son. My father's sniper bullet goes through that guy on the can and then continues through the assassinations of the '60s and '70s until it comes to a moment's rest in my son's question about the odds of Obama's being killed today. He told me that the DC police have arrested many people today who had threatened Obama's life. I told him that people say a lot of things. And what he said was, yes, but some people do them. There, I think, there, with my son saying "some people do them," my father's bullet, and his "Jesus, Donald, I liked it," come to momentary rest.

Chapter 13

Notes on mind control
The malevolent uses of emotion as a dark mirror of the therapeutic process

Ruth Stein

> One man uses his power over another to crush his individuality, his dignity, his capacity to feel deeply (to feel joy, love, and even hate); and ... to stifle the victim's use of his mind—his capacity to think rationally and to test reality.
>
> Leonard Shengold
> *Soul Murder* (1991)

> The methods of establishing control over another person are based upon the systematic, repetitive infliction of psychic trauma. They are the organized techniques of disempowerment and disconnection.
>
> Judith Herman
> *Trauma and Recovery* (1992)

> Regard punishment as a political tactic ... make the technology of power the very principle of both the humanization of the penal system and of the knowledge of man.
>
> Michel Foucault
> *Discipline and Punish* (1995)

INTRODUCTION

This chapter is about malignant processes that use depth-psychological or psychoanalytic knowledge to dominate other people through malevolent uses of emotion. It shows how psychoanalytic theory can be recruited for good or for evil, for harm or for repair. Aligning itself with the present book's theme, it explores the processes at work in mind control methods that rely on brain functioning as well as on relational-transferential factors. While it does this, it makes at the same time some statements about psychoanalysis itself. The symmetrical contrasts as well as the similarities between these techniques and therapeutic methods relying on depth psychology make for an instructive and disturbing subject.

MIND CONTROL

One or another form of mind control, the imposition of thoughts and beliefs and their cultivation in the subjects on whom they are imposed, is a factor that is common to different situations where, in the words of Margaret Singer (Singer & Lalich, 2003), a major researcher of mind control,* "a behavioral change technology [is] applied to cause the learning and adoption of an ideology or set of behaviors." Mind control, according to Singer, is distinguished from other forms of social learning by "the conditions under which it is conducted and by the techniques of environmental and interpersonal manipulation employed to suppress particular behavior and to train others."

Mind control techniques underlie the calm and serene tone of the letter to the 9/11 terrorists,† written as an encouragement for a mission of killing and self-destruction. Mind control had its effect on anthropologist Maya Deren who went in 1949 to Haiti to study local Voodoo dancing, and it explains the overpowering state of trance she entered there that permanently changed her outlook on Voodoo and on her life. Charismatic leaders and agents of so-called coercive persuasion are often capable of achieving far-reaching changes in people. There are regimented "thought reforms," "reeducation programs," "self-strengthening tracks," all names for an assortment of what used to be called brainwashing and is now called mind control. These processes were and are practiced in cults, in religious conversion agendas, under totalitarian governments, in certain therapeutic communities, and in abusive relations of all stripes. They take place in the American Bible Belt, in Israel's religious conversion institutes, in some

---

* In 1983, the American Psychological Association (APA) rejected the research report it had requested of Margaret Singer's chaired Task Force on Deceptive and Indirect Techniques of Persuasion and Control (DIMPAC). Before the task force had completed its final report, the APA submitted a memo, stating that Singer's hypotheses "were uninformed speculations based on skewed data." The APA subsequently withdrew the brief it had submitted, portraying it as premature, but when the report was finally completed in 1986, the APA Board of Social and Ethical Responsibility for Psychology (BSERP) rejected it, stating that the report "lack[ed] the scientific rigor and evenhanded critical approach necessary for APA imprimatur," and stating that the BSERP did "not believe that ... [the DIMPAC has] sufficient information available to guide ... [the APA] in taking a position on this issue." The BSERP board requested that the members of the task force not distribute or publicize the report without indicating that the board found the report unacceptable. Singer's efforts to sue the APA were unsuccessful. I read this report, based on 3,000 cases that were interviewed since 1972, and found it thoughtful, measured, and very informative, even if it did not have exact numbers to buttress its hypotheses. One of the reasons Italian psychologists and theologians Amitrani and Di Marzio adduce to explain the patent volatility and resistance of the APA and of many scholars to the concept and metaphor of "brainwashing" is the possibility it admits "that some form of mind conditioning might be practiced on members within an NRM (New Religious Movements)." It thus seems that the effort at suppressing the concept and the phenomenon of mind control is undertaken in the name of religious freedom that needs to be defended.
† The letter that was found in Mohammed Atta's luggage before he boarded the plane from Boston on September 11 (see Stein, 2009, Appendix A).

Islamic madrassas and mosques, in China, in Korea, and in the USSR. They were part of the experiments conducted by the Central Intelligence Agency (CIA) in the Cold War era, and they are active today, in the American Survival, Evasion, Resistance, and Escape (SERE) plan, a program for learning to withstand torture that is also used as an aid in getting information from terrorists through stress induction and various other methods of torture.*

What stimulated my thinking in this area, in addition to surviving religious and psychoanalytic indoctrination, is the letter that was found in Mohammed Atta's belongings on September 11, 2001. Listening to the prosody of the text, its repetitive, exhortative pulse, one is almost induced into a state of trance. The letter itself, in a kind of *mise-en-abîme*, provides detailed instructions to its addressees on how to induce in themselves a state of calm single-mindedness while preparing to fly the plane into hell, liquidating any person on the way (cf. Stein, 2001, 2009; Kippenberger & Seidensticker, 2004). In this chapter, I address the brain-induced aspects of mind control, trance, and altered states of consciousness. I then proceed to the themes of seduction and emotional enslavement that are implicated in mind control, to eventually reach the intersubjective aspects of the relation captor–captive, as exemplified in Robert Jay Lifton's (1961) study of thought reform in China. This journey takes us from the physiological, through the social, to the individual; from the facilitation of altered brain functioning, to group processes in cults, to the perverse, intimate relationship between indoctrinator and prisoner. Far from exhausting the topic, I am bringing together some fragments to build a picture that will hopefully throw some light on this heterogeneous, historically continuous psychosomatic–relational–cultural phenomenon.

Throughout history a systematic silencing of human consciousness served as an efficient tool for dominating and neutralizing tribe members, citizens, and whole nations. We know that cults, initiation rites, protracted or repetitive religious ceremonies and other forms of mind control existed in ancient Egypt and in other parts of the world. Charismatic experts of human engineering, priests, gurus, elders, kings, and their counselors knew how to psychologically exploit fascination, fear, and guilt for their purposes. One of the proven methods for controlling the minds of people is the use of techniques and rituals to create peak experiences and moments

---

* One of the areas to be investigated by the CIA was mind control. The CIA's human behavior control program was chiefly motivated by perceived Soviet, Chinese, and North Korean use of mind control techniques. The CIA originated its first program in 1950 under the name BLUEBIRD, which in 1951, after Canada and Britain were included, was changed to ARTICHOKE. A declassified CIA document (CIA MORI 190691, p. 1), "Hypnotic Experimentation and Research, 2/10/1954," describes an attempt to hypnotically create unsuspecting assassins (in Mind control Cover-up; updated May 21, 2005).

of illumination. Peak experiences can be rich sources for insight,* healing, and renewal, and can offer vital rest and relaxation. But intense peak experiences can also serve an opposite purpose by undermining the self-feeling of individuals and masses, and feeding the crowds the opium of bread and circus. The more systematic, modern version of mind control addressed here has the form of a preordained plan to influence its objects to adopt a new version of causality and to radically reinterpret their life history while developing dependency on the organization or the leader. The mind control trajectory will further turn its object into an agent of the organization or the leader who runs the thought reform.

SNAPPING

We know that emotional overload, as well as physical techniques such as prolonged rhythmical pounding and drumming, music and dance, chanting and praying, fasting, and sleeplessness, are vehicles for changing brain waves and brain functioning. Under certain circumstances, the brain becomes sensorily deprived or overloaded (Sargant, 1957) and, reaching its limit of normal processing, it "snaps" (Conway & Siegelman, 1995; Winkelman, 2000). At such a point, consciousness changes permanently or for a long time. Trance states—regarded by Denis Wier (1996, 2009) as a "looping of consciousness" that has the capacity to disable critical reflection—are states of mind that are generated by iteration of stimuli, and/or by having the subject flooded with them or deprived of them. Trance states are, for good or for bad, means of saving mental energy by enabling the routinization of many ordinary activities while at the same time disabling higher, more demanding cognitive functions. Trance can be said to be a kind of dissociation that creates the illusion that the images or thoughts that emerge while it lasts are true and valid. We also know that many social contexts and group pressures accelerate and reinforce processes wherein people abdicate personal responsibility, conform to the group's values, and mutilate their individual identities. We know likewise that psychodynamic processes, such as manipulated intimacy and transferences to authority figures, introjective processes (particularly as the term is used by Ferenczi) that are promoted by a certain cunning empathy by the influencer can have tremendous mutative powers. The mutual influences of body, society, and mind that are implicated in these processes reinforce one

---

* Benny Shanon (2002), a cognitive psychologist at the Hebrew University, describes in a compelling document an amazing array of images and insights he acquired traversing altered states of consciousness produced by consuming the Ayahuasca drug in the company of Amazonian shamans (see also Harner, 1990).

another exponentially. Breaking down this multifactorial phenomenon into its components is artificial of course, but I do it in an attempt to clarify the ways in which this social–individual body–mind phenomenon can be orchestrated.

PHYSIOLOGICAL PROCESSES

Military drills, government-organized propaganda and clamorous power parades, socialization rituals, initiation rites, spiritual and religious training are cultic practices within social institutions that manufacture certain physiological conditions. Such conditions interfere with the normal functioning of the brain by inducing trance states and hypnotic changes in consciousness. Trance states can be evoked by rhythmic and acoustical activities, and by disregulating body functioning, such as sleep and food deprivation as well as by powerful emotions. In his study of the neural ecology of consciousness and shamanistic healing, Michael Winkelman (2000) explains the ways in which these activities can induce an altered state of consciousness. Briefly, such activities manipulate, through different venues, the sympathetic nervous system (the activating system of the autonomous nervous system that prepares the body for action) to the point of exhaustion and collapse into a parasympathetically dominant state (the parasympathetic system is responsible for the relaxation response and ultimately for sleep, coma, and death). Prolonged periods of stereotypy (repetition of an invariant pattern of movements with no observable goal, such as in drawn-out breathing, pacing, or drumming) produce an increased rhythmicity and a general slowing of the EEG pattern, slow-wave parasympathetically dominant states (especially when the activities result in collapse), and the release of endogenous opiates, which induce euphoric states.

Powerful emotions, too, can cause brain functioning that disconnects from reality and runs "by itself." Massive sensory assaults flood the capacity of the brain and culminate, according to some researchers, in an ecstatic "snapping." In these states, the brain derails, deviates from its routine processing of reality, and begins running on its own accord, with no external input. The "snapping," chaotic-change moment can be followed by a far-reaching transformation, amounting to a religious conversion or ideological "illumination." The experience of becoming enlightened can be understood in physiological terms as a condition whereby the experience becomes more intense than the ability of the brain to encompass and contain it. Such a neurophysiological state can be experienced as a feeling of being powerfully swept away, of losing control over actions and speech, combined with a sense of loss of self, manifest in such extreme passivity as to feel like a puppet on strings pulled by some exterior power.*

I mentioned a brain and experiential change that is caused by the mechanical facilitation of the functions of consciousness. The common denominator of brainwashing and therapeutic change is *the seeking to effect change through the creation of an intensive experience.* The creation of intense experience is aided by various techniques of assault, flooding, inflammatory rhetoric, powerful visualization techniques, and the like, all of which disrupt the continuity of experience and the regularity of mental functioning (Conway & Siegelman, 1995; Singer & Lalich, 2003; Singer et al., 1986; Singer & Ofshe, 1980, 1990). Many authors and witnesses speak of sudden events of snapping and popping that occur at the moment of an intense experience. Such moments can be short and abrupt, or they can be slow and drawn out. In the wake of such rupture, the information-processing centers of the brain become disorganized and leave the mind open and receptive to new ideas and even to a radical reorganization.

The term *snapping* encompasses different processes and different qualities of consciousness alteration. Michael Winkelman (2000) demonstrates that, although induced by very different means, there is a fundamental physiological similarity underlying different altered states of consciousness.† The abrupt

---

* Trance is a dissociative state in which the mechanism for information processing bifurcates into two separate tracks that run concurrently, a dominant and a latent one. The dissociation can be adaptive, serving as the basis for the automatization of learned skills, which is efficient since it saves energy and aids learning and action. Dissociation can also serve one well in creative or inspirational states; but it can of course be highly pathological. A state of trance is caused by so-called trance-creating loops. Such loops are repetitive or recursive thought-sequences that begin and end with the same thoughts. These loops cause a specific dissociation that is unique to trance in that they lead to the modification of the energy that is needed for the awareness accompanying the processing of thought patterns in new ways. If enough energy is provided to dissociate awareness, and if the pace in which one thought recedes before the thought that follows it relatively slow, a trance is established. In other words, when our awareness is fed by energy that is sufficient into itself so as to become detached from the stream of consciousness and become dissociated, and when the thoughts that are embedded in this awareness fade slowly, leaving a wider place to those succeeding them, we are acting out of trance. An additional element that impacts trance is that of repetition and acceleration: When we execute a task in a routine and repetitive manner, we can get into trance, and the faster we execute the repetitive task, the easier trance will develop. Since trance uses less energy, higher cognitive functions (such as critical thought, decision making, judgment, or precise remembering) become deactivated. In trance the awareness of the body becomes constricted, and so is sensitivity to pain, but self-observation, on the other hand, becomes more acute.

† A variety of conditions can evoke limbic system slow-wave discharges that synchronize and dominate the frontal cortex (Mandell, 1980; Winkelman, 1986, 1992). Certain brain areas (the paleomammalian brain, the hypothalamus, and related areas) that regulate emotions and the balance in the autonomic nervous system, will be activated and will lead to parasympathetic domination of the brain. Parasympathetic dominance can be achieved either by intense activation of the autonomous nervous system until the sympathetic (arousal, fight–flight system) exertion leads to collapse and shift into a parasympathetic mode, or that mode can be more directly accessed through withdrawal, relaxation, and internal focus. Shamanistic and other healing methods use the body's homeostatic balancing acts for its healing potential.

and inexplicable manner in which people on such occasions are launched into a new and intensified level of awareness arouses panic and disorientation. A person whose sense of self has been detonated in this manner will hold on, as in posthypnotic suggestion, to the first interpretation that will cross his way to explain the enormous experience he just went through. If he will be told that the great ecstasy he felt is the Holy Spirit that entered his body and mind, he will believe it. If he is told that the feelings of detachment and distance he is experiencing represent a state of "cosmic union with the All," he will believe it, too. Following the moment of snapping, the brain, whose lucid perception of what has been done to it has been taken away, continues to execute its natural meaning-making activities in its perpetual, wired-in striving to arrive at a modicum of inner organization. In fact, in this state of search and reorganization, the brain is capable of astounding feats of the imagination, write Conway and Siegelman (1995), such as barking like a seal, believing that one is a chicken, or sliding into a variety of regressive states. If the person does not get help in refuting these beliefs, his state may solidify into a (new) personality structure.

In his research on people joining cults, psychoanalyst Stanley Cath (1982) describes physiological mechanisms responsible for changes in cognitive organization and in perceptions of the external world that are designed to diminish anxiety and uncertainty, but that can lead to a collapse in ordinary brain functioning that often eventuates in a mystical state. Cath listened to the narratives of cult survivors and found that the changes they underwent during their cult life (he regards these changes as a shift into "primary, archaic narcissistic states") were almost all produced by "technologies of experience," a blend of Eastern, New Age, consciousness-raising technologies (meditation, visualization, yoga, chanting, body-touch) that have been discovered in the 1960s and '70s and that function as double-edged swords, destroying the mind or enhancing it.

PSYCHOLOGICAL FACTORS

Brain overload is not the only thing that changes awareness. As mentioned, psychological-transferential factors, too, carry momentous importance in processes of "extreme influence." Important in this connection is the finding that when sufficient pain is inflicted on a person, the victim ends by feeling sympathy and affection toward the perpetrator (i.e., Stockholm syndrome*).

---

* The Stockholm syndrome was named after a bank robbery in Stockholm in which a small group of hostages became emotionally attached to their captors and even later defended them. Patty Hearst and, recently, Jaycee Dugard are names of girls abducted, held captive, and thought raped, but who did not try to escape (Dugard) and cooperated with the group's armed robbery (Hearst). "When a person is held in a captive situation and threatened with extreme violence, torture, or death, the immediate instinct is self-preservation. When the captor shows any signs of kindness to the victim, the hostage immediately senses that they can preserve their life by pleasing their captor."

Certain aspects of childrearing bear analogy to these processes. Leonard Shengold (1991) called "soul murder" the "attempt to erase or spoil the separate identity of another person." Sexual abuse, emotional deprivation, and mental torture can end in soul murder. Ordinary parental cruelty or neglect has less far-reaching consequences, but Ferenczi (1932/1955), in his classic "Confusion of Tongues between Adults and the Child" describes how, when a child attains a certain level of anxiety, he becomes submissive to the will of the attacking adult, and will from then on try to guess each of the adult's desires and strive to gratify them. In this situation, the child forgets itself and identifies with the aggressor, at the same time as it succeeds in maintaining the fiction of being loved. The relevant point here is that the abused child's sense of reality is heavily compromised, since the reality of the abusive adult becomes the reality of the child.

The power of affects in all this is central. Intense feelings are not only amplifiers for everything that happens to a person; they are also calls for action. The emotions that are produced under conditions of mind control range from joy and inspiration to anxiety, confusion, fear, guilt, hate, humiliation, and alienation. These feelings will motivate the individual to seek relief from his past or present life stresses, or escape from a sense of emptiness and internal persecutory objects. The fear and the feeling of entrapment and isolation the mind-controlled person feels can create powerful bonds between abductor and captive. The indoctrination can be intimate and passive, as in abduction; or in small groups; or it can take place on a large, public scale, such as role playing, psychodrama, guided imagery and a creation of an intense collective emotional state. Mind control often works through what Ronald D. Laing (1967) called "the politics of experience," which exploit the right that one person assumes to validate or invalidate the other's experience. Laing saw the family, and by extension other social frameworks, as arenas of *the struggle to control behavior by defining experience*. Laing left us penetrating accounts of how he who gets to define experience gets to control it, and the subtle ways this is done in sick families.

There are parallels between totalitarian methods and certain (but not all) models of psychoanalytic training. The infantilization of "candidates," the mistrust toward their independent thinking and competence, the mystification that surrounds senior (training) analysts who occupy positions of power (training analysts were, and in some places still are, elected by votes cast by a small group of older, privileged training analysts in the institute, who promote members who tend to think like them and leave out those who disagree), the intolerance for dissident opinions and their categorization as "nonanalytic" and even psychopathological—all these phenomena have controlled, and still, in part control, the mind of aspiring analysts in training (cf. Reeder, 2004; Kernberg, 1993). Fortunately, psychoanalysis is increasingly liberating itself from these practices, and in this it increases its chances of survival.

GASLIGHTING

Psychoanalysis coined various terms that signify mind-controlling processes on an intimate level, whether analytic or domestic. Terms such as *gaslighting*, *soul murder*, and *driving the other person crazy* have been used to describe procedures of unacknowledged or dissociated domination and psychic exploitation within the imprisonment of the closed quarters of the family or the analytic situation. Gaslighting is a term used by Calef and Weinshel (1981) following the famous film *Gaslight** to describe the process where one person plots to undermine the sanity of an intimate other. They describe the confusion of the victim, who struggles with the feeling that her mind plays tricks on her and that she is losing her sanity. Her anguished self-doubt grows without her suspecting that her husband is deceiving her by deviously staging these distortions. The concept of the effort to drive the other crazy, was developed by analysts who worked with schizophrenics, such as Silvano Arieti (1955), Harold Searles (1959), and Ronald D. Laing (1967). Arieti talks about "externalized psychosis" of people who specialize in creating situations that produce or accelerate psychosis in the other while they themselves remain immune to overt symptoms. Couples therapy works with a solid awareness of phenomena of induced craziness in one or both partners. In such situations the psychotic person may look normal and healthy while his partner will appear confused and mentally ill. Searles describes how the surfacing of a delusion may be a result of the attempt to find relief from the anguish that accompanies states of uncertainty and confusion caused by a malevolent other.

Leon Shengold borrowed the term soul murder from Freud's Schreber case† to describe extreme parental abuse. Erasing the other, writes Shengold, is done via emotional, mental, or physical torture; sexual exploitation; and varieties of parental cruelty and neglect. The massive isolation of affect that is caused by the abuse leads to confusion, denial, identification with the aggressor, or "hypnotic living death." The end result, according to Shengold, is rage, paranoia, and masochistic submission to a cruel morality, fueled by a doglike dedication to the abusive parent. Kaspar Hauser,

---

* Patrick Hamilton's (1939) play, *Angel Street*, was later adapted into *Gaslight* (1944), a movie that shows how a young, newlywed woman is driven to the brink of madness by the machinations of her criminal husband, who attempts to make her doubt her hold on reality in order to commit her to an institution. One of the ways in which he was able to shake his wife's confidence in her own perceptions was to alter the brightness of the gaslights in the house, hence the title of the movie.

† Shengold attributed the origins of the concept to the Nordic dramatists Henrik Ibsen and August Strindberg who wrote about "the attempt to erase or spoil the separate identity of another person." Strindberg, more vociferous than Ibsen, was intensely preoccupied with mental hostility and the battle of minds that happens in the paradigmatic situation of two people who hate each other and strive to drive the other to doom.

the hero of Werner Herzog's eponymous movie,* showed none of these violent emotions. Kaspar Hauser was a soul-murdered, 19th-century German young man who appeared one day in the streets of Nuremberg after having been kept in chains in a cave where he had lived seemingly his whole life with no human contact† and no access to the outer world.‡ This hauntingly beautiful movie, the narrative of a totally isolated youth who was "unspeakably abused" (von Feuerbach & Daumer, 1995) by having his childhood robbed, and his mind and thus his human spark nearly extinguished, is suffused with silent horror and unspeakable grief.§ Kaspar Hauser was kept alive for years under extremely controlled conditions; he grew up to be totally defenseless, bereft of any protective skin over his raw and dangerously exposed psyche; his physical murder was bound to follow his soul murder.

Thus, murder need not be of the body only. In *Discipline and Punish: The Birth of the Prison*, Foucault (1995) writes about the historical shift from the body to the so-called modern soul as the site of control and violence. Foucault sees the invention of the modern "soul" as coincident with the discovery of the prison. Incarceration made it possible to supervise and investigate the person by treating his soul, rather than inflicting on him a physically painful ordeal or killing him. After all, acting on the soul offered many more possibilities of control than the body, limited as the body is in its givenness to pain and torture. Replacing physical torture (and death) as the punishment for those committing regicide, "penalty" became a representation of public morality, and imprisonment the general form of control. Foucault further elaborates how the invention of the prison effects not only the deprivation of liberty, but also the *technical transformation of individuals*. Incarceration involves isolation, grouping and classification, categorization and inspection, and, we may add, arousal and inner emotional pressure with no exit. These new modalities of control, the negative "technologies of care," involve constant, uninterrupted surveillance that is exercised according to a strict codification that partitions time and space. An important point here is that, as with totalitarian reeducation programs, *discipline mostly does not need brute force; it rather operates by a calculated gaze and by "exercise"* (timetables, drills, and other prison rituals). Foucault's descriptions of the organized sadism and the indirect, regimented violence endemic to the incarceration system,

---

* The case of Kaspar Hauser became a paradigmatic and symbolic backcloth for a massive amount of literature, poetry, film, and other art forms over the years.
† It is said that his anonymous captor put opium in his water and while Kaspar slept, he changed his clothes and his pot, cut his hair and nails, and put out fresh food for him.
‡ Feuerbach, Hauser's legal guardian, who wrote a book about his enigmatic case, calls it "crimes against the soul."
§ The grief and dread that overwhelmed me on first seeing the movie eroded my thresholds, which were eroded to begin with due to the analytic phase I was in during that period. I had to leave in the middle; I could see it to the end only the second time.

brings out the affinity with thought reform regimens. Even Foucault's three elements upon which success of disciplinary power depends apply to the mind-control procedures: hierarchical observation, normalizing judgment, and examination. Power does not need to physically attack, or even exclude or repress; instead, power creates the reality and rituals of truth. Kafka's *In the Penal Colony*, too, is a parable about the operators and victims of a torture machine, contacting a person's inner sense of law, guilt, and retribution.

TWO TYPES OF PROCESSES

It seems to me that the literature on varieties of mind control calls for distinguishing two processes of coercive and indirect influence. In the first, love is proffered that later turns into abuse, coercion, and humiliation, whereas in the second, the prisoner/indoctrinator relates to the prisoner with cruelty, then changes his face to become kind, friendly, and warm. In the first type of engineered abuse the redemption seeker (who joins a cult) is lured by the grace, warmth, and gentleness of his new acquaintances. The coercive demands are revealed only gradually, at a point when the follower has already been incrementally caught in the commands of the group. In many ways, this sequence resembles a drug changing its function from positive to negative, from a substance that procures extreme well-being and happiness to the thing that is desperately needed as a salvation from pain, despair, and collapse. In contrast to this prototypical process of becoming addicted, there is an apparently opposite, but basically similar process, in which what comes first is not love, but hate—harsh practices of attack and imprisonment, and mental or physical kidnapping that impose discipline and utter obeisance. Following the victim's "softening" by these attacks, he is granted reprieve and acceptance by the group where he now feels he belongs when he begins to comply. He is hence rewarded in direct proportion to his self-denunciation and immersion in what is dictated to him. None of the coercive-persuasive environments and procedures is a pure type, but the second type of process, hate turning into "loving kindness," seems to be more perverse and chilling than the process where gentleness and seductiveness change into oppression and torture. It seems to me that "love" following persecution and causing of pain makes it possible to gain a deeper glimpse into the thirsts of the soul and into human masochism. Dispensing tenderness, warmth, and friendliness after periods of creating terror, hate, guilt, and shame is a powerful tool for changing a person's character into an obedient, automatic, centerless creature that is extraordinarily pliable and open to receive the pouring into him of contents from the outside. Let me elaborate these two kinds of processes.

"LOVE" THAT TRANSMUTES INTO OPPRESSION AND COERCION (HATE)

The "soft" induction into cults recalls the fairy tales of the witch who poses as a loving mother who suddenly unmasks her monstrous face. This is the luring, pseudobenevolent object that transmutes into the persecutory one.

People who become cult recruits are sedated into euphoric states that encourage the relinquishing of their independent minds. The sense of warm communal solidarity frames the peaceful teachings of the cult leader. The initiates are enthralled by the mind-numbing rituals. Cult followers describe how impressed they are with the cult recruiters, who exude self-confidence and enlightened love, filling up the room with their glowing presence. This first galvanizing encounter is followed by a "love-in," or "love-bombing," a carefully crafted lavishing of nurturance on the potential followers. The leader's love for his followers is incessantly touted, and they are encouraged to visualize it at all times. In this phase, the true purpose of the cult is still concealed under a veneer of social ideals or high religiosity (Jim Jones' anti-racism, Shoko Osahara's Buddhist values, or the Spiritual Inner Awareness of John Rogers are examples). Ideological allegiance is promoted, serving to bond and solidify reciprocal identifications and helping dissolve the bonds to the former self and to its objects.

There is an orthogonal relationship between the power of techniques of "extreme influence" (Amitrani & Di Marzio, 2001) and the intensity of the individual's neediness. According to Cath (1982), many cult adherents have a "black hole," made of the absence of a protective and idealizable paternal presence. Lacking such an internalized object, the person cannot build a cohesive self to initiate adult life. When the time comes for an intimate commitment to a partner, or for anchoring oneself in a long-term pursuit, the anxiety, depression, or the inner void can be calmed by the "love" and inner peace and spiritual strength the group leader proffers. Such love is less personal and therefore less threatening than intimate love at close quarters that would be developmentally appropriate. This "impersonal intimacy" allows, writes Cath, the illusion of having attained safe separation-individuation from the parental figure. It also offers protection from a murderous superego, which is replaced with an external guide who disciplines and lays down rules, but who also "loves" the followers, at least at the beginning.* As long as the followers agree to bypass their individual lives, the leader will contain their inner turmoil, calm their ambivalences, and be available for what is called "filling up the soul."

Heinz Kohut writes that when the parent is unempathic, an idealized but flawed parental imago will be internalized instead of the representation of

---

* This psychic constellation is similar to the concept of "father regression" (cf. Stein, 2006, 2009).

the abilities and skills of the child itself. A glorified, omnipotent parental imago, consisting of the magical expectations based on the child's perceptions of his idealized parents, and a self-representation as impotent and helpless, reflecting the expectations of a tiny child in the face of an unempathic parent, become frozen and cannot integrate with later, more favorable self-representations. Peter Fonagy writes that when there is an internalization of a parent with frightening intentions, there is unbearable pain of an unmodulated "badness" of the self. When this badness is experienced in the mode of "psychic equivalence" (a mode where an inner event and an external happening are perceived as equal, due to an inability to distinguish the two), the sense of badness gets directly translated into "real badness" from which, in extreme cases and in a teleological functioning mode, one can escape only through suicide. All these descriptions emphasize the enormous damage done to the sense of self (not only to the brain) when the person is assaulted by an overwhelming, psychically violent other.

One of the central ideas in psychoanalysis—Jacobson, Winnicott, Fonagy and others elaborated it—is that internalizing the other's representation *before* the boundaries of the self are consolidated undermines and damages a coherent sense of self. In such cases, *the infant is forced to internalize the other not as an internal object, but as a core part of the self.* In other words, if the parent does not connect with the infant's anxieties and does not "digest" and process them and thus creates a mirroring analogue to the infant's self-state, the child (self) will have to adapt to the parent (object), that is, to an alienated presence in its own psyche as if it were itself.* As we shall see in thought reform procedures as well, the process of the colonization of the self by the other is effected through its coerced identification with alien representations of that other.

How is this done in cults? After the initial cementing into the cult comes the indoctrination process. Now things change, either abruptly or, as is more often the case, gradually. Successive stages of harsher work and living conditions and increasingly sinister prophesies of what is to come in the world, and what the new adherents have to do to save themselves and prove worthy, causes the selves of the followers to be incrementally usurped, imprisoned, paralyzed, and enslaved. In extreme cases, they are eliminated or used to eliminate others. I described earlier how sensory overstimulation and physical deprivation create brain perturbations, while at the same time the person's attention and reasoning powers are further narrowed by a series of arcane intellectual pursuits, such as "seminars" or "classes." These "learning sessions" are not real learning sessions, but mind-narrowing, trance-inducing practices, often interspersed with hypnotizing demonstrations of the leader's knowing, healing, and predictive powers. The followers now feel joyful and more

---

* See Leslie Sohn's (1985) "identificate."

alive than ever before. The enormous relief of having their uncertain, fragile, individual ego ideal taken over by the leader fuels a sense of renewal, happiness, even bliss, and a flurry of rebirth fantasies. Ties to family and former friends are cut, and autonomous thinking and planning for the future are relinquished. Human ties are sacrificed for the joy of refinding the powerful parent, who dispenses unconditional love and promises salvation. The participants become increasingly bonded to such a figure, while working harder and harder not to disappoint him and not to be disillusioned themselves.

Dan Shaw (2003), describing people who are, in Singer and Lalich's (2003) words, "naïve, unaffiliated, trusting, and altruistic" (or idealistic), recounts how cult followers "are told to surrender their small and selfish ego, to be magically transformed into pure awareness of the transcendent 'Inner Self,' which is one with the guru and with God" (Shaw, 2003, p. 126). Under these "protected" circumstances, inductees do indeed feel better. Having simplified their lives, they sense a greater harmony within this narrowed world, unaware that the cage has snapped closed. Gradually, however, they become subjected to increasing emotional exploitation and terrorization by the leader's harsh decrees and growing emotional abuse, which in turn are fed by their very compliance, in a circular movement of ever-increasing imprisonment. When the idealized parent figure turns into a sadistic tyrant, the follower comes face to face with his own persecutory internalized objects, for the sake of which self-sacrifice is regarded as a privilege. "But Father reminded me that punishments were deserved ... sacrifice was the role of the chosen few," writes Deborah Layton in *Seductive Poison* (1999, p. 61).

THE PEOPLE'S TEMPLE

Layton (1999) describes the love and the intense, focused attention she received in her first encounter with "Father," Rev. Jim Jones, the leader of the People's Temple. She, like the others, became instantly mesmerized by the concern, devotion, and the penetrating personal knowledge of her he seemed to have.* Like the others around her, she felt deeply grateful to him. From then on, she traces her gradual shifting from the joyous amazement at her finding this person and a new and better life, to her painfully dawning realization of her spiritual "father's" insidiously developing paranoid

---

* Charles Lindholm (1993), in his study of Charisma, notes that the leader is recognized as charismatic only by those whose needs he addresses and whose values he shares. He also notes that he is seen by his followers as possessing the sharpest vision into human affairs. We can link these two observations by realizing that the leader's extraordinary vision accrues from his attunement to his followers.

harshness, disingenuous ruses, and cruel domination, which ended only with the willing death of the whole community she left behind.

Lifton (1961) describes a process that is structurally similar in most cults. As soon as the powerful initial seduction and initiation into the cult had their effects, the relentless nipping in the bud of any manifestation of confidence or self-esteem in the followers began. One way to increase blind faith is to systematically crush the believers' sense that they are good, valuable, effective, and basically "right" in their perceptions. Diminishing them to a small size ensures the adherents' humility and receptivity to the leader's image as superior and infallible, and helps in recasting the horror that is happening as something inevitable, even benign, since it is issued from the leader's superior understanding. In the People's Temple, for instance, the regularly held public rituals of minutely staged humiliation and degradation were followed by the leader's private apologies and "loving" (often sexual) gestures toward those he had publicly humiliated just hours earlier. Layton (1999) describes being publicly mortified by being accused of acts she had not committed (such as allegedly threatening Father with suicide to have sex with her; in reality she had been raped by him). Following such a scene, Father would come into her cabin, apologize to her, stroke her hair, and explain to her that he had to punish her to show the congregation that he did not favor her. Deborah felt deep gratitude for the time and the love Father gave her and for his effort to make her strong. After these rituals, Deborah becomes even more deeply entrenched in the life of the cult. She is now "doublethinking," to use George Orwell's (1949) term, knowing that from now on she has to lie to her family and fabricate another life to keep her biological father from knowing the awful truth, from which she cannot (at least at this stage—she would courageously escape later) extricate herself anymore.

The splitting required for the cultic mode of existence involves a double renaming of both the treacherous attacks of the leader on his followers and the causes of such attacks. The leader's attacks and control of his followers through his projections of his repudiated parts into them are reframed as profound acts of love and caring correction. At the same time, the leader's abusive behaviors are rephrased as his reactions to his disappointments in his followers (rather than the unspeakable and unthinkable disillusionment he causes them). On the positive side, the follower is motivated by the promises he is given of attaining control over what is ordinarily possible. Not only submission that seeks love, but power plays a role here. When submitting to the figure of the father, the follower is endowed with the possibility of being a hero and of fulfilling a triumphant role, which had not been attainable for him in the period of weakness and identity-diffusion preceding his joining the cult.

Often the sense of deep connection the cult followers have with their leader results from his *intimate engagement* with each individual. This makes them interpret the connectedness, which in reality is designed to harm and exploit, as benevolent and caring. The closeness, sometimes achieved by the tight monitoring alone, makes for enhanced suggestibility that makes the followers receptive to the leader's projections of his contemptible parts and unaware of his "sucking" the loving, "selfless" parts from them and donning them himself. The leader and his followers now both perceive the repudiated "bad" parts as residing in the followers, whereas all the goodness is in the leader. They increasingly need the leader's "loving correction" to infuse them with goodness and rightness. Since the leader pretends to love his followers, his hatred posed as love is a full-blown perversion, a false love. Compounding this perversion of emotional truth is the insidious gradualness with which the punishments are engineered. The step-by-step, incremental measures prevent most followers from realizing the deceptions, the malevolence, and the ignominy of being held evil and guilty. At the same time, the very harshness of the punishments very often seems to the followers to point to some higher inscrutable justice. When a follower is made to climb up the hierarchy, she or he is made to inflict abuse on those underneath her or him. When Layton (1999) was later admitted to the 'inner circle' of the privileged, her promotion was accompanied by threats and another turn of the screw on her prison door ("Father made me watch the beatings and had my photo taken holding the rubber hose, which paralyzed my questioning inner voice" [p. 61]). Lifton (1961), too, writes about the self-betrayal mind-controlled people are made to undergo, as we shall see when we turn to the second emotional sequence in mind control.

HOSTILITY THAT TURNS INTO "LOVE"

The idea of harshness followed by "kindness" or "love" was developed by William Sargant in the 1950s in his studies of John Wesley, the charismatic 18th-century preacher who was the main figure to usher in the great Methodist revival in England. Sargant (1957) tells us that Wesley found a powerful method for acquiring control over his audience. In his sermons, Wesley would luxuriate in detailed descriptions of the fire and brimstone of hell that awaited the sinners, pitching up his listeners' fear and terror into a frenzy. His ingeniousness in evoking his audience's terror was matched only by the exquisite skill with which he offered them the means of salvation from it. Wesley's sweet, glowing portrayals of God's mercy that came in the wake of his descriptions of ordeals of torture and annihilation swiftly transformed whole terrified congregations into communities of fierce zealots. England's soul was saved.

Frightening and threatening a person to the point when she is at her wits' end with fear and anxiety, and then introducing a loving, caring, sheltering presence is a powerful technique that induces a radical change where the old personality is reconfigured into a new, docile psychic combination. Such transformation bears unmistakable affinity with so-called psychotic insight. It is the experience of disparate pieces falling into place, of everything suddenly making perfect sense, and it is similar to the production of paranoid delusions that are nonnegotiable and incorrigible. When, in the midst of the most vicious pressures put upon an individual (a cult follower, a congregant, or a religious adherent), her indoctrinator suddenly changes his demeanor to amiability and caring, and he creates a gratitude and an enthrallment that opens and draws the captive person into a powerful bond of receptiveness and loyalty. Somerset Maugham said about the Jesuits' "Spiritual Exercises": "Considering that their effect has been achieved through a constant and ruthless appeal to terror and shame, it is surprising to observe that the last contemplation of all is a contemplation of love" (Sargant, 1957, p. 168).

Like Wesleyan and other religious brainwashing routines, mind control points to a new path to salvation after strong negative emotions have been aroused and amplified in the captives, who are now grateful and eager to do everything they are being told. Lifton's (1961) interviews of people who underwent the thought reform programs in the Chinese Cultural Revolution in the 1950s and 1960s is a compelling study of the psychodynamic and transferential issues involved in this conceit. These programs exploited the effects of systematically administered guilt/anxiety/shame in the context of strong emotional arousal and inescapably shut settings—prisons for the Westerners, "revolutionary universities" for Chinese intellectuals—where people were both forcibly confined and exposed to massive social and psychological pressures (Singer & Ofshe, 1990). After a long process, when the Communist gospel was accepted, "love" and kindness replaced fear. The friendly gestures of the captors and instructors, their tokens of trust and respect, created havoc in the minds of the prisoners. The unexpected and vast gap between the harsh routine and the sudden leniency deeply shook the prisoners whose gratitude and happy servitude now grew exponentially. Thought reform was a psychologically violent and intimate process; the captors were interested not only in the behavior of the prisoners, but also in their innermost motives; they probed into the psychological weaknesses of their subjects with keen observations and great detail, all the while keeping close contact with them. In fact, as Lifton notes, many of the prisoners eventually came to regard thought reform as an ennobling, quasireligious experience, and felt genuinely reborn along with their society; they even experienced a sense of loss when the process was over or when they were expelled from the People's Republic of China.

I chose to focus here on Lifton's work on thought reform not only because it is a well-documented psychological study of the experiences of individuals undergoing these processes,* but also because of the particularly complex object-relational aspects of this human experiment. When the Culture Revolution in China began, Chinese and Western intellectuals living in China were arrested without further notice. Such a person was repeatedly told that he was not the professional he took himself to be, and that everything he thought he was was only a cover and subterfuge for who he really was. Initially, the individual would struggle with himself and with others against the assault on his self-respect and sense of self, but any attempt on his part to reassert his identity was considered arrogant and dishonest and a symptom of his error, and led to renewed pressures on him to capitulate. The incessant denial of personal and professional identity sapped at the person's strength, and his sense of self became amorphous and weak. The external attacks, accompanied by sleep deprivation and harsh conditions, led many people to inner submission. They began to exist in a twilight state between no sleep and no wakefulness, in which they were increasingly susceptible to influence from an ostensibly just and flawless outside and more vulnerable to destructive and self-destructive impulses from within themselves. In parallel to these proceedings, his prisoners persistently researched his psychology and his most personal history. The intimate knowledge they obtained in this way enabled them to access the prisoner's most intensive guilt and shame reservoirs. Gradually and quite unavoidably, the prisoner came to experience himself as evil.

On a deeper level than gratitude for the reprieve offered the prisoner by his captor, are the psychodynamics of the sense of self. One of the most shocking phenomena in the paradigm of hate that turns into "love" is the sequence in which a person is made to believe that he is evil, to be subsequently redeemed from his evil self by extreme means leading to an imposed version of reality, which he is made to perceive as truth. As in Wesley's sermons, the thought reform programs functioned by systematically instilling and amplifying an experience of being evil. The relief from this experience was offered by conversion into ideological fanaticism. The means toward obtaining such relief was the *confession*, in particular, the confession of having committed terrible, vague crimes, which the individual was expected to recognize and acknowledge. Confession played a most crucial role in the indoctrination process, since to confess is to substantiate and to call into existence a hateful, regretful self-condemnation. The pressure to confess, aided by the relentless stream of accusations the cellmates were charged with heaping on the prisoner, was so excruciating as to make him invent

---

* This axial work serves as a target for the criticism leveled by "new religious groups" and other so-called advocates of religious freedom who seek to deny the coercive character of cults and the whole concept of mind control as a delusion. These people, among them members of the APA and other social scientists, occasionally denounce the book, but more often use Lifton's carefully honed qualifications as weapons in their crusade against the "anti-cult" movement, who, on their part, use the book as major evidence.

whatever he thought his captors wanted to hear about himself and others. The prisoner now *wanted* to say whatever he was expected to say—not only in order to forestall further attacks—but *to win the captors' approval and sympathy*. This *self-betrayal* went beyond the betrayal of friends and family which the prisoner was pressured to denounce, since *the more of one's self one was led to betray, the more deeply one became involved with one's captors*. Descending ever deeper, being embedded in a matrix of hostility, emotional isolation, and hopelessness, the prisoner would often spin into dissociative trance states that alternated with primitive anxieties. The dissociation was an attempt to resolve the horrific contradiction between his history and beliefs, and the—at this stage still mystifying—"truths" he was coerced to believe in. When dissociation did not help, annihilation anxieties, suicidal thoughts, and psychotic hallucinations emerged.

On the brink of the point beyond which psychic damage becomes irreversible, the official attitude of his "judges" abruptly changed into an unexpected show of kindness and concern. This about-face notably does not imply any lessening of the demands made on him; the intention of the surveillant is to relax the pressures sufficiently for him to better adapt to the demands and expectations. This so-called leniency (Lifton sees it as the "rebirth" after the "dying" incurred during the assaults phase) has the powerful effect of spurring the prisoner on to a deeper confession. Being offered rest, acceptance, kindness, a renewed identity, and a release from annihilation anxieties, the prisoner *becomes motivated to help the officials achieve their goal*. He now becomes their grateful partner in his own reform. His diffuse sense of guilt becomes specific and detailed, and he learns to see personal evil and destructiveness in his past actions that are now reinterpreted to resonate with the beliefs and inner objects in him that parallel the environment's condemnations of himself. When he accepts the "higher" morality of the group, the harshest judgments join his conscience's most tyrannical parts. Significantly, the mining of the archaic superego by the captors turns a plainly guilty person into a person burdened with total, cosmic guilt, who learns to assess his life up to that point as a series of malicious and shameful acts. The thought reformed person now assumes a so-called negative identity that incorporates his maximal potential for disgrace, the worst image of the kind we each have of ourselves that we do not want to be materialized under any circumstance, the very script that must never come true if we don't want to lose our soul. This is now what these prisoners identify themselves with. Such "negative identity" is assiduously nourished by the reformers and it grows and suffocates the prisoner with ever deeper and more menacing guilt and shame. Psychoanalytically speaking, *mind control is a matter not only of suppressing one's autonomy and rebelliousness, but of tapping into one's inner persecutory objects*, feeding the beasts until they grow and become murderous. Coercion and mind control are not just negative processes of oppression and domination, but positive ones of enhancing one's self-negating parts. The "redemption" and "rebirth," the conversion into religious zealotry or totalitarian enlightenment, a purge from all "bourgeois lies" has all the more power by functioning as a relief from continuous distress. The sense of inner evil can be transcended in the ecstatic experience of being purified and pronounced good.

SELF-LOATHING

From my clinical experience I learned about the toxicity of self-loathing. Whether conscious or unconscious, the sense of self as malicious and unworthy is often the sickest element in one's mental economy that can lead to the most extreme and alienating defenses. The sense of the self as bad, worthless, or repulsive is an attack on one's core and is thus a most dangerous element in a person's psychic economy. Self-loathing is chronic, amplified shame. "Shame is the affect of indignity, of defeat, of transgression, of inferiority, and of alienation. No other affect is closer to the experienced self," writes Gershen Kaufman (1988, p. xxi). The capacity to experience shame is a precious developmental achievement, an acquisition of a sensitive marker of boundaries, warning us against behaviors that will diminish, mortify, or endanger us psychologically. The mature, self-protective version of signal-shame is the dialectical opposite and a good version of the searing, internalized version of traumatic shame that accrues from being found undeserving at one's core. Internalized shame, in contrast to signal-shame, violates our essential dignity as human beings. Self-disgust and self-hatred are often deeply hidden in the recesses of the psyche. It usually takes a long analytic process to uncover these festering emotional judgments against oneself from underneath symptoms, layers of defenses, and various provocations and challenges posed to the analyst. In cases of mind control, the external accusations that one is a criminal fold into interior preexisting feelings of self-hatred and sinfulness. Lifton emphasizes the high value the captors attached to guilt, and their use of guilt feelings as "preferred forms of communication, modes of public competition, and the basis for eventual bonds between the individual and his totalist accusers." I would add shame to guilt, since self-loathing is a proliferation of shame. The mining and cultivation of these affects may have been the most creative insight and grand achievement of the thought reformers. After all, reeducation would be an abstract and quite ineffectual process had the prisoners not been subjected to emotional sloughing of the self. One's exposure and amplification of negative emotions toward the self was aided by the work of the state panopticon searching for degrees of authenticity of these feelings. The surveillance was so thorough that it could detect if prisoners tried to simulate guilt and shame without really experiencing them. The prisoner soon found out that it was highly unsafe to detach inwardly to protect himself and fake these feelings; he was more secure by letting the feelings penetrate his psyche and genuinely feel them.

Put succinctly, the message of the thought reform was that self-abhorrence (which it had manufactured) can be corrected. It could be redeemed by the repentance and self-improvement that come from industriously applying oneself to the doctrine. Grotesquely mirroring the psychoanalytic process, the prisoner was made to analyze the causes of his deficiencies and work through his resistances, until his thinking and feeling complied in terms of the presented doctrinal truths. Having achieved this, he now felt in harmony with surroundings that were not experienced as strange anymore. Surrendering to an all-powerful force in whose might he basked and whose moral righteousness he shared, the prisoner now had the satisfactory feeling of solving all his puzzles and problems, and finally becoming intimate with his peers. At this stage, he is identified as a "progressive" and is permitted to express himself more directly. He achieves a more intimate communication with his reformers and his experience feels more real compared to the feelings of disorientation and derealization that had plagued him in the early stages of this process. Furthermore, at this stage the *prisoner experiences many of his responses as personal discoveries*. He has attained what O'Brien says in Orwell's *1984*: "We are not content with negative obedience … When finally you surrender to me, it must be of your own free will" (1949, p. 76).

Now *the link between the adoption of a negative identity and the learning of the Communist doctrine* deepens, since one depends on the other and both are constantly reinforced through their endless repetition. The prisoner studies the doctrine so as to analyze the reasons for his weakness, work through his resistances, and reach the point when he thinks and feels in terms of a doctrinal truth toward which his whole life converges. The effects of this program were not uniform across people of course. Some of the more intelligent prisoners had the chilling realization that what was important was not the Communist doctrine itself, but *the use of the doctrine and the arguments at its basis for the sake of extending the self-exposure of the prisoners*. In effect, the doctrine functioned here as an explanatory theory for the weaknesses, the guilt, and the truth of inner evil of the prisoner. The doctrine became the tool for the broadening of guilt to the point of an all-encompassing self-condemnation that extended to every aspect of experience.

It can be said that the goal of this multiple-stage process was to *deliver external accusations, and then turn them into self-accusations*. The effort was focused on getting the person to make his confessions out of inner conviction. Getting him to become increasingly involved in his confession procedure made it more and more a matter of his own responsibility. Laplanche (1999), addressing forms in which the other (and Otherness) gets into the self, distinguishes between the everyday, developmentally benign form, called *implantation*, and its coercive variety, *intromission*, a process whereby one is made to host within oneself an alien presence. In contrast to implantation, the transmission from the other/adult to the self/child of messages and signifiers, intromission is done in violent, coercive ways that do not allow translation, repression, or other normal metabolizing, processing, regulating decisions to be made by the receiver. Intromission entails the inescapability Singer and Ofshe (1980) talk about, the oppressive lack of freedom to "metabolize" autonomously what is happening and what one is made to receive. It is the insinuating of the outside into one's body and psyche. The Cultural Revolution made the internalization of self-condemnation into an elaborate art and a craft.

Ronald Fairbairn (1952) believed that every internalization is a measure of coercion that occurs only when the object is tantalizing or withholding, not when it is good and gratifying. When the frustration and misery the object dishes out exceed a certain threshold, the infant "creates" an inner world that mimics and reconstructs the original situation where a person is bonded with bad objects, and hence feels constantly starving, angry, guilty, and deeply anxious.* We (paradoxically) internalize bad objects, writes Fairbairn, because we cannot resign ourselves to their badness and accept it, and thereby differentiate ourselves from them. Thus, we withdraw (sometimes schizoidically) from the people in the external world, but we strive to overcome our helplessness in relation to them by struggling to become their masters, to possess them and to force them to change into good objects in our inner world. *The internalization of the bad object is a key concept in schizoid disturbances and in mind control.* The manifest accounts of mind control downplay the subtler, less conscious nuances of the struggle to be loved that is implicated in the internalization processes. But *the abandonment of the self that is effected in mind control is in many senses equivalent to clinging to the bad object and internalizing it.* Victims of mind control, similarly to soul-murdered children or cult adherents, have their brains washed with emotions, at the same time as they internalize the will of the torturer/perpetrator and identify it with their self. Instead of rejection or defense against an inimical foreign body, there is internalization and identification with it.

MIND CONTROL AND PSYCHOTHERAPY

If we compare processes of mind control to those of psychotherapy, we see a basic, generic pattern that is common to both procedures: (a) seeking to effect a change (b) through the creation of an intense experience (c) that leads to a possible change in the self, whether by subverting and weakening it, or by encouraging it to expand and grow stronger.

---

* And then we are given to the constant lure of seeking momentary relief by its projection back into the external world.

The intentions, the goals, and the effects of the two processes are very different, however. The arousal of strong affect, and often the use of altered states of consciousness,* to get access to the person's inner recesses is a proven way to effect change. At the same time, the primary intersubjectivity and reciprocity that exists in analytic therapy is deliberately and carefully excluded from mind control as well as from authoritative brands of psychoanalysis. In malevolent or power-driven forms of "extreme influence" the "influencer" strives to be maximally impermeable. The effect of this impermeability is that the individual cannot influence the influencer, who is always right. Such an attitude promotes idealization of the analyst as "the one who is supposed to know" (Lacan, 1964) or who knows. One of the supervisors in my analytic training many years ago, advised me never to ask a patient questions, apparently to abstain from intruding on the patient with my, the analyst's, personal interests or trying in any way to influence the analysand, who should be given free reign with his associations. But actually, as I realized to myself, this instruction was a power tactic: when you ask someone a question, you need something from him. Even if you ask for a clarification, you want something from him, at the same time as you are exposing yourself to the other's unpredictability as to how he might choose to respond to you. The analyst asking the analysand a question disrupts some imaginary absolute neutrality; but by asking, the analyst also shows his lack of all-knowingness and perfection, that is, his weakness, and therefore he should not do it. The prisoner, or object to be influenced, shall likewise not be permitted to pose questions, since within a closed system, there is and should be no effective way for the subject to have an impact on the system. The system regards any change in the procedures, any questioning, as threatening its existence. In a system of closed logic, one deals with criticism or complaints by showing the subject that he is the one to be at fault, not the system. Furthermore, perceptions and understandings attained by the individual are distorted and tweaked until they receive a meaning that is the opposite of what he meant in the first place. The system causes the subject to adapt to it in a series of steps, each small enough for the subject not to notice the change that is taking place and not be aware of the goals of the program until a late stage of the process, if at all.

Institutional effects of psychoanalytic training, such as cultivating idealization of senior analysts and the candidates' masochistic tendencies, the transmission of intergenerational hatred in psychoanalytic institutes (see Reeder, 2004), the persecution of a so-called false self, the suspicious searching for pathology in candidates and colleagues; and in particular, the "superego complex" (Reeder, 2004), a hateful and cruel rather than a strong and critical superego, are aspects of totalitarian systems that used to

---

* There is no doubt in my mind that myriad forms of trancelike, hypnotic, and self-hypnotic states intersperse deep-reaching mutative psychotherapy. I hope to be able to show its existence in clinical reports of other writers and of mine in future writing.

characterize a great part of psychoanalytic institutes, and that are still not completely extinct from some quarters to this day.

The differences, however, between mind control and psychoanalysis are enormous and even symmetrical. A thought reform is essentially a technique for changing behavior that is designed to bring about learning and adoption of an ideology or an array of behaviors under conditions of environmental and interpersonal manipulation. Such conditions serve to extinguish certain behaviors and to cultivate others. Mind control, like psychotherapy, use conflicts, though conflicts emerge more organically and are not deliberately created in analytic therapy. Mind control uses conflicts in the service of transforming the subject's self-image into a negative one, or of intensifying an already existing negative self-image through the reinforcement and emphasis on the negative parts that exist in every self-structure. It cultivates and encourages negative affects toward the self and invents ways to justify the negative affects. It also blocks and negates alternative perspectives, which are so vital in any positive therapeutic process, since what heals in therapy is the working through of current experiences in the context of different and new perspectives. Alex Strachey spoke in the 1930s about the mutative value of a benevolent superego, and Roy Schafer elaborated in the 1970s the concept of a loving superego, which he contrasted with the harsh early superego the Kleinians so well described. Psychoanalysis valorizes empathy and compassion toward oneself; it encourages self-acceptance and self-understanding. Mentalization is presented as a process that makes it possible to reach the understanding that the self and the other have their own separate subjectivities that function through emotional processing to differentiate fantasy from reality, self and other. Promotion of the ability to think (about) what one feels stands in contrast to the processes of brainwashing whose aim is to undermine and spoil mentalization, and with it, destroying the possibility of finding one's psychological self through the mind of the other, the secure attachment figure.

The difference between these intentions and the enablement, in psychotherapy, of experiencing the analyst via one's individual lens is vast. In analytic treatment it is accepted that there is a difference between what the analyst offers and what the patient perceives, and the therapeutic process accepts, respects, examines, and interprets this difference. The analyst provides a safe and holding environment so that the analysand can make contact and observe his own subjective emotional contents as one way of representing reality. A safe environment makes it possible for the patient to externalize an inner experience into the analyst or into a transitional space, so that it can serve later for symbolization and self-enhancement. The analyst does not "know ahead of time," but goes after what the patient presents her with spontaneously. She is curious, receptive, and even vulnerable toward the analysand. In using her subjectivity and attuning to the patient's self-needs, the analyst makes it possible for the patient to create a kind of "inner fold," a private drawer, so to speak, where he can store, dream, feel, and create private self-experiences. All this gives the patient a sense and a hope that his innermost thoughts and feelings are sayable and thinkable within the domain of the human. In mind control, by contrast, an opposite process takes place. The subject is robbed of his benevolent and productive self-reflective capacity, and his inner, private life becomes suffused with anguish and a sense of inner badness. This anguish is manufactured and handled publicly, consolidated and fixated by violence and terror that subtend the coercive pressure on the prisoner to internalize the captor's notions about himself with the aid of the "inner prison guard," or "inner captor" (Fairbairn's "inner saboteur" or "anti-libidinal object"), and to *introject* and identify with the captor's ideas about himself. The sense of inner badness that every one of us is afraid of or dislikes or abhors is actively solicited and reinforced. Thus, instead of the metabolizing and processing of painful or terrifying contents into safe and even exciting thoughts, the expressions of self-affirmation are refuted, even vilified. At the same time, values from the outside are implanted violently to interweave with inner structures hostile to the self. Instead of therapeutic affect regulation, mind control seeks to create affect disregulation in order to promote the Fairbairnian internalization of the bad object. In the wake of the subversion of the confidence in self-compassion and self-acceptance, we have the pushing inside and sucking inward of unbearable contents and self-states. The inner and outer siege of these bad objects destroys all attempts to resist them. Instead of enlarging the belonging to the self, mind control is the transferential internalization of an outer conflict with the captors, a persecution whose aim is to turn the person against herself.

REFERENCES

Amitrani, A., & Di Marzio, R. (2001). Blind or just don't want to see? Brainwashing, mystification, and suspicion. *Cult and Society*, 1(1).

Arieti, S. (1955). *Interpretation of schizophrenia*. New York: Robert Brunner.

Calef, V., & Weinshel, E. M. (1981). Some clinical consequences of introjection: Gaslighting. *Psychoanalytic Quarterly*, 50, 44–66.

Cath, S. H. (1982). Adolescence and addiction to alternative belief systems. *Psychoanalytic Inquiry*, 2, 619–675.

Conway, F., & Siegelman, J. (1995). *Snapping: America's epidemy of sudden personality change*. New York: Stillpoint Press.

Fairbairn, W. D. (1952). *Psychoanalysis studies on the personality*. London: Taristocle.

Ferenczi, S. (1955). Confusion of tongues between adults and the child. In *Final contributions to the problems and methods of psychoanalysis* (pp. 156–167). London: Hogarth. (Original work published 1932)

Foucault, M. (1995). *Discipline and punish: The birth of the prison*. New York: Random House, Vintage.

Harner, M. (1990). *The way of the shaman*. San Francisco: Harper & Collins.
Herman, J. (1992). *Trauma and recovery. The aftermath of violence—from domestic abuse to political terror*. New York: Basic Books.
Kaufman, G. (1988). *Shame: The power of caring*. Rochester, VT: Schenkman Books.
Kernberg, O. (1993). Thirty methods to destroy the creativity of psychoanalytic candidates. *International Journal of Psychoanalysis, 77*, 1032–1040.
Kippenberger, H. G., & Seidensticker, T. (Eds.). (2004). *Terror im Dienste Gottes: Die "Geistliche Anleitung" der Attentater des 11 September 2001*. [*Terror in the service of God: The "spiritual instructions" of the September 11, 2001 attackers*]. Frankfurt/New York: Campus Verlag.
Lacan, J. (1964). *Seminar XI: The four fundamental concepts of psychoanalysis*. New York: Norton.
Laing, R. D. (1967). *The politics of experience and the bird of paradise*. Harmondsworth, UK: Penguin.
Laplanche, J. & Fletcher, J. (1999). *Essay on otherness*. New York: Routledge.
Layton, D. (1999). *Seductive poison: A Jonestown survivor's story of life and death in the People's Temple*. New York: Anchor Books.
Lifton, R. J. (1961). *Thought reform and the psychology of totalism*. New York: Norton.
Lifton, R. J. (1987). Doubling: The Faustian bargain. In *The future of immortality and other essays for a nuclear age* (pp. 195–208). New York: Basic Books.
Lindholm, C. (1993). *Charisma*. Hoboken, NJ: Blackwell.
Neuman, J. (1990). Trauma and recovery: The aftermath of violence—From domestic abuse to political terror. New York: Basic Books.
Ofshe, R., & Singer, M. T. (1986). Attacks on peripheral versus central elements of self and the impact of thought reforming techniques. *Cultic Studies Journal, 3*, 3–24.
Orwell, G. (1949). *1984*. New York: Penguin.
Reeder, J. (2004). *Hate and love in psychoanalytical institutes: The dilemma of a profession*. New York: Other Press.
Sargant, W. (1957). *Battle for the mind: A physiology of conversion and brainwashing*. Cambridge, MA: Malor.
Schein, E. H. (1961). *Coercive persuasion*. New York: Norton.
Searles, H. (1959). The effort to drive the other person crazy: An element in the etiology and psychotherapy of schizophrenia. *British Journal of Medical Psychology, 32*, 1–19.
Shanon, B. (2002). *The antipodes of the mind: Charting the phenomenology of the Ayahuasca experience*. Oxford, UK: Oxford University Press.
Shaw, D. (2003). Traumatic abuse in cults: a psychoanalytic perspective. *Cultic Studies Review, 2*, 101–131.
Shengold, L. (1991). *Soul murder: The effect of childhood abuse and depravation*. New York: Ballantine Books.
Singer, M. T., Goldstein, H., Langone, M. D., Miller, J. S., Temerlin, J. K., & West, L. J. (1986). *Report of the Task Force on Deceptive and Indirect Techniques of Persuasion and Control* (DIMPAC report). Retrieved from http://www.rickross.com/reference/apologist/apologist23.html
Singer, M. T., & Lalich, J. (2003). *Cults in our midst: The continuing fight with the hidden menace*. San Francisco: Jossey-Bass (Wiley).

Singer, M. T., & Ofshe, R. (1980). *Thought reform and brainwashing. Document offered as proof of testimony*. Queen's High Court, London, on behalf of the London Daily Mail.

Singer, M. T., & Ofshe, R. (1990). Thought reform programs and the production of psychiatric casualties. *Psychiatric Annals, 20,* 4–21.

Sohn, L. (1985). Narcissistic organization, projective identification, and the formation of the identificate. *International Journal of Psychoanalysis, 66,* 201–213.

Stein, R. (2001). Evil as love and as liberation. *Psychoanalytic Dialogues, 12*(3), 393–420.

Stein, R. (2006), Fundamentalism, father and son, a vertical desire. *Psychoanal. Review, 93,* 201–229.

Stein, R. (2009). *For love of the father: A psychoanalytic study of religious terrorism*. Palo Alto, CA: Stanford University Press.

von Feuerbach, P. J. A., & Daumer, G. F. (1995). *Kaspar Hauser: Beispiel eines Verbrechens am Seelenleben des Menschen* [*Kaspar Hauser: An example of a crime against the psychic life of the person*] (V. J. Mayer & J. M. Masson, Eds.). Frankfurt: Eichborn Verlag.

Wier, D. (1996). *Trance: From magic to technology*. Ann Arbor, MI: Transmedia.

Wier, D. (2009). *The way of trance*. Strategic Book Publishing.

Winkelman, M. (2000). *Shamanism: The neural ecology of consciousness and healing*. New York: Praeger.

Chapter 14

The gendering of human rights

Women and the Latin American terrorist state

Nancy Caro Hollander

INTRODUCTION

The contemporary world has witnessed ethnic and cultural violence among peoples that includes the systematic rape and violation of the female population. Indeed, it is now known that rape has always been by-product of warfare between peoples. However, I should like to point out that during the period of state terror in Latin America in the last quarter of the 20th century, authoritarian regimes waged a brutal war against their civilian populations—women and children as well as men—with the goal of silencing all critical consciousness. When women have spoken out against extreme economic and social inequities, their protests have often been met with the wrath of the torturers' weapons as the terrorist state ruthlessly implements an *ideological cleansing* of the body politic. Often women have been a gender-specific target of misogynist military and paramilitary forces that kidnap, torture, and murder at will. It is my aim in this chapter to analyze the impact on women of state terror in Latin America and to demonstrate that in conditions of extreme political repression women in the civilian population become direct victims of the traumatizing violence organized by the militarized state. My exploration will attempt to explain the complex relationship between gender and political repression in Latin America and the ways in which gender affects one's psychological response to state organized terror.

I first became interested in the psychological impact of state terror in the late 1970s, when many of the progressive political activists I had known in Argentina prior to the military coup in 1976 were forced to flee their country in the years following the implementation of the bloody attack on human rights carried out by the military junta. As if sown by the wind, these activists landed in places as far flung as Paris, Madrid, Caracas, San Francisco, and Toronto. In 1981 I traveled to Madrid, where I planned to interview a number of women refugees about the impact of feminism on their revolutionary political organizations. There, in a café-club an Argentine feminist friend had opened to provide a cultural meeting place for the refugee community. I sat and spoke with dozens of women who recounted their experiences of terror at the hands of the military-inspired repression that had touched every sector of Argentine society. Their stories, painfully difficult to listen to, were testimonies of the most sadistic and brutal physical and psychological torture imaginable and of the anguishing loss of loved ones who had disappeared or been murdered in the violent attack on human rights carried out by the junta.

My contact with these women in exile became a watershed in my professional and political life. I realized that my training as an historian, and familiarity with Marxist and feminist theory had not sufficiently addressed the psychological domain of human experience (i.e., the realm of subjectivity). In the subsequent decade, I was trained as a psychoanalyst and my work was deeply influenced by North American and European feminist psychoanalytic theory. I became involved with the growing movement among politically progressive Latin American psychoanalysts who were dedicating their professional training to the analysis of the psychological impact of political repression and committing their clinical skills to treating its victims. My research and writing have focused on issues related to the psychology of political terror in Latin America and its traumatic impact on civil society.* In this chapter, I want to develop a theoretical approach to understanding the gendered experience of political repression and to explore the extent to which state terror functions to reinforce and challenge the existing gendered arrangements of society.

GENDER AND STATE TERROR: THE PARADOXICAL PROBLEM

Throughout Latin America, women's psychosocial experience is framed by a culture that reflects the sex roles and gender values characteristic of modern industrial society that emerged in the Western world from the 16th century on. It is the specificity of this female role that provides the lens through which we may understand how women's lives are deeply affected when political conditions become so conflictual and tumultuous so as to give rise to repressive state machines that wage economic, social, and political war on civil society. The effects on women are complex and somewhat discrepant, because the strategies as well as the traumatic impact of the terrorist state reinforce traditional female roles in some respects while they simultaneously alter the roles in other respects. The impact of state terror

---

* See for example, "Psychoanalysis Confronts the Politics of Repression: The Case of Argentina," *Social Science and Medicine*, 28(7), 1989; "Psychanalyse et terreur de'Etat en Argentine," *Reuve Internationale D'Histoire de la Psychanalyse*, 5, 1992; "Buenos Aires: Latin Mecca of Psychoanalysis," *Social Research*, 57(4), 1990.

on women is contradictory, in some ways homogenizing and in other ways exacerbating the customarily differentiated experiences of women and men based on their gender identity and roles.

Let me situate the first aspect of the impact of state terror, namely, that it homogenizes customarily gender-differentiated experience. In *Trauma and Recovery* (1992) feminist psychiatrist Judith Lewis Herman analyzes the nature of trauma in a variety of situations, ranging from wife battering to war. Herman writes that combat and rape are "the public and private forms of organized social violence" that might be thought of as "complimentary social rites of initiation into the coercive violence at the foundation of adult society. They are paradigmatic forms of trauma for women and men respectively" (p. 61). As Herman suggests, under "normal" conditions (i.e., in societies with constitutional governments), young men and women are at significant risk of experiencing violence in adolescence and young adulthood. It strikes me that Herman's idea helps to crystallize one important feature of the impact of political repression in Latin America, namely, that in the terrorist state, elements of these gender-specific paradigmatic rites of passage are universalized, so that both men and women are exposed to traumatic experiences typical only of one or the other sex under more "normal" conditions. In other words, there are ways in which the conditions established by the terrorist state tend to impact on women and men by making their social situations more symmetrical so that each is exposed to both paradigmatic rites of passage. I will explore two dimensions of Latin American state terror: the first related to the extreme, chronic stress resulting from the terrorific conditions of everyday life, which come close to the experience typical of combat; and the second related to the acute trauma of torture, which takes on dynamics similar to those of the battering and rape situation.

Second, because the impact of state terror is paradoxical, in certain other respects it reinforces gender differentiated roles and subjective experience. In other words, the Latin American terrorist state strengthens male-dominant institutions and intensifies misogynist ideology, resulting in an increase in the discrepancy between male and female emotional states and interpersonal behavior. The terrorist state strikes out against women simply because they are women. In the generally militarized social environment, as well as in the more specific torture situation, sexual brutalization of women escalates. Women who have dared to move beyond the narrow confines of the dominant culture's acceptable notions of femininity—especially women who have been political activists on behalf of oppositional movements for radical social change—become particularly vulnerable to the reactionary political discourse of the terrorist state. Indeed, such individuals become specific targets of military and paramilitary repression. However, a salient aspect of state terror is that the entire society is permeated by antagonistic attitudes and behavior toward women because the parameters for misogyny widen. When women are disappeared, or are tortured or assassinated in ways that are gender specific, these acts function as a general legitimation of violence against women, in so doing stripping away from male-dominant culture its paternalistic facade. I will thus analyze the ways in which the terrorist state widens the already existing gap in the psychosocial experience of women and that of men. In this context, I will also explore how gender as it is constructed in the male-dominant cultures of Latin America may affect the coping mechanisms employed by male and female victims of state terror.

VIOLENCE AGAINST WOMEN

Male frustration and rage is often expressed through violence directed at women. In fact, male violence against women is an apparently universal phenomenon that functions as a fundamental element in the maintenance of social control of women by men. Indeed, it is in this aspect of male–female bonds that we can see the intimate articulation between the personal and the political domains of gender relations. While the determinants of male violence against women may be complex and multifold, the feminist psychoanalytic account suggests that one important source is linked to men's need to deal with the dread they have experienced from early infancy on of engulfment by the omnipotent mother. Social scientists point to the cross-cultural myths and rituals that reveal the polar opposites of the solution characteristic of the male dread of reengulfment: the idealization and glorification of women, or their devaluation and debasement. In either case, the female's power is neutralized by the cultural rendering of her as either above or below male fear, which is thus contained through the fantasy of omnipotent control over a dichotomized object (Madonna/whore, good/evil, pure/contaminated). However, male domination in the social sphere must be assured (and reassuring) at times through violent abuse of this potentially powerful object. The intrapsychic dynamics of women fashioned within the cultural matrix help to shape their role in gendered relations by preparing them to expect and often endure violent domination by men.

According to recent United Nations reports, domestic violence exists in all regions, classes and cultures worldwide. Existing data demonstrates beyond a doubt that domestic violence is a ubiquitous and serious problem in developed and underdeveloped countries alike. For example, as the Americas Watch and Women's Rights Project (1991) reports, in Brazil wife murder is a common crime, and a man may still be legally absolved on the grounds of honor if he claims his wife was unfaithful. Moreover, physical and sexual abuse in the home represent over 70% of all reported incidents of violence against women in Brazil. And Brazilian law does not generally view rape, marital or otherwise, as a crime against an individual person, but rather as a crime against custom. Thus society, not the female victim, is the offended party. In Nicaragua, 44% of men admit to having beaten their wives, and in Peru, 70% of all reported crimes are of women beaten by their partners (Kaschak,).* In the United States, sexual assault is just as culturally syntonic as in Brazil: Half of all women in the United States are raped at least once in their lifetime, and half of all adult women are battered in their own homes by husbands or lovers (Kaschak, 1992), a situation still legally condoned in some 30 states (Faludi, 1991). Thus, in patriarchal culture—whether in developed or underdeveloped countries—sexual abuse is so endemic that women learn to live in fear that even if they have not yet experienced a violent attack on their person, they may be victimized at any moment. The fear women live with means that they easily identify with feelings of violent victimization even if they have not been personally violently attacked. Indeed, male violence against women is a normative experience of everyday life and is reified in a culture that mirrors, permits, and instructs both men and women in its apparent inevitability.

The fact that women live with this degree of actual violent attack or potential threat of violence has led feminist psychoanalyst Laura S. Brown (1991) to critically reinterpret the etiology and diagnosis of psychic trauma. Among the mental health professions, the generally accepted definition of the traumatic stressors leading to the diagnostic category of posttraumatic stress disorder (PTSD) are those events "outside the range of usual human experience," such as natural disasters or atrocities perpetrated by other human beings. Categories of symptoms that result from traumatic events include reexperiencing symptoms, such as nightmares and flashbacks; avoidant symptoms, which reflect psychic numbing; and symptoms related to heightened physiological arousal, including hypervigilance, disturbed sleep, and distracted mind. However, as Brown argues, the above definition is built on a notion of traumatic stressors being "outside the range of usual human experience," which makes it a paradigm established from the privileged male point of view. This definition of trauma ignores the level of sexual violence that typifies the ordinary daily experience of women. Brown asserts that an accurate definition needs to move beyond the public and male experiences of trauma to the private and female experiences of trauma carried out in secret and within the interpersonal realm. If the hegemonic definition of trauma is left to males of the privileged class, "real" trauma is that experience in which the dominant group is victimized rather than the experience in which the dominant group is the perpetrator or source of the trauma. Brown suggests that the term *insidious trauma* be used to capture the traumatogenic effects of female oppression characteristic of a

---

* A book by Roger N. Lancaster, *Life is Hard: Machismo, Danger, and the Intimacy* of *Power in Nicaragua* (Berkeley: University of California Press, 1994), analyzes in depth gender relations in Nicaragua within the context of revolution and counterrevolution.

male-dominant culture that normalizes and eroticizes sexual assault and does violence in an ongoing way to the female spirit.

LATIN AMERICAN STATE TERROR AND THE HOMOGENIZATION OF GENDER DIFFERENCE

This normative condition of everyday life is the lens through which we can understand the psychodynamics that underlie the Latin American female's experience of extreme complex trauma caused by the terrorist state.* First we shall see how in some respects, ordinarily gender-differentiated experience is made more symmetrical as the terrorist state imposes an ongoing and insidious culture of trauma, subjecting the entire population—women and children as well as men—to a militarization of daily life and to life-endangering conditions customarily associated with combat. In this regard, we have only to recall that 30% of those who disappeared in Argentina during the infamous "Dirty War" were women, many of whom were pregnant. We shall also explore how in its prisons, "secret" detention centers and concentration camps, the state's infliction of torture subjects men as well as women to the complete and total humiliating degradation of psychic and physical integrity customarily associated with battering and rape. We shall then turn to an analysis of how, in a paradoxical sense, there is likely to be an exacerbation of customary gender differences as male-dominant institutions and ideology are reinforced. And finally, we explore how men and women, in gender-specific fashion, struggle to deal psychologically and politically with the insidious trauma of everyday life and the acute trauma of the torture situation, both of which are important dimensions of the terrorist state's assault on its citizens.

State terror has emerged in Latin America as a response to progressive movements for social change. Under conditions such as those prevailing during the military dictatorships of Chile, Argentina, and Uruguay, or throughout the repressive military and civilian governments' confrontation with struggles for national liberation in El Salvador and Guatemala, state terror strikes out at the body politic with the goal of silencing critical consciousness and paralyzing political engagement in the popular classes. Often the only factor uniting the military and other right-wing sectors of

---

* Judith Herman (1992) also criticizes existing definitions of trauma, pointing out that people who live with ongoing traumatic stressors, such as in the instances of battered women or torture victims, both of whom live in captivity, are exposed to severe, prolonged, and massive psychological trauma, which necessitates a diagnostic category reaching beyond that of PTSD. She suggests the diagnosis of "complex posttraumatic stress disorder" to refer to a history of subjection to totalitarian control over a prolonged period, such as that characteristic of hostages, prisoners of war, concentration camp survivors, and totalitarian systems in sexual and domestic life.

society is their own terror of progressive ideology and redistributive policies promoted by the left. State terror develops a strategy that attacks specific individuals and groups who represent an articulate opposition to the status quo, and it also indiscriminately targets union members, intellectuals, relatives of known guerrillas, unarmed political dissidents, or individuals simply suspicious in the eyes of those who randomly detain and secretly kidnap citizens in their homes, their workplaces, and on the streets. The entire population becomes "victims of a doctrine of collective guilt." Social discourse is aimed at paralyzing the population with its contradictory and paradoxical quality. The government's rhetoric often refers to the war between "violence" (attributed to the political forces fighting repressive governments) and "order" (attributed to the repressive state's cleansing offensive against the enemies of Western Christian civilization), in this way masking the relationship between the two that characterizes the military and paramilitary attack on civil society. Society is redefined as a war zone, and the militarization of ideology stresses the couples of chaos/order and enemy/ally, imposing a sense of catastrophic danger and constant unpredictability. While the terrorist state speaks of the need to respect the family and social order, entire families are attacked, destroyed, and violated. The widespread use of policies of disappearing, torturing, and murdering significant numbers of citizens is aimed at imposing a passive consensus within the population.* Everyone knows about the "secret" concentration camps, wakes to see or hear about the disfigured cadavers of torture victims rotting in the streets or floating in the rivers, and knows someone whose relative or friend has disappeared. In such conditions of repression, the population is forced to work out in any manner possible a way of understanding the rules, the cues, and what make a good citizen and what makes an enemy of the state. All are compelled to feel that their homes, their jobs, their loved ones, their own lives are in jeopardy. People come to trust no one, to confide in no one, to seek self-preservation in isolation. Contact with others endangers. Individual behavior is characterized by silence, inexpressiveness, inhibition, and self-censorship, all of which result in depoliticization. In this situation, individuals become obedient and potentially punitive of self and others.

One of the outstanding features of a terrorized population is the compulsion to deny reality, to refuse to bear witness to the sinister drama that oppresses the entire nation. Denial shields the individual from his or her conscience and the internal or external demand to act in defiance of the systematic violation of basic human rights. As parents, for example, alter their vocabulary and their vision of reality in response to their children's questions in order to protect them, they impose denial, repression, and censorship within the family, which contributes to the creation of an apolitical

---

* Unless otherwise indicated, the information in this section comes from Hollander (1992).

and self-censoring generation. The psychology of denial thus functions to sustain the politics of terror. As Salvadoran social psychologist Ignacio Martín-Baró (1988) put it, the psychosocial trauma is one of dehumanization, with the effects of inhibiting four important human capacities: the capacity to think lucidly, the capacity to understand truth, the capacity for empathic response to the suffering of others, and the capacity to maintain hope in the future. Martín-Baró emphasized that the militarization of social life has the effect of stimulating a progressive militarization of the mind, manifested not only in the psychopathologically destructive behavior of members or ex-members of the military, but reflected within the general population as well in the growing preponderance of military forms of thinking, feeling, and acting, which come to be accepted as normal modes of conduct. With social forms of response no longer available to permit the expression of aggression toward the real source of suffering and loss, aggressive attitudes and behavior increase in personal life, being displaced onto less dangerous situations and onto safer objects.

Because it is impossible to live for long in acute pain or permanent collapse, adaptation to the catastrophic conditions of state terror is accompanied by a general impoverishment of psychological resources in adults as well as children. Common psychological responses include the development of dissociative defenses, the pathological regulation of emotional states, and the emergence of a fragmented identity. Politically engaged individuals are forced to make an anguishing choice: If they continue to maintain coherence with their political values and their commitment to struggle against the terrorist state, they pay the psychological price of living with a high degree of constant anticipatory anxiety about their own and their comrades' and family's vulnerability. If they renounce their political commitment and thus abandon a project that has given meaning, significance, and a set of positive values to their lives, they suffer the psychological price of intense guilt and the rejection of the person they once were. As for the perpetrators of state terror, as the report of the Argentine National Commission of the Disappeared (1986) indicates, the original goal of creating a climate of intimidation and fear aimed at silencing a population at some point shifts to "the perverse exhilaration of absolute, uncontrolled dominion over others, which [becomes] an end in itself, a way of life."

This profound level of political repression exists within a deteriorating economic situation for most sectors of society. The Latin American terrorist state imposes neoliberal free market economic policies favoring elite groups at the expense of the popular classes, whose traditional means of struggling for progressive redistributive policies have been smashed. As labor unions are closed down, intervened, or declared to be illegal, male and female workers, employees, intellectuals, and professionals no longer have legitimated means of defending their class and sectoral interests. Rising unemployment rates, forced wage cuts and an escalating cost of living all reflect the assault on the popular classes' quality of life. People's energies are increasingly drained in the frantic effort to defend themselves against the ruthless attack on their material conditions of survival. In such conditions, resistance or opposition threaten to stimulate life-threatening reprisals. With freedom of speech and all forms of organized collective struggle prohibited, this struggle for survival is intensified, physically exhausting the body and demoralizing the psyche.

In this combatlike atmosphere, women and children, as well as men, of the middle and popular classes are exposed during an extended period of time to ongoing political repression and economic crisis. They suffer prolonged repeated trauma that is insidious and progressive, invading and eroding the personality. Women and children, as well as men, become chronically traumatized, and may experience disturbing states of hypervigilance, anxiety, and agitation, or withdrawal into a numbed state of isolation. A sense of tranquility or comfort is no longer accessible as an emotional referent; instead they may suffer a variety of somatic symptoms, including insomnia, tension headaches, gastrointestinal disturbances, stomach pain, back pain, pelvic pain, and cardiovascular disturbances, including rapid heart beat and shortness of breath. Sustained stressful conditions also produce difficulties in affect regulation, including either explosive or extremely inhibited anger, which severely alter relations with others, even—or perhaps especially—in the most intimate of relationships.

It is in the acute trauma of the torture situation, however, that the terrorist state crystallizes its assault on civil society. Torture submits women and men alike to the most heinous crimes against humanity at the hands of "instrumental torturers," whose brutal sadism represents the implementation of a savage policy to do everything possible to kill off opposition to the totalitarian power of the political and economic elite in whose interests the torture is carried out.*

Innovations in torture techniques reveal a calculated and intelligent manipulation of human needs and emotions, including sensory deprivation and overload, interspersed with frightening sounds and images of the victim's loved ones being tortured or murdered, and the injection of chemical substances, which induce a variety of agonizing physiological and psychological states. Careful study of detainees' psychological vulnerabilities aid in the design of personalized psychological torture methods. Conscious attempts to break down the resistance of the victim include the use of the "good torturer" and "bad torturer," a tactic meant to cause confusion in the victim who is trying desperately to maintain his or her moral resistance by clinging to a bifurcated vision of the world into good and bad, in which the torturers represent an evil enemy attacking his or her political project,

---

* For a more extensive discussion of torture, see Argentine National Commission of the Disappeared (1986), Fernando Bendfeldt-Zachrisson (1988), and CODESEDH (1985).

and one's compañeros represent a positive alternative in the world. In addition to beatings, application of electrodes—the infamous *picana*—to every part of the body, slicing and cutting of intimate parts of the body, submersion in water, and so on, the torture victim is often forced to watch the torture and murder of family members, including small children and infants, or to endure being tortured himself or herself in front of family members.

I believe that in the torture situation, men are submitted to an experience more or less equivalent to the pervasively abusive situations many women endure under more normal circumstances. In fact, therapists have noted a similar psychological profile in the victim of battering and rape and the torture victim suffering from extreme trauma. Like women, men in the torture situation suffer the humiliation of being physically brutalized and cruelly taunted for being weak and helpless in their impotent inability to defend themselves. Sexual torture for men emphasizes an attack on their masculine role; they are debased as human beings and as men by being forced to witness the rape and torture of female prisoners—a compañera, a wife, a daughter, a mother—and suffer precisely because of their inability to protect them. As scholar Jean Franco (1992) puts it when commenting on the experience of male prisoners of the military governments of Argentina and Chile,

> abjection often forced male prisoners to live as if they were women, so that, for the first time, they came to understand what it meant to be constantly aware of their bodies, to be ridiculed and battered, and to find comfort in everyday activities like washing clothes or talking to friends. (p. 109)

Mental health practitioners have noted the similarity in the dynamics of psychological domination characteristic of the treatment of women in prostitution, pornography, and battering with those of political prisoners. Central to the enforcement of relations of domination and submission are inconsistent and unpredictable uses of violence and the constant threat of imminent death. In the wake of several bouts of threats of death followed by sudden reprieves, battered women and political prisoners alike often succumb to the paradoxical state of viewing the perpetrator as their savior. The assault on body autonomy thus expands to the psychic domain, with the successful assassination of psychological autonomy in the victim. As in the Holocaust, in the torture situation the terrorist state is often able to secure the complete psychological regression of the victim so that a pathologically symbiotic relationship develops between the torturer and the tortured.* Just as battered women may begin to feel deserving of their abuse because their self-esteem has fallen so low they cease to believe themselves

---

* See, for example, Bettleheim (1943), Herman (1992), and Amati (1976).

worthy of decent relationships, the torture victim may come, as well, to feel he is deserving of abuse and is dependent on the torturer for salvation (Aron, 1987).

The ability to maintain psychic integrity in the face of torture has become more difficult with the increasing sophistication of torture technology supplied by supportive governments of the developed capitalist world and psychologically oriented methods of repression supplied by complicit medical and mental health professionals. While the extent to which the terror, humiliation, loss of dignity, and physical anguish mark the victim is dependent on the structural characteristics of his or her personality and the duration and intensity of the violence experienced, it has been observed by most mental health professionals who treat survivors of torture that they suffer from a variety of psychological disorders. These include extreme forms of anxiety and panic attacks, difficulty and confusion in thinking, loss of self-esteem, social withdrawal, decrease in productivity, abandonment of goals, and even premature death. Torture produces a uniquely global trauma because it constitutes an attack on the self that goes beyond the extraordinary physical suffering entailed. The threat to the victim's psychic structure, the imposition of a sense of total vulnerability and inability to defend oneself or others, the internalization of aggression against oneself due to the impossibility of its expression toward one's torturers, and the guilt related to either breaking down under torture or surviving it when others do not are aspects of its traumatic impact.

LATIN AMERICAN STATE TERROR AND THE EXACERBATION OF GENDER DIFFERENCE

As we have seen, the terrorist state creates conditions that submit both men and women to insidious and acute forms of traumatic violence, exposing both genders to the catastrophic elements inherent in both combat and sexual abuse. We have noted how male and female alike are brutalized and violated by the terrorist state's political and economic policies. But the destructiveness of state terror is contradictory in its impact, so we now examine the ways in which state terror creates conditions that exacerbate gender differences. We can indicate at least two dimensions of state terror in which women are victimized precisely because they are women, resulting in an exaggeration of the discontinuity in the customarily differentiated experience of males and females. Finally, I suggest how within the general range of emotional responses to trauma, men and women cope in characteristically gendered ways with trauma-inducing conditions created by the terrorist state.

State terror intensifies the conventional physical, psychological, and ideological violence directed specifically against women in several ways. Perhaps its most concrete manifestation is in the sexual abuse of women by members of the military, which is endorsed as a method of dealing with the civilian population. Even where rape is officially discouraged by military regulations, its practice is tolerated as a significant aspect of the continual reenactment of power among members of the military because it embodies that institution's association of masculinity, aggression, and depersonalization.*

In the case of El Salvador, for example, the military's standard operating procedures have involved capricious exploitation by soldiers of women in the community. Often a soldier would accuse an individual woman he desired of being politically involved, certain that his word would prevail over hers, so as to secure sexual compliance from her. Women have been kept in bondage under such conditions, afraid for their lives should they refuse unwanted sexual advances. In many cases, women have been known to "give" themselves to individual soldiers as their private sexual property so as to prevent becoming the collective property of a whole battalion. In other situations, women have traded their own person in a quid pro quo for the release of young men targeted for death by the military. A woman's body becomes a commodity in a market controlled by military officers: "Thus, while men may trade in cigarettes or male prestige when seeking favors, women more often must resort to the coin of the flesh" (Aron, Corne, Fursland, & Zelwer, 1991).

In Guatemala, rape by army personnel and even by male members of the civil patrols has been one of the painful factors in the military assault on the Indian population. A rare occurrence among the Indians in the past, rape has increased due to outside influence from ladino (urban White) culture and from the occupation of Indian communities by military forces for extended periods of time (Manz, 1988). In the Quiche province, it has been reported that the army's practice of raping young women has been so widespread that it is nearly impossible to find women between 11 and 15 years of age who have not been subjected to sexual assault by the army (Aron et al., 1991). Even foreign nationals are not beyond the military's war against dissent. For example, in its publication *Rape and Sexual Abuse: Torture and Ill Treatment of Women in Detention* (1992), Amnesty International reports that a Roman Catholic nun from the United States, who had been working in an indigenous community in Guatemala for several years, was abducted in 1989 by uniformed police officers. She was blindfolded and taken to a warehouse near Guatemala City, where for 12 hours she was beaten, burned with cigarettes, repeatedly raped, and sexually abused in ways she referred to as "too horrible to describe."

In Argentina, during the military dictatorship's Dirty War, bored junior officers who were members of the torture squads would cruise the streets in

---

* Because of the widespread practice of rape by soldiers, many feminists have come to think of war as intimately linked with violence against women; see Brownmiller (1976) and Stiehm (1983).

their infamous Ford Falcons looking for pretty young women to sequester and take back to camp to rape, torture, and then kill. Such activity represents a kind of ritual sport, which has the important function of strengthening the male bonds of military domination through the enactment of traditional forms of patriarchal control over and willful violation of women's bodies (Simpson & Bennett, 1985).

Under state terror the insidious form of trauma suffered by women in normal conditions due to the widespread incidence of sexual abuse is clearly intensified. Rape in this context receives overt political sanction and thus moves beyond the image of an isolated criminal act to one of a normative act of social control carried out on behalf of a collective goal. As Amnesty International (1991) points out, because of their cultural or social circumstances women are often isolated by the human rights violations they suffer, afraid that any public denunciation of their humiliating assaults will bring reprisals from their own families, social isolation from their communities, or additional attacks by the government or military. Women from disadvantaged social or economic sectors, as well as their counterparts from the more privileged classes, often find that they have no recourse in official channels. And those women who speak out, denounce political repression and organize politically against it, do so at extreme personal risk, as they often themselves become targets of further repression. Such has been the case of the Guatemalan Mutual Support Group for the Appearance of Our Relatives Alive (GAM), several of whose leaders were abducted and murdered in 1985.

The state's political attack on women resonates within the society at large as aggression in daily life increases. Men who are unable to direct their rage at the state that is responsible for their victimization are apt to displace these emotions onto safer and traditionally socially acceptable objects. Wife battering and rape is a likely arena where male aggression is enacted, given the fact that the customary relationship between machismo and violence has been made even more acceptable by the ideology and practice of the state. In Argentina, for example, according to a University of Buenos Aires team of feminist social scientists interviewed in 1990, during the military dictatorship, violence in daily life was so ubiquitous that the increasing incidences of wife battering, rape, and child abuse did not attract the attention of progressive and feminist forces. It was only with the abolition of state terror and the emergence of democratic government that these "invisible" insidious forms of violence became recognizable and were addressed as social concerns.

Directly related to this insidious form of daily assault is the increased pressure on women related to their division of labor in the family. Women suffer the consequences of the terrorist state's economic war on the popular classes more than men because it is women who are accorded the responsibility of caring for the members of the family.

The identity of the Latin American woman is derived from her position in the family and especially from her "sacred" mothering role, where she is overprotective in her nurturing, absolutely devoted to her children, and willing to sacrifice her own desires to please her family—especially the male members. Her love is sacrifice and selflessness personified. (Bunster-Burotto, 1994, p. 304)

So the woman suffers the strains of trying to make ends meet, it is she whose labor is increased by the endless search for ways to economize and maintain decent nutritional levels for her children, it is she whose wages are the first to drop with rising unemployment. It is she who must fight with her husband not to drink up the little money there is for food, and it is she who cannot abandon her children when her husband has fled the intolerable pressures. Indeed, as it is put in terms of the Salvadoran situation, "hunger is the other war, a bomb exploding amid children, heaving up broken bodies and the ragdoll living. The women live this war, rock the starving babies too weak to howl, carry the tiny white coffins out to the fields" (Golden, 1991, p. 41). During the Salvadoran Revolution, when thousands of men left their homes to participate in the armed struggle, they left women to care for the children. Families were so disrupted that today almost 38% of Salvadoran families are headed by women.* It is they who have been on the front lines of the economic war for survival that accompanied the military conflict in that country for over a decade.† For those who themselves participated directly in the revolutionary struggle,‡ much of their work entailed confronting the "victims of misery and despair" who inhabit the slums, "dejected, isolated individuals, families plagued with alcoholism, domestic violence, prostitution, violent machismo, and malnourishment" (Golden, 1991, p. 31).

---

* It is worth noting that the upheaval represented by civil war and revolution has added to other cultural and economic factors that have contributed to high numbers of female-headed households in Central America. In Nicaragua, for example, for decades, chronic unemployment drove men into a migratory labor force searching for work far away from families and home. The resulting instability was aggravated by cultural traditions in which men have enacted their machismo through the simultaneous establishment of several families, one of which usually loses the man's function as the male head of household. By the time the Sandinista Revolution triumphed in 1979, over one-third of the families in the capital city of Managua were female headed. Argentine psychoanalyst Marie Langer, a consultant to the Sandinista government, has noted some of the problematic aspects of this phenomenon in dealing with families in Nicaragua. According to Langer, this pattern of female-centered and male-absent families results in profound female resentment and anger toward men and an unhealthy degree of emotional investment in children, who then have a difficult time separating and becoming independent of their mothers without a great deal of guilt. This problem was taken up by the Association of Nicaragua Women (AMLEA.) See Hollander's introduction (1988) in Langer's *From Vienna to Managua*.
† Interview with Maria Ester Chamorro, September 14, 1992; see also Thomson (1986).
‡ Women represented 50% of the rank-and-file of the popular organizations, 30% of the Sandimista Front of National Liberation (FMLN) combatants, and 40% of the revolutionary leadership (Golden, 1991, p. 75).

In all Latin American countries ruled by state terror, the combination of worsening economic conditions and political repression reinforce one another to intensify the difficulties women have in fulfilling their roles as maternal providers, whether or not this responsibility includes working for pay outside the home. In the case of Guatemala, while the genocidal attacks by the military on the indigenous population include murder and torture of women, its effects are felt in less obvious ways as well. Thousands of women have had husbands disappear or murdered, and they are left with the psychological sequelae of traumatic loss in addition to the responsibility of overseeing the survival of the family. They usually are the sole providers, and they must try to secure aid from male relatives to help them work the land. In many instances women are forced to abandon the land until their sons are old enough to farm; in the meantime many leave for the cities to search for work, where their customary fate is to become part of a huge network of domestic laborers superexploited in the homes of the wealthy urban elite (Manz, 1988).

In the case of Argentina, the military dictatorship's Dirty War had a profound impact on women of the popular classes and the extensive middle class as well. Interdisciplinary research teams that studied the psychological, social, and economic conditions of families throughout the country during the military dictatorship found that among the popular classes in the northern and southern provinces, men manifested their stress through increased alcoholism and emotional withdrawal from the family. Growing unemployment among men, either because they were fired or because their demoralization made them incapable of sustaining a job, meant that women had to become auxiliary or sole providers of the family. It was noted that in response to the repressive political climate and harsher economic conditions, men who were not physically disappeared by the state often disappeared in other ways. For example, it was not unusual for men to psychologically abandon their wives or to express increased hostility toward family members, and in such cases, women were left to struggle alone to make ends meet and to care for their children. Women, as well, suffered a number of psychological difficulties, including nonspecific anxiety, demoralization, agitation, and a variety of psychosomatic disorders. However, it was noted that women seemed able to develop multiple strategies permitting them to continue to function adequately in their maternal role. In my opinion, the observations of these researchers are a concrete demonstration that women are motivated to sustain their familial functions because of their own infantile experience with their mother in the male-dominated but female-centered family. Female embeddedness in the caretaking relationship to others and the adequate fulfillment of this role is a fundamental ingredient in women's self-esteem. This perspective is shared by Maria Ester Chamorro, who believes that political repression is more difficult on women than on men because the latter are more distant emotionally from their children and women's emotional connections are so intense that they experience their children's deprivation as if it were their own.

Among the middle class, especially in Buenos Aires, the significant number of women who were professionals were hard hit by the military dictatorship. The majority of these women were in the liberal arts and social sciences, the traditional sector of professional female employment in the public sphere. They were direct targets of the state's policy of closing or intervening entire academic departments at the universities, including psychology, sociology, anthropology, and history. Many state-funded projects that had employed hundreds of female professionals were closed as well. These women felt themselves endangered simply because of their membership in specific professions that were identified by the state as subversive. Many chose to abandon or to deny their professional identity in preference for officially defining themselves as housewives with only a high school education. In many cases, Porteño middle-class women were forced by political circumstance for the first time in their lives to embrace a traditional role in the home, and many found themselves devoting their energy to housekeeping, cooking, and childcare. Their professional and political projects liquidated, women in their late 30s and early 40s, with children now grown, became pregnant and mothers of young children again, feeling as if the nuclear family and their role in it could buffer them from the brutal political and social conditions created by the military dictatorship.*

This pattern was reinforced in Argentina by the official ideology of the state, which through the mass media permeated public discourse with reactionary visions of the family, presenting an ideal feminine model, usually embodied in the first lady of the nation. The terrorist state in general projects bourgeois notions of femininity as models of appropriate womanhood applicable to women of all classes, and the traditional role of wife-mother is romanticized and even associated with the glory of the nation. Attributes such as passivity, self-abnegation, and obedient adoration of the male sex are emphasized and presented as the complimentary traits to those of male glory and heroic dominance. A constant ideological assault aims at creating a sense of inadequacy and self-blame in women whose lives do not mirror this ideal. In Argentina, this ideology was used in a highly manipulative and sinister fashion. For example, when newspaper articles reported the appearance in city streets and along country highways of corpses of young people, disfigured by bullets and unspeakable torture, the text claimed

---

* Interview with members of a multidisciplinary team of postgraduate students in Feminist Studies at the University of Buenos Aires, August 1990. These women traveled throughout Argentina during the military dictatorship as part of multidisciplinary teams, including mental health professionals, to study the family situation in the context of political repression. The results of their study were published only when they became part of the archival data of the Asamblea Permanente de Derechos Humanos.

that they were the bodies of hoodlums who had carried out illegal actions, forcing confrontations with the police. The deceased would be named as common criminals or subversives who had broken the laws of the nation and whose violent attacks on society had been legitimately responded to by the authorities. Responsibility for their deaths would be displaced onto the family, and parents would be depicted as guilty of inadequate supervision, accused of failing at the important task of socializing their children to respect authority and private property. The media's cruel message was meant to mystify the reality of the criminal acts of the state by according blame to its victims for the violent deaths of their children. Given women's maternal role, this ideological tactic was intended to reinforce the guilt, however unrealistic, that mothers feel when they blame themselves for not having been able to prevent a traumatic loss of a loved child.* This type of ideological torture is part of the repressive state's exaggeration of the gender differences of ordinary life.

In addition to this ideological assault, state terror also imposes on women a particular kind of sexual abuse and violence reserved mainly for them because they are women. This uncontained aggression, which serves to exacerbate conventional gender difference, occurs in many contexts, including the torture rituals practiced by the militarized state.

Thus, in spite of the fact that the torture situation subjects men to psychological experiences similar to those customarily suffered by sexually abused or battered women, there appear to be gender-based differences in the practice of torture as well. There are indications that female torture victims suffer greater sexual violence than their male counterparts. In a sample of 28 persecuted or tortured women from Latin America living in Canada, 64% had been sexually abused, 44% of whom had been violently raped (Chester, n.d.). A mental health professional in Los Angeles who treats refugees from El Salvador also notes that sexual assault is more a part of the torture of women than of men, recalling that in her personal experience as a therapist treating victims of political repression, between 15 and 20 female patients and only 2 male patients reported sexual violation as part of the torture experience (Deutsch, Personal Communication, August 29, 1992). In El Salvador, the army reserved specific torture techniques for women, including forced nudity and rape. In contrast, the 10 forms of torture aimed exclusively at men did not relate to the prisoners' sexuality (Aron et al., 1991). Among the long list of torture techniques used on women, the special ones that represent a focused assault on the sexual organs include the use of electric shock on nipples and vaginal areas; sharp instruments to cut and slice the breasts and abdomen; and the insertion of objects, animals, and

---

* See the Latin American and Caribbean Women's Collective (1980) and Huneeus and Busto (1989).

insects into the vagina that bite and scratch and rip away at the insides of the victim (Bunster-Burotto, 1994). Mass rape by security officers or torturers, or forced rape by fellow prisoners, is often part of the repertoire of torture.

International Human Rights organizations have documented that military and paramilitary assaults on civilians suspected of supporting dissidents or revolutionary movements often include leaving massacred bodies at the scene of an attack in terrorific conditions meant to psychologically and politically immobilize those who have survived. An especially sadistic example is that of the mutilated body of a dead pregnant woman, whose fetus has been slashed out of the abdomen and stuck in her mouth. This particular method of terror is a brutal symbolic representation of the intersection between psychological and political processes: It reflects both the unconscious male dread and hatred of the fertile omnipotent mother who, it is feared, consumes rather than feeds the baby, as well as the conscious sadistic rage aimed at women who, through her political opposition and her capacity to bring forth from the womb more rebels, is seen as a challenge to male authoritarian power.

Assault on the female psyche is a central aspect of the torture of women. Their identity as women is viciously ridiculed. While being gang raped, the female torture victim is accused of being a whore and is thus blamed for her own dishonor. The torturers also use the female psychological connection with others as a form of torture. This may happen to female political activists or to male activists' wives, mothers, or daughters, who themselves have not been politically involved. Women are forced to witness the torture of comrades, friends, relatives, especially children, or else they are tortured in front of their loved ones. If the victim is a political activist, she is forced to betray either her comrades or her family, either the political cause or her personal relationships. The torturers force her to choose between her political links or her maternal links. The prisoner's knowledge that her decision inevitably imposes brutal torture on people she loves is frequently experienced as worse than being tortured herself.

A special category of torture relates to pregnancy. Three percent of the disappeared in Argentina, for example, were pregnant women. These prisoners, and their female counterparts made pregnant as a result of rape by their torturers, faced the cruel fate of having their fetuses suffer the pathogenic effects of torture and their newborns stolen from them to be distributed illegally among military families and those complicit with the military state. Literature from Argentina, Bolivia, Chile, Brazil, Guatemala, and El Salvador points to the profound agony of pregnant women whose tortured bodies cannot shield the infant growing inside them and of women who go through labor surrounded by taunting and cursing soldiers. Such women endure an acutely traumatic torture that is truly gender specific (CODESEDH, 1985).

GENDER AND THE PSYCHOLOGICAL RESPONSE TO LATIN AMERICAN STATE TERROR

I would like to turn now to the question of the ability to resist torture, which is clearly aimed at the demolition of the psychic apparatus. Following upon the classic studies of concentration camp behavior by psychoanalyst Bruno Bettleheim (1943), much has been written about the fact that individuals with a history of political consciousness and commitment are better prepared than those with no political awareness or sophistication to withstand the multiple levels of paradox and confusion induced by the torture experience. However, there are also accounts in the Latin American literature to indicate that a deeply held personal morality and connection to community or family can function as well among nonpoliticized torture victims to strengthen their capacity to resist complete psychic destruction. One Argentine female ex-prisoner, for example, reports that while many political activists broke under torture, she witnessed a number of individuals with no political history who were able to resist. It was her impression that people with a strong moral sense and a deeply felt connection to the lives of other human beings and to the community could maintain a clear sense of good and bad, right and wrong, in opposition to the torture strategy aimed at eliminating this capacity by causing extreme confusion and then identification with the torturers (Amati, 1976). One Chilean female ex-prisoner, an activist in the Movimiento de la Izquierda Revolucionaria (MIR), indicates that her ability to withstand the inhumane tortures to which she was subjected was because, in addition to her political convictions, she felt strong links to the people—family as well as comrades—whom she loved and on whose behalf she had been politically involved (Diaz, 1976).

This capacity to sustain oneself in the face of the acute torture situation on the basis of love for others, as well as or instead of the commitment to a set of abstract political principles, seems to resonate particularly with the gendered attributes of women in patriarchal society. This phenomenon has been observed among nonactivists as well. For example, in an investigation carried out by a team of mental health professionals of the psychosocial effects on families that had had one or more of their members abducted and tortured, it was suggested that as a result of such an experience, the families lived fearfully and in isolation, their members having little opportunity to work through the sequelae of the traumatic event. The study pointed out that in many cases, family ties were broken under the stress, with the formation of new couple relationships and new familial nuclei. Investigators observed that the inability to psychologically elaborate the torture experience was more marked among the men than the women and that the most frequently noted behavior among fathers was introversion, isolation, emotional armor, and psychological absence in relation to other family members. The men were inclined to express rage and rigidity in situations that demanded flexibility and dialogue. In contrast, the mothers—and women in general—seemed to have more internal resources that permitted psychological elaboration of the trauma. It was noted that the results of this study confirmed other such investigations into the impact on families of extreme political repression (CODESEDH, 1985).

LATIN AMERICAN WOMEN FIGHT THE TERRORIST STATE

In spite of all the attempts of the terrorist state to eliminate political opposition and to paralyze dissent, and in spite of the profound traumatic impact of its policies, there are multiple examples in Latin America to indicate that even in the environment of extreme political repression women and men retain the capacity to fight for a society that honors human rights and social justice. I would like to end this paper with some remarks about the relationship between gender, intrapsychic life, and political struggle.

In Latin America, from the late 1970s on, a specific form of political practice emerged in response to the policies of state terror that depended on the mobilization of the capacities inherent in female-gendered intrapsychic life. From Argentina to Guatemala, mothers and grandmothers' organizations sprouted like flowers in a deadly draught to uphold the value of life and to demand the reappearance of the hundreds of thousands of sons, daughters, and grandchildren disappeared by the military dictatorships dominating the continent.

In Argentina, the Mothers and the Grandmothers of the Plaza de Mayo were the first organized visible opposition to the terrorist state during that country's Dirty War. The women came from diverse backgrounds, and the thread that united them was their profound commitment to their children and grandchildren and their unswerving refusal to passively accept their disappearance (and torture or murder) perpetuated by an illegitimate government. On behalf of their loved ones they tenaciously fought the terrorist state at tremendous risk to their own lives. Participation in the mothers and grandmothers' organizations has been extraordinarily positive for the participants. Mental health professionals who have assisted the Mothers and the Grandmothers of the Plaza de Mayo argue that its significance must be understood within the context of the intended goal of state terror, which is the imposition of silence, a policy representing the demand for total social disavowal. They describe the pathological effects of complying with what they call "the silence rule," including extreme dissociative phenomena, the denial of reality, learning problems, the impoverishment of fantasy life and so on. The value of the Mothers and the Grandmothers is its refusal to accept the terms of the terrorist state, their articulation of the existence of a responsible party that has kidnapped their children, and their demand of an accounting from those responsible. In their coming together to act, these women have given up the narcissistic dyad "I and my child/I and my grandchild" in the interest of the common concern for all the disappeared children and grandchildren. They do not seek revenge; their action is beyond the realm of retaliation. They demand the return of law and its ability to contain arbitrariness within the social order.

This experience, of course, is not unique to Argentina. In El Salvador, for example, in the midst of the military's onslaught against the civilian population, the Congregation of Christian Mothers for Peace and Justice gave birth to literacy brigades, health projects, and cooperatives. Women who had previously spent their lives going "from the grinding stone to the well" became activists fighting for their rights, confronting the army and often winning. Several mass associations of women were constructed in the war zones, and the Committees of Mothers of the Disappeared have maintained a presence of defiance and protest even when the military outlawed demonstrations throughout El Salvador and severely punished those who were brave enough to continue manifesting open opposition. Their ethos of life has challenged the military's ethos of death (see Golden, 1996).

It should be noted that the many ways women organize to struggle against the terrorist state sheds light on the relationship between political resistance and mental health. It is postulated that women such as these have been able to work through the traumatic loss of their loved ones in part because for them their activist group has become a new privileged object, and they have been able to give a new meaning to group links beyond the family. "The struggle for peace and justice means another type of relationship with the object ... [and] the ego ideal system has been altered: to be a mother now means 'to fight for all our children' or 'to fight for life'" (Kordon & Edelman, 1986, p. 63).

Mental health professionals who have worked with the mothers and grandmothers organizations relate that these activist women tended as a group to be more able to elaborate the trauma they suffered and to deal with the general impact of state terror than their husbands and those mothers who were not engaged in an organized confrontation with the political forces responsible for their traumatic loss (Kordon & Lagos, Personal Communication, August 14, 1989). It is argued that this experience is healing precisely because the individual who has known traumatic loss is no longer cut off from the group. Her loss is no longer individualized, detached from its historical context and from the collective process, but is now part of the political struggle which produced it and can now potentiate its reparation. It is this transcendence of isolation and this commitment to act as historical agents that many Argentine mental health professionals believe to be essential to the resolution of the pathological effects of state-induced trauma. This membership in a group that acts on behalf of the participants' loved ones and of political justice in general performs the function of, in Jessica

Benjamin's terms, the identificatory subject of desire customarily beyond the reach of women in patriarchal culture. In this case, women achieve subjectivity through their identification with the political cause that permits them to actively assert their desire and to demand of society a response.

REFERENCES

Amati, S. (1976). Alunas reflexionces sobre la tortura para introducir una discusion psychoanalytical. Published in English in *International Psychoanalytical Studies Organization Newsletter*, 2 (April 1976).

Americas Watch and Women's Rights Project. (1991). *Criminal justice, violence against women in Brazil*. New York: Human Rights Watch.

Amnesty International. (1991). *Women in the front line: Human rights violations against women*. New York: Amnesty International.

Amnesty International. (1992). *Rape and sexual abuse: Torture and ill treatment of women in detention*. New York: Amnesty International.

Argentine National Commission of the Disappeared. (1986). *Nunca Mas: The report of the Argentine National Commission on the Disappeared*. New York: Farrar, Straus & Giroux.

Aron, A. (1987). Problemas psicologicos de los refugiados Salvadorenos en California. *Boletin de Psicologia*, 4(23), 7–20.

Aron, A., Corne, S., Fursland, A., & Zelwer, B. (1991). The gender-specific terror of El Salvador and Guatemala. *Women's Studies International Forum*, 14(1/2), 37–47.

Bendfeldt-Zachrisson, F. (1988). Torture as intensive repression in Latin America: The psychology of its methods and practice. *International Journal of Health Services*, 18(2), 301–310.

Benjamin, J. (1997). *Bonds of love*. New York: Pantheon Books.

Bettleheim, B. (1943). Individual and mass behavior in extreme situations. *Journal of Abnormal and Social Psychology*, 38, 417–452.

Brown, L. S. (1991). One feminist perspective. *American Imago*, 48(1), 119–133.

Brownmiller, S. (1976). *Against our will*. New York: Penguin.

Bunster-Burotto, X. (1994). Surviving beyond fear: Women and torture in Latin America. In M. Davies (Ed.), *Women and violence* (pp. 156–176). London: Zed Books.

Chamorro, M. E. (1999). Personal interview. Los Angeles, CA, May 1995.

Chester, S. (n.d.). Women and political torture: Work with refuse survivors in exile. (Unpublished).

Comité para la Defensa de la Salud, la Ética Professional y los Derechos Humanos (CODESEDH) (1985). *Seminario Internacional: La Tortura en America Latina*. Buenos Aires, Argentina.

Diaz, G. (1976). *Arbeiterkampf*. 17 January, 6–9.

Faludi, S. (1991). *Backlash: The undeclared war against American women*. New York: Crown Publishers.

Franco, J. (1992). Gender, death, and resistance: Facing the ethical. In J. E. Corradi, P. W. Fagan, & M. A. Garretón (Eds.), *Fear at the edge* (pp. 104–118). Berkeley, CA: University of California Press.

Golden, R. (1991). *The hour of the poor, the hour of women: Salvadoran women speak*. New York: Crossroad.
Herman, J. L. (1992). *Trauma and recovery*. New York: Basic Books.
Hollander, N. C. (1988). Introduction. In M. Langer, *From Vienna to Managua: Journey of a psychoanalyst* (pp. 1–21). London: Free Association Books.
Hollander, N. C. (1992). Psychoanalysis and state terror in Argentina. *American Journal of Psychoanalysis, 52*(3), 273–289, discussion 291–292.
Huneeus, T., & Busto, M. (1989, October). *Women and violence in a violent society*. Paper presented at the annual conference of Society for Traumatic Stress Studies, San Francisco, CA.
Kordon, D., & Edelman, L. I. (1986). *Efectos psicologicos de la represion politica*. Buenos Aires: Sudamericana-Planeta.
Kaschak, E. (1992). Engendered lives: A new psychology of women's experience. New York: Basic Books.
Lancaster, R. N. (1992). Life is hard: Machismo, danger, and the intimacy of power in Nicaragua. Berkeley and Los Angeles: University of California Press.
Latin American and Caribbean Women's Collective, The. (1980). *Slaves of slaves: The challenge of Latin American women*. London: Zed Press.
Manz, B. (1988). *Refugees of a hidden war: The aftermath of counterinsurgency in Guatemala*. New York: SUNY Press.
Martin-Baró, I. (1988). La violencia politica y la guerra como causas del trauma psicosocial en El Salvador. *Revista de Psicologia de El Salvador, 7*(28), 123–141.
Simpson, J., & Bennett, J. (1985). *The disappeared and the mothers of the plaza*. New York: St. Martin's Press.
Stiehm, J. H. (1983). The protected, the protector, the defender. In *Women and men's wars* (pp. 367–375). Oxford, UK: Pergamon.
Thomson, M. (1986). *Women of El Salvador: The price of freedom*. Philadelphia: The Institute for the Study of Human Issues.

Part 4

Resistance

Eyal Rozmarin's chapter (Chapter 15) suggests that social and historical forces play an unconscious yet decisive role in our lives. Telling the story of a conversation between Israeli parents about the prospect of their children becoming soldiers, and of an analytic relationship between two Israelis, the chapter aims to bring to light a hidden balance of power between family bonds and collective attachments. The chapter uses ideas developed by Michel Foucault, and Gilles Deleuze and Felix Guattari in the field of critical theory to examine the ways in which families function as social agents, that is, as socializing institutions. Rozmarin examines the effect of the collective trauma underlying the dominant discourse of the Israeli society, and the hypercollectivity of the Israeli kibbutz, in generating powerful unconscious conflicts that haunt subjective and family life. Arguing that collective affiliations and, consequently, collective politics are inseparable from individual psychology and interpersonal relations, Rozmarin suggests that political awareness and political exchange can play a crucial, liberating role in the therapeutic relationship and life in general.

In Chapter 16, Steven Botticelli explores processes of identification as they shape political resistance to the Israeli occupation of the West Bank and Gaza. Drawing on the narrative accounts of a Palestinian woman living under the strictures of occupation in Ramallah, and of Israeli soldiers who have refused to serve in the occupied territories, he traces the role of identification in giving form to these individuals' resistance.

Adrienne Harris (Chapter 17) explores the slow rise to resistance and consciousness of Israeli soldiers coming to terms with the 1982 Lebanon War through an extended examination of Ari Folman's film *Waltz with Bashir*. In the film, through the use of animation, symbols, and images packed with the trauma and losses of half a century's experience with war and genocide, the burden of history imbricates the burden of soldiering. Looking at the film, at the critical and popular reception of the film and at the discussion in 2009 with psychoanalysts and filmmakers in Tel Aviv, along with interview material Harris made in 1985 with women in the peace movement in Israel, the chapter charts the appearance and disappearance of areas of discursive freedom in which to resist war and to acknowledge its costs.

Lynne Layton's chapter (Chapter 18) seeks to understand contemporary social and psychic obstacles to political resistance. It looks, in particular, at the difficulty sustaining resistance in a period marked by neoliberal free market principles that leave people on their own to fend for themselves. Taking up Freud's connection between disavowal, perversion, and resistance, the chapter describes a perverse pact forged between an abandoned population and political and economic structures committed to lying. An analysis of the film, *In the Valley of Elah*, and personal experiences of political resistance illustrate the role sexist, gender norms, and norms that separate the psychic from the social, play in sustaining the fetish structure of social perversion.

Chapter 15

Living in the plural

Eyal Rozmarin

I

A while ago, a few of us, Israelis who live in New York and others on a visit from Israel, gathered for an evening at the Brooklyn house of a friend. Over dinner and wine, a conversation developed about the extent to which parents have influence on the choices their children make. We spoke about influence in general, and since there was a teenage boy among us, in particular about whether Israeli parents can or should try to prevent their children from serving in the Israeli army. The question came up because all of us around the table were in agreement that the Israeli army has been for a long time the tool of highly questionable if not outright criminal repression and violence. All of us, either from personal experience or from exposure to the stories of others, had knowledge of the practical and ethical catastrophes inflicted by, and upon, young Israelis during their army service. Those of us who are psychologists could tell more about the insidious trauma, forever unsettling the efforts of soldiers to live their lives after, or as is often the case in Israel, between periods of enlistment. The 2006 summer war between Israel and Hezbollah, so careless of both Lebanese and Israeli lives and to the opinion of most a costly failure, was still fresh in our memory. The Gaza offensive of January 2009 was still to come. After years of scrutiny over the way it wages wars, reigns over civilian populations, and protects Jewish fundamentalism, the Israeli Army is no longer a sacred cow. Yet there were deep doubts among us as to whether our generation could or in fact should try to protect the next generation from the daunting prospect of army service and its traumatic consequences.

The conversation was serious, often tense, and complicated. But in the end there were two major reasons given as to why parents may eventually stand back. The first was in the register of collective identity. As Israelis, some argued, we are responsible for our collective survival and may have to accept the necessity of such sacrifice. We may have to do so in recognition that, despite our misgivings about our collective actions and despite the risks we all know about, our children's service is justified by the common cause of sustaining our existence in a hostile environment. It is simply, if painfully so, that our national existence, as perhaps any national existence, exacts a necessary price. This price is often to raise our children to be soldiers and risk their lives.

The second reason given around the dinner table had to do with the relationships between parents and their children; more specifically, with the experience of parents whose children grow up and out of their sphere of influence. By the time they reach their late teens, some argued, parents hardly have any power over their children. They make their own choices, they do what they think proper, and they consider the opinions of friends and cohorts much more than they value those of their parents. Social belonging and group identity take precedence over family loyalties and parents' wishes. There is not much parents can do to stem this socializing tide. But even if there was, my friends reflected, we would hesitate to do so. We all need to belong, both parents and children, and belonging involves compromise. In Israel, we might have to admit, belonging still means adopting a certain national ethos, and as part of it being a soldier, independent of any particular circumstance or politics.

As the conversation drew to an end there was among us a palpable sense of melancholy. As if we all recognized, our discourse exhausted, the inevitability of loss and heartbreak. Our children will go on to the army under our conflicted blessings. Their fate, as indeed our own, is out of our hands. There was also a sense of apologia, a clear if inexpressible repentance. As if while we spoke with each other there has been building between the words a shared need to recognize a tragedy in the making; to recognize and apologize to each other for facing it paralyzed and helpless.

After dinner, we all went out to the garden for a breath of fresh air and to discuss plans for the future. I cannot recall how it came about, but all of a sudden we found ourselves, children and adults together, playing a game called Holocaust alphabet.* You go over the alphabet, letter by letter, and for each letter you recall something that has to do with the Holocaust. A is easy—Auschwitz, B—Bergen Belsen, C—crematorium, D—Dachau. E—Eichmann, of course, F—the Führer, G—the Gestapo, And so on and so forth. Once you're done you can start all over. A is also Akzion, the name given by the SS to the gathering of Jews in the village or ghetto for the purpose of transportation to the Ukrainian death forests or the Polish death camps; not to mention *Arbeit Macht Frei*—work will set you free—the sign that welcomed those arriving to Auschwitz and B—Birkenau. C—*Caput*! Someone shouts triumphantly; *Caput* being the word that often came during those years after *Juden, Juden Caput*. No! Someone else argues with equal triumph, *Kaput* is in K. C is difficult. We can't wait to get back to G

---

* In Hebrew: *Alef Bet Shoah*.

and H, with Göring and Göbbels, Hitler and Himler, waiting to be named in a cheerful chorus ...

A stranger will not understand it, but for us this game is great fun. We play it excited and joyful, undoing the weight of our dinner conversation with the lightness of our peculiar version of remembering things past. It brings out the mix of irony and tragedy that is such a familiar aspect of who we are as Jewish Israelis. It recalls the Holocaust jokes we used to tell as children: How can you fit 10 Jews in a Volkswagen? Two in the front, two in the back, and the rest in the ashtray. It reminds us of that old, always-successful technique for avoiding laughter in the wrong moment—invoke a Holocaust image in your mind's eye. For me it was always a row of naked men standing in front of a ditch they have just dug in a shadowy, black-and-white forest, about to be shot by a group of uniformed men who are already aiming their guns.

But, of course, this game did much more than recall the lost time of our ancestors. Following a conversation in which we concluded that our children will be sacrificed, it summoned from the recesses of our collective psyche the rationale for this sacrifice. Generation after generation, dating back to Abraham, Sarah, and Isaac, our families have been pried open by the commands of god and the decrees of pharaohs, kings, and executioners. Having just experienced our helplessness as history asserts itself over our own families, we dreamt up the long, horrible moment in which this helplessness emerged most traumatically in our recent collective past. And between our dinner conversation about the present and our after-dinner memory tournament, a tragic story told itself as if inevitable. If then we were a scattered, peaceful minority and now we have a nuclear bomb, it still, somehow, remains that we are victims, unable to resist the fatal course onto which the next generation will now march.

II

My wish here is to reflect about this somehow, to look at that which drives, and often makes the passage of such binding legacy across the generations appear inevitable. I use the term *binding legacy* where some would use trauma, since I believe the notion of trauma is far too narrow and too negative to describe what passes between us. The idea of trauma is also often already subsumed in its own rationalization. It has become in some domains, the Israeli national narrative included, the first half of an equation of which the second half is self-justification. To say that we give our children to the army because of our collective trauma is both true and a truth that must be deconstructed. It is precisely one of the tragedies of our binding legacy that what our trauma does, we think it *should* do to us. Our task is to identify the full range of what passes through the generations, the spheres in which it travels, the many ways in which it becomes meaningful and animates our lives. What is it that flows through and comes between parents and their children? How does it use the family, and drives parents to forfeit what may be their deepest instinct and most profound responsibility? How does it make it so that parents hand their children over to the collective and its life-defying projects?

These questions are, obviously, significant in general, but they are of particular significance to psychoanalysis. Psychoanalysis is a discipline wholly invested in the discourse of the family. From its origins to the present, psychoanalysis conceived of the family as *the* unit into which the individual is born and through which he becomes a meaningful subject. Yet we are all born into societies and cultures, into norms, customs, rituals, and myths. Our personal and familial stories are never insulated from the direct effects of social forces, of politics, economics, and history. Psychoanalysis, however, for the most part treated the effects of the social indirectly. True to its focus, it conceived it mainly as a set of constraints on subjective experience established through family dynamics. The process by which the social comes into the family in traditional psychoanalysis is called the Oedipal complex. Through this process, the social crosses the boundary between the family and society, gets played *in* the family, and becomes a sovereign principle of the subject's psyche. Classic Oedipus involves a rich set of characters, taking part in an unfolding drama where desire, shame, and anxiety become wedded to social norms. It concludes in the establishment of a powerful social agent, a superego, at the very heart of the subject. It results in the transformation of a rebel-child of many potentials into a well-socialized girl or boy. If the social causes any further trouble, this trouble is now internal to the subject, a matter of personal suffering and responsibility. It is the ultimate accomplishment of the Oedipal process that once concluded, social reality and its vicissitudes can only figure as a psychological affair.

Since this originary thesis, psychoanalysis continued to see socialization and conformity as one of the most important developmental tasks of both families and individuals. It accepted that it is the role of parents to raise well-adjusted children. In both theory and method it suggested that those who failed to develop properly were failed by their families of origin.

It largely remains true for psychoanalysis that whatever drama, conflict, or vicissitude it hypothesizes, it looks for and finds within the family and not between the family and its surroundings. Psychoanalysis generally assumes confluence, if not harmony, between family and society. It does not attend as a matter of general theory to the possibility of diverging interests, or conflicts, between the family and society. As if it always holds true that parents raise their children to embrace a social order with which they fully identify. As if we all wholeheartedly accept the social reality around us. But how can psychoanalysis address the relations between family and collective, when the collective asks the family, as in the old Biblical story

God asked Abraham, for the sacrifice of one's child? When there emerges a potential conflict, as it does, consciously or unconsciously, in many Israeli families, what is the psychoanalytic stance? What should it be?

That psychoanalysis does not question the relations between family and society has made it subject to fierce criticism, leveled most forcefully by Michel Foucault, Gilles Deleuze, and Felix Guatarri. Foucault (1999/2003, 2003/2006) critiqued psychoanalysis as a discourse at the service of the power structures that characterized the social context in which it came into being. The point in time, the turn of the 19th century, as in Europe societies were coming firmly under the rule of bourgeois hegemony, increasingly identified along the lines of statehood and nationality. One of the crucial aspects of this historical process, Foucault argued, was the organization of society around the idea of normalcy and the development of practices designed to define, regulate, and enforce norms. The nuclear family, itself consolidating through this process as a fundamental social unit, came to have a crucial role in this new organization. It became the focal point of multiple discourses, received the status of a primary social institution, and was delegated a formative social responsibility. The family was entrusted with the role of raising normal, that is, productive and well-disciplined children, who grow up to assume their given place in the social order.

In this, the family has taken upon itself the functions previously served by priests and princes, the responsibility held up until then by the church and the aristocracy. The family was now called up to make subjects believe what they must believe, fear what they must fear, so as to conform to the new sovereign order. But in transition, the function had been readjusted to the needs of its new holder. The power previously wielded through religious dogma and princely violence came into the hands of families transformed. It came in the form of reasoned knowledge based on norms.

The central premise underlying this arrangement is that there can be no conflict between the family and society. Both family and society aim to raise productive, law-abiding citizens. Both family and society are best served when the family takes it upon itself to function as a formative social institution, and to propagate the social norm. It is in this context that Foucault identified the tenets of psychoanalysis as reproducing the demands of social power. Psychoanalysis, Foucault argued, abided by these premises. It joined other disciplines governed by an enlightened bureaucratic ethos to give these premises the status of reasoned knowledge, and developed intricate methods to regulate and enforce them. It did so by establishing a discourse that renders all conflict familial and mental, and that confines its drama and potential resolution to the interior of the subject. This drive of psychoanalysis to interiorize all conflict is what made Foucault famously conclude that psychoanalysis is nothing but a subtle, more insidious extension of the 19th-century asylum.

Deleuze and Guatarri (1972/1983) made the psychoanalytic notion of the family the focus of their joint oeuvre. They paid in this context close attention to psychoanalysis' archnotion, the unconscious. Since psychoanalysis sees family relations as formative of the subject's very psychic structure, they reflected, its notion of the unconscious is conceived along the same lines. Simply put, for psychoanalysis the unconscious is structured around given constellations of mother, father, and child. The remainder of the meaningful world, all sociality and reality, materialize through the internalized representations and the intricate relations between these three characters. Everything there is outside the family psychoanalysis reduces to this fundamental, mythical triangle. Everything that may happen to the subject is already inscribed in its tensions and angles. But the idea that subjectivity and its unconscious aspects are defined by family relations is a device, Deleuze and Guatarri go further than Foucault to argue, for maintaining a hegemonic social order. It is that order that, precisely, makes a certain kind of family its cornerstone. Its aim is to push to the background, in fact to repress into a forever unreachable unconscious, the conditions required to sustain its power.

Yet, Deleuze and Guatarri (1972/1983) argue:

> the family does not engender its own ruptures. Families are filled with gaps and transected by breaks that are not familial: the Commune, the Dreyfus Affair, religion and atheism, the Spanish Civil war, the rise of fascism, Stalinism, the Vietnam War, May '68—all these things form complexes of the unconscious ... If in fact there are structures, they do not exist in the mind, in the shadow of a fantastic phallus distributing the lacunae, the passages and the articulations. Structures exist in the immediate impossible real. (pp. 96–97)

There is in the unconscious far more than the internal affairs of the family. Families live in the world and the world lives in them; the world as it is, as it is thought or unthought, as it has been inscribed in and between the lines of history. The unconscious is filled with all matters, present and past, with stories and myths, culture and society. It holds and is held by more than one person or one family. There is, in other words, a collective unconscious in and between all of us. In positing an unconscious confined to the inside of an assumed self-contained subject, and to the effects of a solitary drama alleged to have taken place in his family of origin, psychoanalysis is complicit with the repression of social history and reality. According to Deleuze and Guatarri (1972/1983), psychoanalysis should recognize this, its basic fault, and address it. If it cannot, it should be substituted by a more critically enlightened, more liberatory discourse.

But a psychoanalyst for whom this critique rings valid still faces an irresolvable conundrum. Unless he rejects the psychoanalytic endeavor in favor of other forms of meaningful engagement with others, he must think and practice within an order that he finds deeply problematic. Moreover, if he chooses to engage in psychoanalytic discourse as a means of addressing the dilemmas forced on individuals by collective power, such as to compromise or sacrifice their lives for corrupt causes or risk painful alienation, he will find himself in the position of a tortured parent. Quite like a parent who cannot remove his family from the collective and its ominous claims over his children, the psychoanalyst will be able to contemplate the possibilities only from within the repressive order. Quite like a parent, blinded and trapped by the collective's narratives, it will be very difficult for him to see the dilemmas forced upon him vis-à-vis his patients as anything but the trouble of intimate relationships.

This is because the forces by which the social represses, the forces that push all subjective matter into the family and render families inwardly bound islands, these same forces make the psychoanalytic relationship itself a family organization. And so what confounds families in general, confounds also analyst and patient. All problems appear as family problems and are configured in terms of family relations. The psychoanalytic notion of the transference reveals this effect most powerfully. When we are consigned to seek all formative meaning in the words and deeds of our parents, any intimate agent of meaning becomes a parentlike figure. On the other hand, any subject in search of guidance and care is perceived, and perceives himself as an infant. Caregiving and caretaking are cast as the drama of child and parent. Psychoanalysis is forever locked in the dilemmas of intimate desire and personal identity. Thus, for better and for worse, the psychoanalytic relationship reproduces the family's binding paradoxes. Like parents, we cannot take those in our care out of the world in which we all live, nor out of the simulacrum of this world that we recreate in the analytic relationship. Yet we are called, as crisis perpetually looms, to address their destiny and save them.

At this point, where the family and psychoanalysis each find themselves in a bind, the discourse offered by Foucault, and Deleuze and Guatarri reaches a limit. This is because, as psychoanalysis most fundamentally offers, subjective life is not driven by any single force or reason. The unconscious is not only a depository of discursive canons, not even if we conceive of them, as Foucault did, as a complex set of contradictions. Subjectivity is not only an effect of objective forces, consciousness is not only a charade, the subject is not simply a symptom. The subject is always also an agent who inhabits a self-conscious, willing position. As lost in the effects of social forces as he might be, he stills desires, seeks meaning, and makes decisions. And more often than not he has the idea of freedom. If our goal is to help change the course of any singular life, to help liberate the subject from what binds the most basic yet unconscious tenets of his being, we must understand not only how social power binds the subject, but also how the subject takes this power, owns it, and resists it. We have, in other words, to meet the subject where he is an effect but also agent of his destiny.

In this junction, of social power, subjective life, and agency, the question becomes this: How do we as analysts keep our hold where parents lose their grip? How do we make it so that those in our care do not fall prey to that which makes them want to be soldiers? How do we save them, with and against their will, from the fate they seek to evade in asking for our learned protection? If it is true that families raise soldiers because they are agents of collective forces, then the answer is as simple as it is formidable to realize—we must untie the knots that bind the family to the collective and its causes. We must—in our domain—challenge the process that Foucault, and Deleuze and Guatarri expose so clearly, whereby society has organized to form the family as a normativity-producing institution. We must conceive the psychoanalytic relationship as a locus of critical resistance, as a practice of an exploration for alternatives. The families we create with our patients should be different. That Brooklyn evening described above reveals what vortex of present and past, pain and desire, intimate and national sentiment; what storm of memory, guilt, and delirium will rise to face us.

III

Asaf is a 35-year-old man whom I have been seeing for several years in my New York office. He is, like me, an immigrant from Israel. Asaf grew up in a kibbutz, that extreme form of collective life that took over childcare at birth, replacing parents with night-nannies and siblings with cohorts, reducing the nuclear family to an almost abstract association, responsible neither for feeding nor for soothing. Asaf remembers crying alone at night, quietly, so as to not wake up the other children. He remembers also a period of nightly escapes from the children's house to where his parents lived, a few hundred yards away. Scary voyages in the dark, shadows lurking behind trees, wolves howling in the distance; journeys ending in his being taken back, reproached and shameful. His mother tells him that his escapes began when he was about two, a toddler. At some point his parents yielded and found a way to let him sleep with them. Later in childhood he went back to sleeping in the children's house without protest. He could not know it then, but many kibbutz children went through similar periods of nightly escapes in search of home, most of them thwarted. Parents could not allow their children home without risking the collective's wrath for the transgression.

In the afternoons, Asaf often went to visit his grandfather. He was his grandfather's favorite grandson and, except for special occasions, the only one to visit him. They would spend hours together, preparing food, playing games, and telling stories. Asaf's grandfather came to Israel as a young man, soon after the war in Europe ended, and was one of the kibbutz's founders. He lived through his teenage years during the war, most of them in a Polish ghetto, then in a concentration camp. It remains an open question how he survived there as long as he did.

Asaf was a strong, handsome boy, a friendly and popular athlete. As was expected of boys like him, when his time came he enlisted in an elite unit of the Israeli army. He spent a great deal of his service in the then occupied, dangerous southern Lebanon, and the still occupied, rising Palestinian territories. There were frightening moments and cruel reality all around him, but he still does not like to talk about it. It seems to him irrelevant. There is a strong policy, a "don't ask, don't tell me how terrible it is what you go through in the army" between parents and their soldier children. Have a meal, go out, have fun, they tell them when they come home for a break, but don't tell me what you're doing. This policy is so powerful that even those men who lived through harrowing times together hardly ever talk about it. It is strong enough to force a silence years and continents away, even between patient and therapist.

It is part of the Israeli social contract, perhaps not only the Israeli, that trauma is acknowledged collectively but forbidden recognition between individuals. There is a day a year when the nation stops to remember its dead and wounded, and to retell stories of heroism. But otherwise, unsolicited personal memories of hardship or proclamations of sacrifice are considered unpatriotic whining. The same is true for the Jewish-Israeli archtrauma, the Holocaust. There is a day a year dedicated to sacred collective commemoration. On the side of the profane, Israeli politicians never hesitate to enlist the Holocaust in the service of their agenda, grand or miniscule. But those who came to Palestine in the years immediately after this unimaginable, always personal catastrophe, those who Tom Segev (1994/2000) called "the 7th million," were consigned to nightmares and silence. The nascent Jewish hegemony took upon itself the task of absorbing the trauma, transforming it into a collective drive for national revival. This task is still its nonnegotiable raison d'etre. But the individuals to whom it all happened were left possessed by intolerable, shameful secrets; a repression second only, and perhaps a mirror image, of that applied to the stories of the exiled Palestinians whose land they settled.

On this wasteland of doubly forfeited past and the disavowal of two diasporas, a kibbutz was founded, and children were raised in an experiment where the family was deconstructed. But weakened as it was, pried open to collective will, the family remained vital. In the dark of night, children who could barely walk or talk rebelled and went in search of home. Family dramas still unfolded; stories were told, secrets were whispered. Yariv's grandfather did not tell him until later, just a sliver even then, of what he had gone through in the ghetto and in Auschwitz. But what connections were drawn in the mind of a child between the silence of his most intimate kin, the loud ethos of a new kind of Jewish man, and the collective's ritual forgetfulness and telling? What pain did he absorb, what lessons did he take then? What filled the void left by this lethal, double-edged absence—no past except that hinted to, with shame, between the lines, no parents to rest by unless he risked the dark of night and took upon himself the burden of transgression?

From a very young age, Asaf had his own shameful secret: he was attracted to the other boys. He had to keep his secret in an environment without privacy, and he became very good at being secretive. Yet there was a stubborn anxiety that his desire would be revealed and mocked, and he did much to counter that prospect. It was not until he came to New York that it felt possible for him to explore his sexuality, but coming out made for new secrets. While, on the one hand, he found men he loved and lived with, on the other, he pursued quick, transitory liaisons, steeped in a deep sense of danger. Repeatedly he was overcome with fear and shame. He loathed himself for giving over to what felt to him a dark force, yet for a long time while he could not but return to this precipice. As if, indeed, to find sense of peace he needed to engage in, and survive transgression and its shamefulness. As if he needed to prove himself immune to his own fears, able to travel and return in one piece from a place of hunting abjection. He lived, in sex, so many aspects of memory and repression.

There was another area of trouble for Asaf: he was constantly concerned about his physical well-being. He went through long periods where he experienced bodily sensations that felt like symptoms of unknown illnesses. Yet despite close medical attention no illness was ever diagnosed and his symptoms remained mysterious. There was a period of strange feelings in the abdomen, another with breathing difficulties. There was also a long period of inexplicable vertigo that could not be accounted for by any medical examination. All of these went away as incidentally as they came. Under their sway he feared cancer, a heart condition, a brain tumor. He was afraid of dying. But, in fact, he was all along, and remains to this day a thick-haired, towering man, a modern day Samson.

Asaf came to me by an unorthodox route. We met at a dinner party. We spoke quite a bit, my professional hat was off. We had, it turned out, mutual friends, and we grew up a few miles away from each other. A few days later he called and said that he'd like me to be his therapist. I said it felt, to me, complicated. We had met socially and might meet socially again. One of our mutual acquaintances, in particular, was a close friend and may have already blown any professional cover I might need to create a therapeutic setting. If this is a problem for you, I'll understand, he said. But for me, knowing you and your friends actually feels safer. I decided to go with his instinct. Perhaps, I am thinking, with the years of our working together since then in my mind, he needed precisely that—someone whose secrets were already apparent. Knowing my friends meant he could easily imagine what my life in New York was like and, if he wished so, learn many more details. He could also imagine that growing up by the same rolling hills, covered with orange and avocado groves, overlooking the sea, we had been all along also treading the same conscious and unconscious landscape. We grew up with the taste of the same foods, heard the same songs, learned to laugh at the same kind of humor. We had also beneath us the same ghostly terrain, rife with the hushed secrets of grandparents, the untold pain of parents, and the lies that surround the erased ruins of Arab villages.

We have gone through a great deal in our work together. Far more than any narrative could do justice. But it all led, in some fashion, to one moment of recognition that transpired between us recently. Asaf has been going through a particularly difficult time, with both love and business relations in upheaval. It is a time of serious challenges; important decisions need to be made, but it is also a time of great promise. So much is in flux precisely because he has been able to begin lifting himself out of some fundamentally bad compromises. Yet it is infinitely difficult for him to see the forest from the trees. He sees danger all around him. He sees, particularly, risk, disappointment, and betrayal in his relationships. He is not deluded. January 2009 is an anxiety-filled time for everyone in New York, the epicenter of an ominous global recession. It is also true that some of the people around him have been profoundly disappointing, which is why he is breaking with them. But Asaf is trapped in a compulsive funk that won't let go, as if he lives under a spell that cannot be lifted. The spell decrees life as a paradox, a zero-sum game between mindful depression and dissociative resilience. He should finally, we both agree, be able to experience his difficulties. He can allow himself to feel, and to protest as much as he needs to, that he is scared and tormented. Strong and conscious as he became after years of hard work on himself, he no longer needs to keep his feelings secret. But once he starts, his fear and torment seem to have no end, and at a certain point he feels completely overwhelmed and utterly defeated. He then begins to dread that if he cannot put it all behind him, once and for all, he will go crazy.

The theme has been between us for a long time, that this dilemma harkens to when he was a toddler crying all alone in vain or gathering himself to march night after night in the dark to find his parents. There lives in him, brave and inconsolable, a little boy torn between desperate longing and courageous self-assertion that only reaffirms his loneliness. The theme has also been between us that his closeness with his grandfather left him with an untold burden. The horrors that his grandfather had to overcome so he could live, the secrets that he had to live with on his own, sparing his children and grandchildren. Did they not all live inside Asaf as fear and shame and passion constantly at war, as guilt for living and forgetting? Did they not stand there, begging him to live and leave them all behind while haunting him in the form of a ceaseless, faceless illness?

But all the insight gained between us was not helping him these days. He was compelled beyond the reach of our carefully pulled together reason. There were only two options, drown in despair or move beyond it, collapse or prove yourself invincible. And since neither was fully possible he was angry and exhausted. "Can you imagine life outside these options?" I asked (by far it was not the first time I had done so). But on that day he had a new idea: "Maybe," he said, "I should decide on a specific day in which I dwell on all of this and mourn the person I could never be. Maybe if I had a day like that, I could live differently on all the others."

This felt like an epiphany, a way to put the past in its place. But in the following session he was worse. I am tired, he said, I can't go on like this. Maybe I should just pack up, leave this awful place, go back to Israel. This was not a new fantasy. By now he knew it to be born of a nostalgia for something that no longer existed. The kibbutz, like most of them, has been privatized. He was no longer in his 20s. Israel would not be for him what it has been, which in any case drove him across the ocean. But then, on that day, as was true for the most recent two weeks, there was another element, yet unmentioned between us, in the picture. In this, the first half of January 2009, Israel was in the midst of an all-out war on Gaza. He had been spending a lot of time following the news, and the topic came up between us. It came up and soon thereafter became a heated argument. He said it was regrettable that Israel had to do what it was doing, but it had to. How could it go on tolerating those Qassam rockets? I said it was a criminal war waged by a cynical government in control of a vast army, against a helpless civilian population trapped by an equally cynical government in a starved ghetto.

But they are shooting at us.

But we have been oppressing them for decades, and have put them under siege for the past two years without any reasonable chance to live the kind of life that you and I take for granted.

We have no choice.

Yes we do.

What do you want us to do?

Talk with them.

There is no one to talk with, they want to destroy us.

Did we not negotiate a truce that held until recently?

They broke it.

It went on and on. At some point the hour ended and we were already standing up. Asaf is always conscious not to overstay my hospitality. "You are crazy," he said, walking out of the room. "See you next time," I answered.

After he left I was upset and confused. Upset, because the war in Gaza had stirred in me an almost unbearable mix of helplessness and outrage, and I had just had another opportunity to experience it. Confused, because I literally found myself in a fight with a patient, a fight that did not seem to leave either of us well. It was not that I simply could not help myself. There was a moment of hesitation, and in that moment I had an instinct that it would be good for us to talk about the war, no matter where it took us, and I allowed myself to follow that instinct. I did not feel regret after he left, but some of my clinical alarms were ringing. Did I stray too far? Did I lose my clinical bearing? Did I forget what this relationship was all about? One alarm, indeed, had to do with my role in our relationship. Was it my job to be political? But it was not about being political per se. I wouldn't have been less political if I had nodded and agreed with him. I would have simply conveyed to him, had I done so, that we held the same views, that we had the same politics. What I had done was to break an assumption that since we were alike, of the same collective, we had the same outlook on what had been going on in Israel. How much our alliance depended on this assumption now became the question that bothered me. To what extent should I have made an effort to maintain this assumption alive, so as to be able to keep my analytic responsibility, to be for him what he needed me to be.

Another alarm rang a more general note, that in this fight I abandoned any hint of what we now call thirdness (J. Benjamin, 1995, 2006). It was true, once the moment of hesitation concluded in my surrendering to the rising conflict, neither in thought nor in deed have I been able to move out of my position. I kept fighting to the session's bitter end, even as we were already standing. There have been many moments like this before, with Asaf and with other patients. Debates, heated arguments where each of us had taken a sharp angle and stayed there. But on this occasion the slippage felt critical, and the lack of perspective threatening. Did I relinquish my most basic duty to him? Did I betray my deepest responsibility? Have I deserted him when he needed me to hold the complexity of his, of our, situation in mind, to be sane in the face of insane circumstances? Have I not, instead, materialized in front of his eyes as yet another person driven by these circumstances into madness? He said that much, did he not? When he told me "You are crazy." Crazy is not how he needs me.

A third alarm rang more viscerally than the first two. It rang thus: Right as you may think you are, you have just taken yourself out of the consensus. On this day, when almost the entire Jewish Israeli collective aligns behind the war, as mad and criminal as you think it all is, you are exposing yourself to your patient as a member of a marginal ideological minority. Not only marginal, but impotent as well, and unpatriotic if not traitorous. You, against all these brave, beautiful soldiers and the families who worry about them. Questioning what they are doing for you, and in questioning making it more dangerous for them, breaking their spirit, offending their service.

If he had been one of those you know would have agreed with you, then, well, you belong in that same unfortunate boat and can safely commiserate together. But this one identifies with most of them in that extended kibbutz, he feels warm and safe in their company, he longs for it. You, on the other hand, don't. You are different and have always been. You never felt part of their fraternity; it always scared you. Now he knows how lost you are, how hopelessly disturbed and frustrated. Now he knows the secret of your loneliness—you do not belong. Not a good place to be, the psychoanalyst exposed as a stranger among his own.

IV

There it was—the specter of a parent's fate in the family that he and I had created. He wanted to go away, to join the other boys both literally and symbolically. To enlist again in the extended military that Israel sometimes is. In a time of fear, his life in turmoil, he wanted to go back where he belonged, where he could feel again safe and oriented. But I became alarmed, remembering the cost this belonging had already exacted from him. What he saw as self-evident and reassuring, I saw as surrendering again to the particular rendering of the collective story that haunted his life. A great fear came up in me that he was compelled against his best interest, to perpetually seek what in my mind was a dangerous destiny. A destiny that was first inscribed in him during his nightly escapes as a toddler, and wound up manifesting in his tortured desire and inexplicable physical pains. The destiny of a child held by forces stronger than any family bond on a stake, an Isaac without any mother or god to save him. My own child in transference, slouching helplessly toward his own demise like a lamb to the slaughter, in a macabre echo of the way his grandfather's generation was so cruelly derided by the new uber-Jews of Zion to have done. Only now the slaughter was self-perpetrated, masking as a choice of the sort that marked his life since he was an infant spending his nights alone: falling asleep in order to keep on living, psychic death as a condition for finding home. So I resisted, I tried to protect him, to argue against going back and its false premises. He looked at me as many boys surely look at their parents. "You are crazy," he said, as he walked out the door. I was left feeling a wayward, shamed analytic parent. Here I was, exposed in my futile resistance and its incomprehensible reason. Exposed and expelled—an exile of the collective that should be his home.

Could this be what parents face, what they try to avoid before it all becomes conscious? To be seen by their children as pitiful losers in the triangular balance of power between family, child, and collective? To be seen in their weakness, in their helpless protest, clueless outcasts in a collective that basks safely in its ideology? Could it be that parents let their children go in order to avoid, for their own sake and for the sake of their children, a crisis of identity without resolution? Could it be then, on the other hand, that to stand up and resist, to persist without letting your child go, means to keep one's child away from ever belonging without conflict, or worse, from ever belonging at all? I surely feared that much, instinctively if unknowingly, in the moments following my argument with Asaf. In the family that he and I had created on the east bank of the Hudson, to do so would have meant—if he had listened—to risk making him an eternal exile, even if he eventually went back to Israel. Who in their right mind would want that for their child? Who in their right analytic mind would want that for their patient? If in the end the dilemma is between oblivious yet reassuring belonging and awareness that beckons only alienation and solitude, can we make a choice for the latter with a clear conscience?

And in fact, as was evident in our Brooklyn conversation, a frayed conscience is part of the experience of parents facing such dilemmas. If we are shocked into a paradox of emotions, we are also thrown into a confusion of desires and obligations, perhaps even crueler and more traumatic than the confusion of tongues that Ferenczi (1932/1955) wrote about. In this confusion of tongues, which is between the discourse of citizens, accountable to a given collective and social order, and the discourse of parents responsible for the well-being of their children, no choice is free of trouble. Where the well-being of the subject for whom we care and the demands of the collective are in conflict, our choices, whether all along or at singular, dramatic moments, are always also choices between loyalty to this loved one and loyalty to the collective, which is also an object of love. Whose interests should prevail, to whom do we own our allegiance? What are we more accountable to when our obligations seem to be in conflict? Who is more deserving of our love? Underneath the inevitable clash between different kinds of caring and belonging rests a no less irresolvable problem in the realm of ethics. The question is always also what is right.

In this realm, any decision for is also against a responsibility. Any choice we make is also a betrayal of something or someone we love. To choose the good of one's self or of loved ones is to betray the common good, the well-being of the many, those we depend on and identify with. Yet, in surrender to the collective's hunger for good members and good soldiers there always persists a trace of suspicion and alienation. One can never be loyal to a principle without loosening attachments to concrete objects. One can never abide by an ideology without willingness to sacrifice individuals or instances that oppose it. A world in which parents resign themselves to sacrificing their children will always be haunted by the shadow of doubt about the limits of love and responsibility. Children will never fully trust their parents; friends and neighbors will always sense the distance that can grow between them if their relationships came into conflict with social normativity. On the horizon of each and every turn there would be lurking a responsibility betrayed, a love forsaken, and an immense loss of one or another kind. In the clash between the discourse of society and the discourse of parenting, there might always be a place where, letting social desire take hold where parental holding is wished for is as bad as meeting the plea of a love-needing child with adult sexuality. In this place we are all destined to be both abused and abusers. And as is the case with the abuse Ferenczi discovered, the trauma could not but double under the denial forced by the social desire that was its perpetrator.

"There is no document of civilization which is not at the same time a document of barbarism," Walter Benjamin (1940/1968, p. 256) famously argued. This may be equally true about the way the forces of civilization, of social power and history telling, come in and between the subject and the family. Foucault, and Deleuze and Guatarri have told some of the story of this barbarism, and theorized about the means by which it circles and invades the family. They exposed structures and processes that have come to exert over our lives immense power, precisely because that power is devised to keep unconscious through the configuration of the family. They rejected, in my view rightly, much of the psychoanalytic canon in this context, as siding with and enhancing the repression affected by this configuration rather than critiquing it. But they did not, perhaps by the nature of their discourse they could not, offer recourse other than alternative structural principles and ideologies. If their critique does much to enlighten us as to the social conditioning of our living, and in some instances suggests new ways to think and engage socially, it cannot, by itself, fully address the cause of freedom as a concern and desire in concrete subjective life. Psychoanalysis on the other hand, with its deeply thought and exercised capacity to encounter and make sense of subjective and intersubjective living *in practice*, can do much to drive their thought into the domain of lived experience, and to animate its prospects.

It is in this vein that I introduced Foucault, and Deleuze and Guatarri into the midst of actual family relations; those of my friends with their children, those that Asaf and I created in our relationship. What I have tried to show is how the forces they conceptualized and their attendant barbarism transpire during critical or prolonged moments of conflict, as heartbreaking, irresolvable dilemmas between opposing loyalties. How they are suffered, but also negotiated and resisted, how they overwhelm but also reach a limit beyond which something in the subject rebels. I tried to show regarding the collective's stirrings what psychoanalysis knows best, how the repressed keeps manifesting and works both for and against its own purpose. I tried to show how we enlist but still refuse to be blind soldiers; where we are conscious agents with a sense of self and need for meaning that never goes away completely. In doing so I hoped to find a new angle vis-à-vis the question of social repression, and the potential for it to be met, appropriated, and transformed in and between us. My wish was to deliberate new means for how to address the plight of those who ask for our assistance. But it was also to find ways to use the power of psychoanalytic exploration beyond its normatively circumscribed mandate. My wish was to address subjective and family life not only as the effects of social forces, but also as sites of power and signification themselves—hubs of meaning and will that could be expanded outward, transform the way we think about the social and its effects, and perhaps drive actual change in our collective living.

What I have come upon, and where such change perhaps could be imagined, is one of the most traitorous crime scenes where the collective reaches in and grasps at the core of subjectivity. This crime of theft and fraud, sometimes of murder, where subjectivity becomes the hostage of sovereign ideology, is perpetrated at the point where one's sense of self, one's own identity draw meaning from collective renderings of life and history. It is the point where the subject can conceive himself only as one of many. Where identity is both consciously and unconsciously collective, and where being depends on belonging with others across time and territory. In this definitive terrain personal memory meets official history, subjective loss and fear go to work for the collective shame-terror propaganda industry. The objects of one's desire and guilt become deceptively hybrid, both singular and plural, both personal and social. One's need and duty toward those he loves become inseparable from his attachment to the ethos by which the collective aims to govern the very possibility of living and loving.

In this space, which is neither personal nor impersonal, Asaf's love for his grandfather is indistinguishable from the belief instilled in him that a strong Israel saved his grandfather from the Holocaust and continues to make his life possible.* American immigration officers get mixed up with Nazis, and an imperial Israel becomes a ghetto in grave danger. His ancient longing to be a boy whose parents cared for him masks as nostalgia to a past whose pain is muted and as a loyalty to national ethos. The anxiety of daily life and the anxiety of exile become a single catastrophe. The possibility of happiness comes under the sway of old, unquestioned normativity. To be means to belong, to live in identity. In this space, though, repressed trauma becomes physical trouble, the dissociation of a toddler lost in the dark becomes the delirium of adult sexual adventures, and a home that

---

* During his speech on the eve of the recent Holocaust commemoration day in Israel, April 20, 2009, Israel's Prime Minister Benjamin Netanyahu made sure to warn Iran's President Mahmoud Ahmadinejad thus: "We will not allow the Holocaust deniers to carry out another Holocaust against the Jewish people. This is the supreme duty of the state of Israel. This is my supreme duty as prime minister of Israel." The speech became a headline in the daily *Haaretz:* "Netanyahu to Iran: We will not allow a second Holocaust." Such rhetoric, which among other things clearly cheapens and trivializes the memory of the real tragedy that met the Jews in Europe, is very common in Israeli politics. But curiously, it hardly ever meets criticism. As if it actually soothes, rather than alarms, the audience.

never was is substituted by a country that never made it possible. To be one's self means not to be alone, to be harmoniously together. In this space, precisely, a parent's instinct to keep his child safe comes under a deceitful identification of the child's well being with that of the collective. Love and rationality both become servants of state ideology.

It is the vortex of semicollective, semisubjective forces operating in this space that surfaced between the two acts of the Brooklyn evening I reported. The memory of our Holocaust came up, casting a spell with the charm of a child's game to remind us who we are and where we come from. It came to make us forget any objection to the fatal cost demanded of our families by our shared identity. All differences of opinion silenced, our shame and sadness melting away, the joy of our belonging with each other took over. And it was fueled by years of teaching that instilled in and between us the unquestioned knowledge that it is that horrible trauma that binds us. And it is true, with all these years bearing down on where subjective and collective meet in each of us, this trauma as told and silenced, numbed and celebrated does in fact bind us. We love each other for having it in common. It adds to our relationships a deep layer of intimacy and meaning. It might even be that we love our trauma for giving us identity and pulling us so close across decades and oceans. To give it up will not be possible without a great sense of loss and disorientation.

But giving up the trauma and its spell is precisely what is needed. For Asaf to live a good life he must somehow untie the knot the makes him long for parents who could not be there and suffer for the ghosts that haunt the reason behind their absence. He must resign himself to knowing that in some ways all of them have been betrayed by a collective ethos that made their pain its rationale and weapon, and the betrayals have caused damage. His life, he must concede, should not be lived as a commemoration. It should not hinge upon a melancholic fantasy that refuses recognition of what has been forever lost to history. His loss, moreover, he will have to realize, is not what the collective's sanctioned narratives and veiled threats made it out to be. It is, more likely, of the psychic reproduction of those gray-concrete wartime shelters we all ran to when, once every few years, the sirens sounded; where we trembled for our lives but also knew for certain who was bad and who was good. There will be in this realization both relief and sadness. There might also be a new kind of anger. In the end he might feel that he does not want to go back to Israel, or he may make his home there as a form of compromise, as staying in the United States would also be. But in either case his choice will be one driven by consciousness and a will for self-realization rather than by an unconscious fantasy governed by the collective's design for him. And in this vein, he would need to allow himself to think that in offering all that to him I am not crazy; that my position is neither thoughtless nor disloyal or antagonistic. As for my part, I will have to step out of the role I took, or rather fell into, during our enactment of the subject–family–collective drama, to indeed be more thoughtful and less antagonistic to the ghosts of our shared history. I will have to better recognize my own conflicts related to belonging and identity. I will have to admit more than I'd like to that this tortured and deceitful ethos is my beloved burden also.

That all of this came up between us in the form of a political debate about the wars of Israel is not that surprising. The wars of Israel are present at the core of both our family histories. My childhood, like his, is marked by them. In 1967, me at 5 years old and my mother 9 months pregnant, my father went to fight the Six-Day War and stayed away for months. He was away when my sister was born, but the war was present. For a long time my sister's nickname was Kalkilia, given to her by my father's army buddies after the town they took together in the West Bank. Six years later he was gone again for months during the October war. This time, while on vacation he told stories about Ismailia, on the far side of the Suez Canal in Egypt. During these months I spent my days looking at maps and drawing battle plans to take over the entire Middle East, like him. I also made with my mother's help a war diary combining newspaper clippings with my own personal comments. To this day, my relationship with my father consists of heated political debates about war and peace, and in them our entire history of intimacy and conflict is enacted. We live in a triangle, he and I, quite different than the Oedipal. In this triangle of identity and love, the third party is Israel. His Israel and mine are not the same, but tearing them apart has been like breaking up with him. The Israel he believes in I cannot accept, the Israel that I want he thinks impossible. The father he was to me is also the soldier he was in the Israeli army. The sister that came into our home during the war he made, with tender love and pride, a token of his conquest. Israel and home, our love and politics, our ability to trust each other and our trust in our collective's story telling are inseparable.

The same is now loudly true between Asaf and me. It is my point precisely that it has always been so, but now the ghosts are materializing and we can hope to deal with them. By ghosts I mean not only those of the dead, but also those, far more insidious, that carved the texts on their memorials. These ghosts of history and politics—history *as* politics, nationalized their death, made it collective property, and told us how it should become our purpose. These ghosts need to be exposed and exorcized, or they will keep the two of us paralyzed together. We need to make their presence matter for conscious experience, to undo the repression that hides their manipulations. We need to see and feel how they arise and gravitate between us, how they contrive to make us blind perpetrators of their brutal message. We need to tease apart those aspects of our being that have been pushed together beyond recognition under their pressure. We need to claim what we wish for ourselves, and send the draft letters handed to us by the nationalized dead, themselves held hostage and deceived to guilt us into battle, back into the purgatory of history's sins. And as we do so we could begin to ask ourselves the questions that herald the possibility of greater freedom. Can we be brothers without being brothers in arms? What other kinds of brotherhood are possible? Can we be our fathers' sons while not abiding by the truths they stood for? Can we belong in exile? What kind of parents can we be? Will I accept a son who wants to go? Will he accept a son that fears and clings to him? What does a life look like when its purpose is living? Many more questions, most asked and answered without words, by getting in and out of conflict, strong and vulnerable together, looking for answers other than those given to us. And step by step we will begin to dig our way out of the burial site where our souls have been co-opted.

What vision of ourselves can we hold in our minds as we begin this upward journey? If we are to be free of the demand to make collective history our purpose we need, beyond knowing as means of resistance, a sense of what we want for ourselves and for our loved ones. What should I wish for him, what should I wish myself, what should parents wish for their children? Herein lies the challenge: in opening a space of exploration between an understanding of how things are and a desire for what could be, in knowing where we come from *while* imagining a different future.

Adorno (1951/1999) warns us, in a different context, of the trouble inherent in such ambition. To work against the ruling reason, against its laws of right and wrong, principle and actuality, madness and civilization, means to face the specter of unreason. It means to sometimes lose one's sense of grounding and identity, to at some point not being able to avoid feeling all wrong or sick or crazy. Asaf and I can both testify to the deep truth of this insight. He often asked himself, and me, if he was going mad; most urgently as a war was raging in him between dueling truths and during his bouts of inarticulate physical trouble. I surely felt both wrong and lost in my anxiety and rage when he said he might go back to Israel. It is, in fact, a formative dynamic of our relationship, of all of my analytic relationships, that we swim and float through waves of reason and unreason, mutual understanding, and confused togetherness. These waves often manifest in our making or not making sense, yet not making sense is sometimes better, more hopeful. Another patient told me once, and it seems true often: When we make sense together I feel buried. Perhaps we all feel buried when we strive to understand each other using a sick reason. Perhaps, despite the fact that it is sick, we all feel crazy when this reason threatens to abandon us, like a cruel warden, locked behind the thick walls of an existential Alcatraz.

If psychoanalysis has always searched for reason in the seeming unreason of the troubled subject, it has also for the most part made its business to mark it as unreason and to escort the subject back into the safety of social consensus. In this, psychoanalysis has perpetrated its own crime. Aligning with governing power, it relegated a great deal of what subjective and intersubjective life could be to perpetual alienation. If psychoanalysis wishes to serve the subject in his effort to challenge the destiny charted for him by the unconscious machinations of collective power, it should remain loyal to his distressed unreason as a form of truth rather than repress it. It should expose the madness of the collective and the sickness of its compulsory story telling. It should reveal how social forces circulate in an unconscious that drives the subject's most intimate experiences and willful action.

But, and this has been my main point, even if it takes a critical stance vis-à-vis the collective's truth, psychoanalysis is left with the dilemma of the family it replicates. For psychoanalysis, as for the family, the task is to raise an individual who is neither a helpless soldier of the common sickness nor a lonely prophet destined to living in a social wasteland, an individual who can be both critical and desiring. In other words, the task of psychoanalysis cannot be only negative. Psychoanalysis cannot conclude in answering the question how not to live. It has to offer, both in general and in each analytic dyad, some ideas about how to strive for and enhance the prospect of good living. This particular responsibility and cause is what makes psychoanalysis a unique site of inquiry and potential source of insight for other forms of discourse. If it aims to answer to this cause, psychoanalysis cannot but offer a third pole to the two of the critical dialectic—a positive of desire and imagination to inhabit the clearing carved out of the domains buried under social repression by the double negative of critical dialectics. And the positive perhaps comes back to this: What does it mean to be a good parent, how can a good parent fulfill his infinite responsibility? On this, psychoanalysis has already said a great deal and could say much more, if it partakes in forms of sense-making other than the common.

Perhaps, to evoke again Ferenczi's insight, our first responsibility is to recognize the trauma that is commonly being denied. If it is true, as I argue, quite in line with Freud's thinking, that civilization bears down on the subject and governs his ability to think and love, we should recognize the trauma that this social government inflicts on his life. We should register in protest what psychoanalysis too often accepts as a sign of well-being, that the forces of collectivity introduce limits to the subject's most intimate attachments and to the loyalty of those whose care he most relies on. We should recognize that even as the existence of these limits is both repressed and rationalized, it stirs suspicion and causes malaise in each of us. To deny these limits is to perpetuate the trauma, and the unacknowledged isolation and unintelligible self-doubt that are its mark. Perhaps therefore, in this domain, the first responsibility of the good parent and the good analyst is to find a way to say, "I know that I am hurting you when I'm asking you to be normal."

I hope that I was able to show how difficult it is to do otherwise. We identify *with*, so we could be identified *by* others. Identity is a fundamental relationship between the subject and the collective, and until there is another way of living in the plural, identity is a necessary self-construct. But parents do hurt their children as they side with indifferent collective forces and in doing so bring to life a complex equation of loyalty and love. Even if they cannot help it, parents still make love conditioned on identification, not only within the family, but also between the subject and a multiplicity of collective narratives. The stakes are always high, but in some circumstances it is a question of death and life. As I have tried to show, even if in such extreme circumstances most of us have no simple answers, there is great potential in raising the questions and not giving up.

This chapter is dedicated to Asaf with love and gratitude.

REFERENCES

Adorno, T. (1999). *Minima moralia* (E. F. N. Jephcott, Trans.). New York: Verso. (Original work published 1951)

Benjamin, J. (1995). *Like subjects, love objects.* New Haven, CT: Yale University Press.

Benjamin, J. (2006). Two-way streets: Recognition of difference and the intersubjective third. *Differences, 17*(1), 116–146.

Benjamin, W. (1968). Theses on the philosophy of history. In *Illuminations* (H. Zohn, Trans.). New York: Schocken Books. (Original work published 1940)

Deleuze, G., & Guattari, F. (1983). *Anti Oedipus: Capitalism and schizophrenia* (R. Hurley, M. Seem, & H. R. Lane, Trans.). Minneapolis, MN: University of Minnesota Press. (Original work published 1972)

Ferenczi, S. (1955). Confusion of tongues between adults and the child. In *Final contributions to the problems and methods of psychoanalysis* (pp. 156–167). London: Hogarth Press. (Original work published 1932)

Foucault, M. (2003). *Abnormal: Lectures at the College de France 1974–1975* (G. Burchell, Trans.). New York: Picador. (Original work published 1999)

Foucault, M. (2006). *Psychiatric power: Lectures at the College de France 1973–1974* (G. Burchell, Trans.). New York: Picador. (Original work published 2003)

Segev, T. (2000). *The seventh million: The Israelis and the Holocaust* (H. Watzman, Trans.). New York: Macmillan. (Original work published 1994)

Chapter 16

The politics of identification
Resistance to the Israeli occupation of Palestine

Steven Botticelli

Identification has always been a promising candidate, among psychoanalytic concepts, for theorizing the elusive link between the psychic and the social, traversing as it does the boundary between the two. In beginning to think about identification in its political significance, I have in mind a person's inhabiting a sense of self as like or unlike some other individual or group, as this becomes the basis for acting in the world in a manner that intends or enables an alteration in power relations. As Judith Butler (2004) reminds us, identification depends on difference: "The one with whom I identify is not me, and that "not being me" is the condition of the identification. Otherwise ... identification collapses into identity, which spells the death of identification itself" (pp. 145–146).

In her monograph *Identification Papers*, Diana Fuss (1995) puts the question precisely: "What ... is political about identification? What role does identification play in the world of social interaction we call politics?" (p. 8). In an incisive introduction she spells out the promise and pitfalls of attempting to harness the power of identification for politics. On the one hand, identification with the struggle of oppressed or exploited groups has provided the basis for coalition building, which (contra the assumption of identity politics, that only those who have directly experienced a particular form of oppression can effectively fight against it) has arguably been responsible for whatever successes have been achieved by movements for the civil rights of women and minorities. For instance, in the summer of 2008 supporters of Barack Obama took to wearing t-shirts announcing "My middle name is Hussein" in an effort to counter the Islamophobic baiting of the candidate. These individuals could be seen to be trying to recover, even promote, a vilified identification with Islam from which the candidate had gone to some lengths to distance himself. Fuss cites other academics and activists, among them Barbara Harlow, Douglas Crimp, and Elin Diamond, who have placed political hopes in the fostering of identifications across lines of race, gender, sexuality, or citizenship. For instance, Crimp (1993) cites his experience in ACT UP to argue for the power of politically productive identifications to be made across identities, as he notes the unpredictability with which such identifications are made:

> A white, middle-class, HIV-negative lesbian might form an identification with a poor, black mother with AIDS, and through that identification might be inclined to work on pediatric health care issues; or, outraged by attention to the needs of babies at the expense of the needs of the women who bear them, she might decide to fight against clinical trials whose sole purpose is to examine the effects of an antiviral drug on perinatal transmission and thus ignores effects on the mother's body. She might form an identification with a gay male friend with AIDS and work for faster testing of new treatments for opportunistic infections, but then, through her understanding that her friend would be able to afford such treatments while others would not, she might shift her attention to health care access issues. An HIV-positive, gay Latino might fight homophobia in the Latin community and racism in ACT UP. (pp. 316–317)

Nevertheless, grounding political hopes in the possibilities for expanded identifications has its limits. For one thing, there is the danger of succumbing to an overly voluntarist notion of identification, which after all is a largely unconscious process. Butler adds a further cautionary note, noting that any identification is assumed "through a set of constitutive and formative exclusions" (cited in Fuss, 1995, p. 9); any identification implies and includes disidentifications, those things that I am not. The potentially regressive political effects of such exclusions are evident, for example, in considering what is implied by identifying oneself as "an American," as this designation implicitly defines one apart from and in a sense against all other "nationalities." Finally, Fuss is attuned to the sense in which identification represents an appropriation of the other, noting that Freud's very theorization of the concept took place in the historical context of colonialism, a colonialism that Freud as a subject of the Austro-Hungarian empire identified with (notwithstanding and in some important ways at odds with his Jewish identity; see Editor's Introduction, this volume). She maintains that

> every identification involves a degree of symbolic violence, a measure of temporary mastery and possession. ... identification operates on one level as an endless process of violent negation, a process of killing off the other in fantasy in order to usurp the other's place, a place where the subject desires to be. (Fuss, 1995, p. 9)

Noah Glassman (personal communication, November 2008) wonders whether this formulation of Fuss's might contain an unintended appreciation of Winnicott's (1969) clinical concept of object usage, even as the aim or outcome appears to be somewhat different from what Winnicott had in mind. Where Fuss sees a desire to usurp the other's place, Winnicott posits destruction of the (m)other in fantasy as a necessary step in the acceptance of the other's independent existence and the creation of a shared sense of reality. This latter is to be regarded as a developmental achievement and a prerequisite to the ability to identify with another subject. Perhaps Fuss's more malign rendering of the process reflects a disciplinary difference, reflecting the gap between the abiding concerns and preoccupations of cultural studies and clinical psychoanalysis.

In her recent work Butler (2003) vests significant hope in the political possibilities of identification, construed in the very broadest terms, as she posits the fact of humans' common bodily vulnerability as the basis for "a normative reorientation for politics" (p. 17). With the impending Iraq war clearly on her mind, and with the conflict in Israel/Palestine as an important reference point, Butler writes movingly of how we are "implicate(d) … in lives not our own" (p. 14), of how we are "impressed upon by others" (p. 16). The recognition of this vulnerability we share with each other "makes me work to forge new ties of identification" (p. 26). If "part of what I am is the enigmatic traces of others" (p. 32), I am enjoined to take responsibility and have concern for the welfare of these perhaps not fully identified people who have left their traces, which is to say, everyone. Clarifying what some may have misconstrued as a strong anti-identitarian position in her earlier work, Butler here suggests rather the importance of "confounding identity" (p. 36) as a means of crossing boundaries between individuals and nations—and makes clear that it is the process of identification itself that permits such confounding.

In this chapter I explore processes of identification as they shape political resistance to the Israeli occupation of the West Bank and Gaza (where, since the 2005 removal of Israeli troops and settlers, the occupation has continued in the form of control of Gaza's airspace, sea, and land entrances and exits by Israel). My choice to focus attention here is multidetermined. As an example of what Khalidi (2006) describes as "settler colonialism," in which an indigenous population is replaced by or simply subordinated to a new one while being denied meaningful control over their own governance, the Israeli occupation of Palestinian territory allows an engagement with a rich and developing literature in colonial and postcolonial studies for its (here, of course, only partial) edification. The Israel–Palestine conflict also compels my interest, apropos of the subject of this essay, due to the powerful identifications many of us who are not ourselves direct parties to the struggle nevertheless have with one side or the other, or in some cases both. Clearly one does not need to be Israeli or Palestinian, or even Jewish or Arab, to have strong allegiances and identifications here. (Of course, no American can consider himself a disinterested party; with the promise early in 2008 to provide Israel with $30 billion in military aid over the next 10 years, the American government has taken sides on behalf of all American taxpayers, continuing the longstanding support that has made the occupation possible.) Further, such identifications are not neatly correlated with national or religious identity. For instance, many Jews, more identified with the social justice orientation of Judaism than with the national interests of the state of Israel, have created organizations to defend Palestinian rights (e.g., in Israel, the Israeli Committee to Prevent Home Demolitions; in the United States, Jews Against the Occupation).

The historical narrative of Israel–Palestine is a highly contested one embracing diverse perspectives on the nature of the conflict, some diametrically opposed to each other. While acknowledging the existence of other viewpoints, I state my own at the outset, as it shapes everything that follows: the occupation itself* is at the root of this conflict, which has created so much suffering on both sides

I. THE PALESTINIAN PROFESSOR AND HER ISRAELI INTERROGATOR

> You treat us all like terrorists, so we might as well behave like them. Give me my hawiyyeh, do you hear me, Captain Yossi?
>
> Suad Amiry
> *Sharon and My Mother-in-Law: Ramallah Diaries,* 2003, p. 46

So says Suad Amiry, Palestinian woman, professor of architecture, resident of Ramallah, to her Israeli interrogator. Having lived in her town for most of the past 7 years "illegally" after becoming exhausted of the effort of keeping current a burdensomely obtained and renewed visitor's permit, Amiry decides to take matters into her own hands and appears unannounced at the office of an Israeli civil administrator to demand she be issued her residency card (i.e., *hawiyyeh*). There is a relationship; this man in the past has periodically summoned her to his office for meetings, the purpose of which is never entirely clear. Lacking a residency card, Amiry has been unable to travel outside of Ramallah, as she would not be allowed to reenter upon her return. Permitted entry to his office by a secretary flummoxed by the appearance of a demanding Palestinian woman at an unusual hour, Amiry proceeds to turn the tables:

> "Get me a cup of coffee and a cigarette," I ordered Captain Yossi as I sat down cross-legged on the chair next to his desk. Not knowing what to make of this reversal in our relationship, Captain Yossi went out and, a few minutes later, came back with a Marlboro cigarette and a cup of muddy Israeli Army coffee. (2003, p. 44)

---

* For my current purposes, I am referring to Israel's seizure and subsequent military occupation of the West Bank and Gaza following its June 1967 war against Jordan and Egypt.

Having gotten his attention, Amiry proceeds to make her plea:

> "Listen, Captain Yossi, I have been dealing with your nonsense for too long now. Do you have any clue what I have been through in the last seven years, waiting for my stupid *hawiyyeh*? Do you have any idea why every single Palestinian man, woman and child is participating in this uprising? [The year is 1988, one year after the start of the first *intifada*.] It is because we can no longer take your baloney (a polite word for shit). Do you know what it means for a wife to live away from her family, her husband and her children? Do you want to know why Palestinian men have been freaking out and running around stabbing Israelis in the back on Jaffa Road? (At that time there were no suicide bombers.) Ask me. I know. I know exactly how it feels to be driven to the edge, of doing mad things. Look at me, Captain Yossi. Do I look like a criminal to you? Tell me." My voice was getting louder with every word.
>
> Yossi was totally stunned. He didn't want to answer yes or no. He did not know what to do with a female professor who was losing it.
>
> "Here—I have packed my suitcase, ready to go to prison after the trial," I added.
>
> "What trial?" He was trying to calm me down.
>
> "You claim to be the only democracy in the Middle East. You claim to have courts. Here I am. Put me on trial, charge me for the crimes I have committed (so far). Here is my bag."
>
> I opened it and started pulling things out of it, one at a time.
>
> "Here is my toothpaste, here are my slippers, here are my books, my t-shirt, my ... my ..."
>
> I was taking things out of my suitcase and throwing them all over the floor of his office.
>
> "I am not leaving." I stopped for breath and continued. "Put me on trial if you think I am a terrorist. Why not imprison me? You treat us all like terrorists, so we might as well behave like them. Give me my *hawiyyeh*, do you hear me, Captain Yossi?"
>
> I was screaming my head off.
>
> "Seven years waiting for my stupid *hawiyyeh*, does that make you happy? Do you want us to freak out and break down? Look at me. Does that make you happier?"
>
> I burst into tears.
>
> Yossi stood still; like all men he didn't know what to do with a crying woman.
>
> I could see that he was capable of handling Palestinian demonstrators, rebels, stabbers, terrorists. He could handle bombs, dynamite, tanks, fighter planes, and submarines. He was trained to handle them all.

BUT NOT A CRYING WOMAN.
NOT A WOMAN FREAKING OUT.

> I watched a stunned Yossi walk out of his office. Soon after, he came back with another Marlboro cigarette, another mud (which I drank this time) and a piece of paper with Hebrew scribbles on it, which he claimed said, "Give this (crazy) woman her hawiyyeh." (pp. 45–47)

Resistance, in the political sense, can take many forms. For the purpose of this chapter, I would like to consider Amiry's challenge to her oppressor an example of resistance and attempt to evaluate it in terms of its political and personal effects. We might first note in Amiry's encounter with Yossi her apparently successful refusal of what Fuss (1995) refers to as the colonizer's "prescribed identifications" (p. 148), whether that be passive subject of Israeli occupation or would-be terrorist. She defies the implicit injunction by which "(t)he colonized are constrained to impersonate the image the colonizer offers them of themselves; they are commanded to imitate the colonizer's version of their essential difference" (p. 146). In her usurpation of Yossi's authority in this encounter, she takes advantage of what Fuss, in an extraction of Fanon, posits as a point of vulnerability for the colonizer, who "inadvertently places himself in the perilous position of object—object of the Other's aggressive, hostile, and rivalrous acts of incorporation" (p. 146). And, in so doing, she makes plain the purely arbitrary manner in which this Israeli administrator deploys his power over the local population; if a residency permit can be granted when a woman throws a fit in his office, clearly there has been no justifiable "security" or other reason for withholding it.

Even as Amiry refuses to "imitate the colonizer's version of (her) essential difference" (Fuss, 1995, p. 146), she does engage in imitation of another sort, adopting the part of the "hysterical woman." Her impersonation throws her interaction with Yossi into a different register: No longer the Palestinian subject to her Israeli interrogator, Amiry through her performance has made herself the hysterical woman to the stolid, uncomprehending man. Perhaps to call this an "impersonation" assumes too much: It is not clear from Amiry's account whether her performance was a form of mimicry, or masquerade. As Fuss explicates the difference between the two, was Amiry deliberately taking on this role in a parodic way (mimicry), or unconsciously inhabiting it, with no irony felt or intended (masquerade)? According to Fuss, the distinction rests on "the degree and readability of (the) excess" (1995, p. 146) of the performance. While we don't know what Amiry felt or intended, clearly Yossi found the performance convincing. Apparently perceiving nothing ironic in Amiry's behavior, he responds to her just as a man (at least a certain kind of man, at certain times) does when confronted by a hysterical woman: he accedes to her demands.

The question of mimicry, its meanings and political effects, has been an important focus of theoretical interest within postcolonial theory. According to Homi Bhabha (1994), the significance of mimicry in the colonial context comes from the desire on the part of the colonizing subject "for a reformed, recognizable Other, as a subject of a difference that is almost the same, but not quite" (p. 86). The "but not quite" is a crucial piece of the equation for Bhabha, because "in order to be effective, mimicry must continually produce its slippage, its excess, its difference" (p. 86). In enjoining the person under colonial control to become like him, the colonizer enacts a "complex strategy of reform, regulation and discipline" (p. 86), producing "a particularly appropriate form of colonial subjectivity" (p. 87), the better to dominate her by. Yet even as Bhabha maintains that "mimicry is one of the most elusive and effective strategies of colonial power and knowledge" (p. 85), he is alert to the possibilities of its subversion, through as Fuss (1995) puts it, "the ever-present possibility of slippage—from mimicry into mockery, from performativity into parody" (p. 147). It can become difficult to distinguish a mimicry of subjugation from a mimicry of subversion. As Fuss points out, "Given the various and continually changing cultural coordinates that locate identity at the site of both fantasy and power, one would have to acknowledge, at the very least, that the same mimetic act can be disruptive and reversionary at once" (p. 148).

It is here, in the effort to evaluate Amiry's actions with Yossi, that the distinction between the Israeli occupation of Palestine and "classical" colonial occupations becomes discernible and pertinent: there is no "civilizing mission," no apparent desire to create "a reformed, recognizable Other" (Bhabha, 1994, p. 86) of Palestinians on the part of Israel. As Amiry points out, and as other observers have noted, Israeli authorities have preferred to regard Palestinians as "a terrorist population" (Warschawski, 2004, p. 26), seeing no need to attempt to fashion them into mimic Israelis. Lacking whatever subversive possibilities her colonially mandated imitation of Israeliness might have held, Amiry seems intuitively to understand the need first to establish a sense of sameness with Yossi in order to advance her purposes. Her inhabitation of the role of hysterical woman provides a successful basis for doing so: the codes that shape gender performance in Palestinian and Israeli Jewish society apparently are not so very different, and Yossi responds to Amiry's "hysteria" in a way that shows that it is culturally intelligible to him. In this sense her actions, though unwelcome, read to him as "I am not so very different from you"; she has established an identification with him.

Amiry's (2003) adoption of the part of the "woman freaking out" additionally seems well chosen (however unconsciously: "when I had left home that afternoon, I had no idea what I was up to," she maintains (p. 47), if not inevitable, in light of recent feminist-influenced psychoanalytic understandings of hysteria. Interpreting the hysterical symptoms of the women who were

Freud's patients a century before Amiry presented herself to Yossi, Dimen and Harris (2001) conclude that "hysteria was political or social protest in the only form available to women" (p. 26) at the time, and note the conflict of freedom and agency contained therein. In Dimen and Harris's brief renderings of the stories of these women, we hear echoes of the drama of female assertion in the face of recalcitrant male power that structures Amiry's encounter with Yossi: Anna O. and Frau Emmy "insisted on their analysts listening to them" (p. 7); Elisabeth von R. "assert(ed) an independence of which…Freud could make no clinical use" (p. 7). Summarizing the emphases of other authors in their edited volume, they note the "distortions in reaction to prohibited forms of … identity" (p. 27) to be found in the personalities of Freud's hysterical women.

It is in regard to this question of identity that Amiry's actions to some degree can be judged to have been successful. Her challenging actions have set off a play of identificatory processes within and between herself and Yossi that (as we will see) ultimately seem to alter each of their identities. For her part, she has shown both Yossi and herself that her identity exceeds whatever determinations they each might have taken to define it: whoever else Amiry may be, she also apparently is this: a "woman freaking out." In this, she has successfully violated what Homi Bhabha (1994) has identified as one of the central features of colonial discourse, "its dependence on the concept of 'fixity' in the ideological construction of otherness" (p. 66). Amiry has gained more space for herself, psychically and (with her procurement of her *hawiyyeh*) within and beyond the physical space of her occupied West Bank town. In her identification with the "woman freaking out" and in a sense with Yossi himself, Amiry has offered a lovely example of Butler's conception of identification as a process that "confounds identity" (2003, p. 36), not permitting to rest the illusion of Amiry's identity as a thing fixed, total, fully known.

Amiry provides this intriguing coda to her encounter with Yossi:

> Five years later, the phone rang.
>
> "Hello, Dr. Amiry, this is journalist Yossi speaking. I would like to conduct an interview with you about the last round of Israeli-Palestinian negotiations in Washington for my article in *Ha'aretz*. Can we have an appointment?
>
> No cup of coffee this time? This was the first thing that came to my mind, but I said nothing.
>
> I could not help but shudder when I heard the voice of journalist Yossi (my interrogator). "Sorry, Captain Yossi. I think this time I am in a position to say NO. You may have changed jobs, become a human rights activist and a freelance journalist, but for me you remain Captain Yossi.
>
> "By the way, I never thanked you for the *hawiyyeh*." (2003, p. 47)

Who has the power to confer identity? Such power is not necessarily proportional to one's social power. As identity is always constructed in relation to some real or imagined other, even those in a position of greater social power—colonizer over colonized, husband over wife, and so on—depend on their complement to sustain their identity. This realization allows the recognition that the seemingly normalizing concept of identity has the potential to provide what Harris (2009) has described as a "crack" in the "agendas and process of power hierarchies" (p. 145). Identity may function as a mode of discipline and social administration by which people are subjugated, as Foucault (1976) argued, but in addition it may operate on the powerful whose identity investments may at times provide a potential site of resistance for those they subjugate. As the writer Jesse Green (1999) has put it in a much different context, "A queen is owned by her subjects, without whom she has no one to rule. Without ruling, she has no one to be" (p. 31).

As we see with Amiry and Yossi, although the power imbalance between them as Palestinian and Israeli remains, Yossi now has relinquished the personal power he once wielded over Amiry. We might wish to believe that Yossi's encounter with Amiry has altered his identifications, perhaps even playing a role in his career change. In his solicitation of her for an interview, he recognizes her explicitly as a doctor and implicitly as someone whose views are worth inquiring about. In return, Amiry refuses to confer recognition on Yossi's new identity, taking the power he has offered her to "fix" him ("for me you remain Captain Yossi") in the same way he for so long did with her, maintaining the reversible complementarity (Benjamin, 1990) of the relationship. In the context of the power imbalance between these two, mutual recognition might seem an unlikely possibility, although that imbalance does not in all cases preclude it. Consider, for example, the Parents Circle, a group of Palestinian and Israeli parents whose children have been killed in the conflict between their peoples. In recognizing each others' grief, these parents function as a bereavement support group as well as political advocates for peaceful resolution of the conflict.

In considering the limitations of Amiry's act of resistance, as I have chosen to construe it, we note that her actions did not alter the terms of the occupation itself, for herself or anyone else. We might also wonder at the fate of her hysterical identification, born as it was out of her fundamentally disempowered position, and how it has lived among her other identifications as professor, Palestinian, spouse, and the like. Even so, we see that a true intersubjective exchange has taken place between two people, one that appears to have left each of them altered in some way. The remainder of Amiry's diary entries, which extend from prior to her encounter with Yossi in 1988 through late 2003, trace the increasing brutality of the occupation: the ever-growing obstacles to physical mobility through the imposition of curfews, the construction of bypass roads, the restrictions of checkpoints; the invasion of the West Bank by the Israeli army in 2002; the construction of the separation wall. Amiry ceases to have any kind of human contact with Israelis, encountering them only on the other end of the guns and tanks through which they impose the occupation. In Butler's (2003) language, this situation would seem to represent "a refusal of discourse that produces dehumanization as a result" (p. 24). Denied the opportunity to speak with Israelis, Amiry is deprived of a relationally constituted part of her identity, a piece of what makes up her personhood.

Fuss (1995) detects in Fanon's *Black Skin, White Masks* (1952) two separate readings of racial "otherness" under colonialism, which together could be taken to provide a periodization of the Israeli occupation of Palestine. In the first reading, the colonized person functions as an Other to the colonizing subject, the repository of repressed or disavowed characteristics. For Fanon, this meant that "not only must the black man be black; he must be black in relation to the white man" (cited in Fuss, p. 142). Alternatively, and in the case of the Palestinians, subsequently, the colonized function for the colonizer not as Other but as object, in the common rather than the psychoanalytic sense of that term; the "in relation to" drops out of the equation. "(E)ven otherness may be appropriated exclusively by white subjects ... the black man under colonial rule finds himself relegated to a position other than the Other" (pp. 142–143). Likewise, separated from contact with Israeli people by figurative and literal walls, Palestinians become thingafied, made invisible and irrelevant, even as a repository of disavowed characteristics. "(W)hen subjectivity becomes the exclusive property of 'the master,' the colonizer can claim a sovereign right to personhood by purchasing interiority over and against the representation of the colonial other as pure exteriority" (p. 145).

Fuss goes on to strike a Winnicottian note, in appreciating the intersubjective origins of subjectivity. "If psychoanalysis is right to claim that 'I is an Other,' then otherness constitutes the very entry into subjectivity; subjectivity names the detour through the Other that provides access to a fictive sense of self. ... (O)bjecthood, substituting for true alterity, blocks the migration through the Other necessary for subjectivity to take place" (1995, p. 143). Denied the possibility of encounters like Amiry's and Yossi's, in which there is an opportunity to destroy the other in fantasy (see Winnicott, 1969), Palestinians are deprived of the very basis for personhood. Perhaps this is overstated: certainly there are other forms of "otherness" within Palestinian society itself that could serve as the basis for "the migration through the other necessary for subjectivity to take place." Nevertheless one is left to wonder at how people denied the opportunity of human encounter with the people who essentially control their lives and livelihoods may thereby be driven to actual destruction, destruction which (contra to Winnicott's best-case developmental scenario) the object may not survive.

In speaking as I do of the trauma of nonrecognition, the foreclosure of identification, I refer to processes taking place in an individual mind, as well as collectively in many minds. I thereby evoke, without more fully resolving, the vexed problem of the relationship between the psychic and the social. I hope nevertheless to have here offered the concept of identification itself as a possible mode of trying to think the psychic and social at once.

Amiry, by virtue of the denial of her rights as a Palestinian, cannot but live in a state of conflict with the state of Israel and its representatives. Then there are Israelis themselves who choose to oppose the occupation, even though they have the option of acquiescing to it.

II. ISRAELI SOLDIERS WHO REFUSE TO SERVE

> What happens to a soldier, decent people, in the occupation is that power takes over, power poisons you. You can do anything. I was witness to beatings, roadblocks, curfew, going in the middle of the night to get people. And I thought it was OK because we were all decent people. ... We did this because this is what we were taught. The feeling was so strong that "we are the victims, we can do anything—we have a moral right."
>
> Ofer Shorr, *Israeli refusenik*
> Powell, 2003, p. 4

On January 25, 2002, a quarter-page letter appeared in the Israeli daily newspaper *Ha'aretz*. Fifty-two soldiers and officers of the Israeli army reserves, including captains, called on their comrades-in-arms to join their act of refusal to serve in the occupied Palestinian territories of the West Bank and the Gaza Strip. This refusal followed the reservists' realization that, instead of bringing an end to the occupation, the 1993 Oslo Accords between Israel and the Palestinians were being violated by Israel. Israel had withdrawn only partially from the occupied territories, and, although the settlements in the West Bank and Gaza Strip were supposed to be dismantled, they were in fact being expanded. The reservists were particularly disturbed by the force they were expected to use as soldiers against a civilian population collectively penalized for its uprising against the occupation, long before Palestinian suicide bombers started to explode in the Israeli streets. (Chacham, 2003, p. 1)

Later in 2002, Israeli writer and cultural critic Ronit Chacham interviewed nine of these "refuseniks"—Guy Grossman, Assaf Oron, Shamai Leibowitz, Ramie Kaplan, Yaniv Iczkovitz, Ishay Rosen-Zvi, Tal Belo, David Chacham-Herson, and Yuval Lotem—about their decision to (as stated in their formal declaration) "not continue to fight this war of the settlements ... not continue to fight beyond the 1967 borders in order to dominate, expel, starve, and humiliate an entire people" (Chacham, 2003,

p. 2). In taking this action, these men drew inspiration from earlier movements of military refusal, such as Yesh Gvul in 1982; at that time 168 soldiers went to prison for refusing to participate in the invasion of Lebanon.

Most of these nine men described the impetus for their refusal as based in their inflicting or witnessing the infliction of some brutality against Palestinians in the occupied territories. In Iczkovitz's case, the decision followed on the simple experience of watching through binoculars a Palestinian family going about their daily routines: "I saw every detail of their daily lives, and it pained me" (Chacham, 2003, p. 62). For others, travel provided a new perspective through which to view their actions, after which it felt impossible to continue to serve as before. Such was the case for Lotem:

> In Gaza, I saw some kids who were afraid of me. They moved to the side when I passed them by. I smiled at one of them. Much to my surprise, he didn't smile back. It took me a while to see myself through his eyes. Then I saw South African kids with the same look. That's when I started to comprehend that I'm on the ruling side. (Chacham, 2003, p. 116)

Given the centrality of military service to the consolidation of Israeli identity (as Chacham puts it, "the army seals the relation between the citizens and the state,") the men's act of refusal "amounted to questioning an essential part of their identity" (Chacham, 2003, p. 8). Though not technically a representative sample of all those who refused to serve in the territories, Chacham's nine interviewees are characteristic of that larger group as middle class, professional men, typically raised and currently living in Tel Aviv or Jerusalem, mostly Ashkenzi. Oron believes it is precisely this privileged status within Israeli society that made it possible for him to refuse:

> Maybe I can allow myself to rebel because I take it for granted that I live comfortably. Perhaps the ability to rebel is related to one's proximity to Zionism. ... The people who can make a bold move at a given moment are those who are not constrained economically, culturally, or socially. (Chacham, 2003, p. 30)

Mizrahi Jews (of Middle Eastern origin), who typically serve in nonelite units, are less likely to come to a position of refusal. As Chacham-Herson observes, "the Middle-Eastern Jew must demonstrate his hatred for Arabs, otherwise he will be confused with them" (Chacham, 2003, p. 148).

Wanting their actions and declaration of refusal to serve as a wake-up call to Israeli society, the refuseniks were met with a public reception that Chacham (2003) characterizes as "turmoil followed by rapid disappearance" (p. 7). After a period of widespread media attention and public debate conducted within Israeli media, the army declared the movement irrelevant and refused to further engage with the issues the refuseniks had raised. After a suicide bombing on Passover night in March 2002, and the retaliatory launching of Operation Defensive Shield, the refuseniks found themselves marginalized, their position having become in many Israelis' eyes difficult to support. On a personal level, many of the men found themselves ostracized, scorned, accused of being traitors or victims of propaganda. The army's response to the men's insubordination was in some cases to transfer them for duty out of the territories into Israel proper; more typically, they faced legal consequences. Usually this consisted of being sent to prison for 28 days, as part of a disciplinary procedure, rather then facing a formal court martial and trial. The state has gone to some lengths to avoid prosecution by trial, as this would bring publicity to the refuseniks' cause (and, not incidentally, would allow the defendants to raise the Geneva Conventions in their defense). At least one refusenik, David Zonsheine, petitioned to be given a full military trial, was refused this by the army, and was refused again on appeal by the Israeli High Court of Justice (Chacham, 2003).

These men's conscious self-understandings as expressed in interviews provide insight into their taking their brave and (numerically speaking) unusual public stance, and I will examine them in some detail next. Nevertheless as psychoanalysts we are naturally curious about the underlying, perhaps unconscious factors that shaped their decision. One effort to tease out the possible psychodynamics undergirding pro-Palestinian activism among Jewish Israelis was a 2006 study conducted by Israeli researcher Aner Govrin. His qualitative empirical study aimed to understand "radical leftist Jewish Israelis'" eschewal of a "preference for (their) in-group of Jewish Zionists," in favor of sympathiz(ing) with a hostile out-group" (p. 624). He conducted interviews with 40 Israeli men and women "active in left-wing radical groups such as Taayush, Indymedia, Women in Black, Bat Shalom, Gush Shalom, and Green Action" (p. 633). This rather eclectic sample embraced activists with a specific focus on the Israeli–Palestine conflict, as well as others with a broader "revolutionist agenda" or who worked for feminist or environmental causes. While their specific actions as activists are not delineated, Govrin provides the example of one female journalist

> active in a radical leftist group that supplies food and medical aid to Palestinian villages under siege. This political group also demonstrates against the Israeli army's and government's violation of Palestinian human rights. Rivka affirms the right of every Palestinian to return to his or her homeland, even if that means that Palestinians will be the majority in Israel. (p. 623)

Based on his interviews, Govrin (2006) proposes that this set of sympathies and actions reflects the domination of an "underdog schema" over a "we-ness schema," an emotional configuration born in at least some cases of a childhood experience of self as a stranger to one's caretaker. For example, one study participant recalled feeling as a child that "people on the street or in the grocery store evoked more empathy [from his mother] than I did" (Govrin, 2006, p. 640). Dynamically, Govrin speculates, such experiences led to

> disidentification with and rejection of many aspects of the original bad object (parent or other caretaker), as well as identification with their wounded self. The deficit in positive object experiences resulted in a lack of receptivity to any positive aspects of the external bad object. On the other hand, it led to an idealized image of any "other" that either hates the extended bad object or is oppressed by it. (p. 643)

Further analysis led Govrin (2006) to distinguish between the "furious provocative radical" and the "compassionate radical" (p. 641), based on the balance in feelings between anger at Zionism and Judaism on the one hand, and compassion for Palestinian suffering on the other. Govrin acknowledges that 11 of his subjects did not fit within the classificatory scheme he developed here.

While important differences between Chacham's refuseniks and Govrin's study participants (whose current military status is not reported, but among whom no military refusers are identified) give reason for caution in making direct comparisons between them, it is striking how little Govrin's descriptions of and conclusions about his "radical leftists" seem to apply to the refuseniks. For one thing, although clearly aware of the political meaning of their actions, the refuseniks do not refer to themselves in political terms—such identity categories as "radical" or "leftist" do not appear in any of their interviews. Lotem goes further in this direction than the others in explicitly disavowing a political identity, construing his actions rather in psychological terms: "Refusal is connecting with one's self" (Chacham, 2003, p. 116). (For that matter we cannot be certain that Govrin's study participants, either, consider themselves to be "radical leftists"; perhaps this is simply Govrin's designation for them. In any case it is one of the limitations of his work that he does not attempt to understand the social processes by which comes to be defined within the Israeli polity a supposedly moderate "center," against which a position that includes a willingness to act out of human concern for Palestinians is regarded as "radical leftist.") His very framing of his research question assumes what needs to be questioned and better understood, that is, the process by which political resistance comes to be viewed as aberrant, even pathological, with attendant consequences for the task of identity making.

The refuseniks constitute their resistance in part by locating themselves in relation to a number of identity coordinates, many of which recur throughout the interviews: identification with Palestinian suffering, and in many cases Palestinian resistance; Israeli; Zionist; Jewish; victim. While the particular configuration of identifications and disidentifications with each of these positions varies across this group of men, in general their identity construction appears to have proceeded along a model of both/and, rather than (as with Govrin's we-ness versus underdog schemas) either/or. Though not without some salient disidentifications, the refuseniks seem to have followed an indentificatory logic of expansion, rather than exclusion. Speaking of the Israeli public he and his fellow refuseniks hope to influence, but seemingly also of a process he himself has undergone en route to his act of refusal, Oron captures the spirit of this identificatory inclusiveness when he says "My patriotism is for both peoples. ... we must be ready to break through mental and emotional barriers" (Chacham, 2003, pp. 33–34). Lotem expresses a similar sentiment in speaking of his daughter: "I want her to understand that she belongs to a bigger tribe than the one that defines itself according to racial criteria" (Chacham, 2003, p. 117). Grossman also evokes the sense of expansion of self entailed in his act of refusal when he says "Calling myself Lieutenant Grossman was a big part of my identity. In time, I understood that that was not all of who I am" (Chacham, 2003, p. 79).

These men's identification with Palestinians is in each case based in their awareness of Palestinian oppression and suffering. Kaplan expresses this sentiment in its simplest form when he says, "I couldn't remain indifferent to the Palestinians' suffering and degraded living conditions. I didn't feel good manning a checkpoint where I'd detain or interrogate poor Palestinians heading to work at 4 in the morning" (Chacham, 2003, p. 38). For some, this identification springs from their awareness that subject to the same conditions, they would behave as the Palestinians have, in their acts of resistance. Speaking of Palestinians jailed for resisting Israeli soldiers' invasion of their homes, Lotem says "If my home were subject to an occupying army, I'd do exactly what they did" (Chacham, 2003, pp. 108–109); "I would also resist my occupier" (Chacham, 2003, p. 115). (Notably, this identification does not preclude Lotem from fellow-feeling with the Israeli settlers on the opposite side of the conflict: "I think about the fact that they'll have to be evacuated [in some imagined eventual resolution to the conflict], and I feel empathy for them. Their pain is real" [Chacham, 2003, p. 117]). Rosen-Zvi, a religious Zionist, responded to his rabbis' and teachers' criticism of his refusal to serve with these questions: "When was the last time you were hungry? When was the last time you roasted in the sun for hours on end at a checkpoint? When was the last time a foreign army invaded your home and locked you up, together with your children, in a single room? (Chacham, 2003, p. 128)" For Chacham-Herson, identification with Palestinians serves as an escape from an unbearable identification with the role of perpetrator: "Figuratively speaking, I know that I wield the stick that beats them. There have been times when I wanted to be the one who got the beating instead. Then I wouldn't have to bear the guilt" (Chacham, 2003, p. 143). Belo, of

Sephardic background, expresses his identification in his own idiom: "I am a kind of Arab!" (Chacham, 2003, p. 73).

The men seem more mixed, and in some cases rather conflicted, in their identities as Israelis. On the one hand, they can be patriotic. Chacham-Herson states flatly, "I want to serve my country" (Chacham, 2003, p. 138). Rosen-Zvi sees his refusal as a patriotic act: "The best way to serve your society is by refusing to perpetuate its injustices" (Chacham, 2003, p. 127). In other cases apparent contradictions in their narratives suggest the presence of some inner conflict. Thus, Grossman can declare "I love the state" and a moment later refer to Israel as "this militaristic state of never-ending wars" (Chacham, 2003, p. 85). Leibowitz says he is "very worried about Israel's fate," shortly after having offered what amounts to an immanent critique of the Jewish nationalism (Zionism) that was the basis for the very creation of the state of Israel:

> While in exile, the Jews concerned themselves with social, ethical, and spiritual questions such as charity, caring for the sick and aged, and studying the Torah. Once we gained our independence as a nation, our educational system neglected these issues in favor of nationalism—and, dare I say it, the sanctity of arms. (Chacham, 2003, p. 92)

Two others are at least implicitly critical of Zionism: Kaplan in his avowal of support for a one-state solution (generally taken to refer to a state for all its people, without distinctions made with regard to religion), and Lotem in his joking, offhanded reference to "three states: a Muslim one, a Jewish one, and a normal one" (Chacham, 2003, p. 115). Rosen-Zvi, on the other hand, derives the justification for his refusal from his Zionism: "refusal is practically the only Zionist position you can assume today" (Chacham, 2003, p. 133).

For three of the men, their Jewish identity is mentioned as related to their decision to refuse. Leibowitz, himself the grandson of a well-known Jewish religious thinker who publicly opposed the occupation as soon as it was imposed in 1967 (Chacham, 2003), maintains that "a religious Jew must ... adhere to the precepts of not oppressing non-Jews. ... Precisely because I'm religious, I must oppose the occupation" (p. 89). He criticizes Israeli judges who order the destruction of Palestinian homes and the detention of people without trial by comparing them with Ahab's judges in the Bible. Chacham-Herson avers that his "position arises from the feeling that you cannot be a Jew, the son of a refugee people, and oppress refugees" (Chacham, 2003, p. 137). For Rosen-Zvi, the occupation itself has been justified for some through an inappropriate invocation of certain Biblical precepts:

> It's important to remember where the "You have chosen us" actually comes from. ... The feeling that we are superior began as the rhetoric

of consolation ... [from the time] when the people of Israel were downtrodden under Roman rule and had nothing of their own. ... They fantasize about their imaginary might. It's a form of consolation for the weak, and as such not very dangerous. Today, of course, the situation is totally different, because the weak have become the strong, but they continue to employ the same old rhetoric and myths. (Chacham, 2003, p. 136)

Implicit in Rosen-Zvi's description here is his refusal of the position of victim, and this refusal is perhaps the most salient common feature that emerges in examining the refuseniks' identity construction. (This is true with one exception; Belo claims victim status for the group that includes himself when he asserts that "(b)oth sides are victims, and I don't want to start measuring who is worse off" [Chacham, 2003, p. 75].) For Leibowitz, this refused identification is matched with a recognition of Israel's victimization of Palestinians: "there can be no peace between occupier and occupied; it's almost like asking for a peace agreement between rapist and victim while the rape is being carried out" (Chacham, 2003, p. 90). Chacham-Herson maintains twice in the course of his interview, "I am not a victim" (Chacham, 2003, p. 138, p. 146), even as he allows that he "understand(s) why (his) friends see themselves as victims after so many suicide bombings" (Chacham, 2003, p. 142). He extends victimhood as well to Israeli soldiers serving in the occupied territories, as "They're the ones who will have to face their children one day and tell them, 'What can I say; I thought we didn't have a choice.' They'll be victims of their own guilt" (Chacham, 2003, p. 146). Expressing his view in a totalizing way, he complains that "the Jews view themselves ... as ... perpetual victims" (Chacham, 2003, p. 141).

In a provocative essay, Adi Ophir (2000), a professor of philosophy at Tel Aviv University and himself a refuser of military service during the first *intifada*, attempts to pry apart the history of Jewish persecution from what he views as the inappropriate use to which this history has been put in the construction of Jewish identity. A self-described "post-Zionist," he believes that Zionism has lost whatever usefulness it may once have had in addressing the political and cultural problems of Israeli and diaspora Jews. In keeping with the postmodern emphasis of post-Zionism, he sets out to critique essentialist conceptions of Jewish identity that developed during the Zionist period and that in many corners continue to hold sway. Toward that end, Ophir attempts to question not only particular constructions of identity among Israeli Jews, but "the forms and practices of identification" themselves. For instance, Ophir asks the question of what allows one "to be entitled to an identity and to become authorized to represent it or bestow it on others" (p. 184). In particular, he wants to understand "how the victim position functions in Israeli culture; in the state's ideology ...

and in the construction of the constituting narrative of the Jewish state" (p. 182).

Making a subtle but crucial distinction, Ophir (2000) maintains that "(b)eing a victim is not a ready-made identity but a position with which one identifies, a strategic position to be attained and maintained in various cultural fields" (p. 183). He describes modern Israeli Jews as "fascinated by the victim position, in love with their losses" (p. 178), although he does not speculate on why this should be the case. For Ophir, the effects of this fascination are far-reaching: being "hooked" (p. 178) to the position of victim as they are leads Israeli Jews to assimilate any and all losses they may experience—including and for Ophir perhaps especially "the loss of a coherent Jewish identity" (p. 178) attendant to the rapid pace of social change—to an already-existing idea of themselves as victims. This manner of construing themselves leads to a particular reading of Jewish history, in which "the many, variegated, and quite different cases in which Jews fell victim to non-Jewish aggression throughout history boil down to one long, repetitive story of victimization" (p. 178), as well as to a refusal to recognize the victimhood of others, including in instances in which they are doing the victimizing.

Unfortunately Ophir does not take up the question of what allows one to move beyond an attachment to the identity of victim. Certainly other soldiers (and civilians) must have encounters with Palestinians similar to those described by the refuseniks, without coming to share the feelings or take the actions of the refuseniks. Further, while Ophir is alert to the function that "victim" status may play as a justification for aggression and in this way serve the expansionist aims of the Israeli state (the development of a settler movement, now more than half a million strong in occupied Palestine;* the inclusion of swaths of territory, including aquifers and other natural resources, beyond the Green Line in the building of the separation wall), he does not consider that an insistence on one's victimhood may also stand as a claim for acknowledgment of injury. In his focus on the discursive, Ophir pays insufficient attention to the psychological, including the psychological reality of trauma. For instance, when Ophir refers to Israeli Jews' "traumatic preoccupation with the Holocaust" (2000, p. 179), he seems to imply there is no reasonable psychological basis for this "preoccupation." Yet we have reason to believe that the traumatic impact of the Holocaust has not been adequately absorbed or processed in individual psyches or by Israeli society as a whole.

In *The Seventh Million*, Tom Segev's (1993) exhaustive survey of the relationship of Israelis to the Holocaust, he documents the at best ambivalent reception Holocaust survivors received from the *yishuv*, the existing Jewish

---

* Including the Jewish population of East Jerusalem, itself illegally annexed to Israel; see Judt (2009).

community living in pre- and early-state Israel: "they were despised for having gone to the death camps 'like lambs to the slaughter,' instead of defending themselves, and because in Israel people refused to believe their stories. ... Their first contact with a country that could only respect dead heroes was a trauma" (p. 466). He notes "the great silence that surrounded the Holocaust through the 1950's ... almost to the point of denial" (p. 513). In his view this attitude followed from the "regret and shame" (p. 513) felt by the *yishuv* toward Holocaust refugees, whose experience could not easily be assimilated to the triumphalist narrative of Zionism that had animated their own immigration to the land that would become Israel. Not until the trial of Eichmann in 1961 did this attitude begin to shift, allowing a "process of identification with the suffering of the victims and survivors" (p. 361) to take place. By the 1980s, the Holocaust had become "a constant and intense preoccupation" (p. 513) in Israel, and was pressed into service by some as a kind of retroactive justification for Israel's existence. (Some defensiveness had been roused on this matter in light of the work of the Israeli "new historians," among them Segev himself, who brought to light some of the less palatable aspects surrounding the creation of the state [e.g., the forcible expulsion of Palestinians]). One imagines that neither of these attitudes—silence and denial, followed by "preoccupation" harnessed to an ideological program—is conducive to coming to terms with the experience as trauma.

Further, we know from work such as Davoine and Gaudillière's (2004) *History Beyond Trauma* and Faimberg's *The Telescoping of Generations* (2005) of how the nonrecognition of traumatic experience leads to its symptomatic appearance in the children and grandchildren of the trauma survivor, as it is transmitted through the psyches of subsequent generations. Finally, there is the generating of new traumatic experience among the young in a country that compels military service among its (Jewish) population, when that country has existed in state of declared or de facto war for most of its 60 years. One recent observer (Bronner, 2008) has referred to Israel as a "country that has tended to dismiss the psychic damage that can result from being a soldier in war" (p. 22). Danny Brom, an Israeli psychologist specializing in work with trauma survivors, has said, "Our kids serve in the army for three years and go to combat units being hyper-alert and looking for enemies, and then they are supposed to just forget about it. On the last day of army service they don't ease you out. They teach you how to write your c.v." (cited in Bronner, 2008, p. 24). Their service in the army being a condition of upward mobility in Israeli society, most Israeli youth would not want to jeopardize their future opportunities by appearing to not have survived their military service "intact."

If psychoanalysis has some contribution to make toward the understanding and resolution of political conflicts in our time, it is surely in its appreciation of the psychic consequences of unprocessed individual and collective trauma. One such consequence is the role that trauma may play as the often unrecognized impetus for the infliction of (more) violence. Jessica Benjamin (2004; personal communication, June 2008) has emphasized the power of acknowledgment of injury to potentially interrupt cycles of violence that are perpetuated through unrecognized trauma, and to provide a measure of healing. This conviction is instantiated in her work with groups of Israeli and Palestinian mental health workers that attempt to create space for the expression and acknowledgment of injury. I see Benjamin's practical and theoretical work in this area as part of an effort to articulate a specifically psychoanalytic morality, one in which the good is achieved through the attempt to transcend the tendency to "take sides" (that is built into the structure of identification: if I am this, then I must not be that) by striving to create a "thirdness" that allows a voice for both positions.

I see the refuseniks as having within themselves attained a degree of thirdness. They have engaged in sometimes painful reconsideration and reconstrual of their identities in order to achieve a more expansive identification that includes recognition of the suffering of both Israelis and Palestinians. Notably, this recognition does not express itself through a perfectly balanced articulation of the interests and grievances of each side. Conscious as they are of the imbalance of power between the parties to the conflict, their thirdness moves them to take a side—against the dictates of the Israeli state, and on behalf of the suffering of Palestinians. This juncture is where psychoanalytic morality meets socialist morality, by whose lights the good is achieved by taking the side of the exploited and the oppressed, by identifying with them. The American socialist Eugene Debs gave eloquent expression to this position, speaking at his trial for sedition in 1918: "While there is a lower class I am in it, while there is a criminal element I am of it; while there is a soul in prison I am not free" (cited in Bartlett, 2002, p. 606).

The refuseniks demonstrate that such "partisanship" does not preclude consideration for the well-being of Israeli Jews, and in fact in the most important ways (in imagining the possibility of a better future for both Israeli Jews and Palestinians, and the actions that need to be taken now to help create that future) act to promote such consideration.

REFERENCES

Amiry, S. (2003). *Sharon and my mother-in-law: Ramallah diaries*. New York: Pantheon.
Bartlett, J. (2002). *Bartlett's familiar quotations* (17th ed.). Boston: Little, Brown and Company.
Benjamin, J. (1990). Recognition and destruction: An outline of intersubjectivity. *Psychoanalytic Psychology, 7*(Suppl.), 33–47.
Benjamin, J. (2004). Beyond doer and done to: An intersubjective view of thirdness. *Psychoanalytic Quarterly, 73*, 5–46.

Bhabha, H. (1994). *The location of culture*. London: Routledge.
Bronner, E. (2008, December 14). In search of the soldier in his past. *New York Times*, pp. 22, 24.
Butler, J. (2003). Violence, mourning, politics. *Studies in Gender and Sexuality*, 4(1), 9–37.
Butler, J. (2004). *Precarious life: The powers of mourning and violence*. London: Verso.
Chacham, R. (2003). *Breaking ranks: Refusing to serve in the West Bank and Gaza Strip*. New York: Other Press.
Crimp, D. (1993). Right on, girlfriend! In M. Warner (Ed.), *Fear of a queer planet: Queer politics and social theory* (pp. 300–319). Minneapolis: University of Minnesota Press.
Davoine, F., & Gaudillière, J. M. (2004). *History beyond trauma*. New York: Other Press.
Dimen, M., & Harris, A. (2001). *Storms in her head: Freud and the construction of hysteria*. New York: Other Press.
Faimberg, H. (2005). *The telescoping of generations: Listening to the narcissistic links between generations*. London: Routledge.
Foucault, M. (1976). *The history of sexuality: Vol. 1. An introduction* (R. Hurley, Trans.). New York: Vintage.
Fuss, D. (1995). *Identification papers*. New York: Routledge.
Govrin, A. (2006). When the underdog schema dominates the we-ness schema: The case of radical left-wing Jewish Israelis. *Psychoanalytic Review*, 93(4), 623–654.
Green, J. (1999). *The velveteen father: An unexpected journey to parenthood*. New York: Ballantine Books.
Harris, A. (2009). The socio-political recruitment of identities: Commentary on paper by Lynne Layton. *Psychoanalytic Dialogues*, 19, 138–147.
Judt, T. (2009, June 22). Fictions on the ground. *New York Times*, p. A21.
Khalidi, R. (2006). *The iron cage: The story of the Palestinian struggle for statehood*. Boston: Beacon Press.
Ophir, A. (2000). The identity of the victims and the victims of identity: A critique of Zionist ideology for a post-Zionist age. In L. Silberstein (Ed.), *Mapping Jewish identities* (pp. 174–200). New York: NYU Press.
Powell, B. (2003, March 13). Refuseniks: Three Israeli soldiers tell why they will not serve in the occupied territories. *Public Affairs*, pp. 1–5.
Segev, T. (1993). *The seventh million: The Israelis and the Holocaust*. New York: Hill and Wang.
Warschawski, M. (2004). *Toward an open tomb: The crisis of Israeli society*. New York: Monthly Review Press.
Winnicott, D. W. (1969). The use of an object. *International Journal of Psychoanalysis*, 50, 711–716.

Chapter 17

Dread is just memory in the future tense*

Adrienne Harris

Waltz with Bashir (Folman, 2008) is a "haunting" film. Many reviewers said this. Traumatic to watch, I would also say. The filmmaker, Ari Folman, draws the viewer quite deliberately into an experience of ghosts and dangers simmering just beneath a chilly, dark surface. Through its structure, through its brilliant use of animation, and through its elegant and deeply satisfying color palette—the ranges of white, grey, black, ochre, and an occasional hit of yellowy-orange—the viewer is brought slowly but inexorably into full consciousness, alongside the narrator, of the long shadow of the Lebanon War and its deep bite into the psyches of soldiers. Somber, moody, anxious: the whole ambience of the film contributes to one central idea: once repression has been unsettled, there is a drive to know, to see, and to bear witness, but progress to this end can also feel like sleepwalking through the dark smoke of a dream.

It is one of the conundrums of antiwar films that it's hard to make such a film without making war exciting. The pornographic impact of violence and the didactic lessons of wars' dangers don't always work together. In this film, Folman keeps a lot of control through pacing, through the often low energy, faintly depressive style of conversation and through the use of animation. As we watch this animation we can and we cannot quite find a mimetic tie to these drawn figures. The animation itself seems slightly off pace, off center. It is a perfect way to describe and construct the experience of depersonalization, in the character and in the viewer. Film scholars are just beginning to have a theoretical language to talk about animation and its effects. Animation links us to other experiences with animation, particularly from our own childhoods. We are led into a regressive state, conjured by the particular magic and unreality of animation. Animation engages us in a very particular relation to bodies. They are contested, unreal and magical, changeable, funny and violently

---

* A quote I find so useful clinically, that Margaret Little (1985) takes from Winnicott. A version of this essay was given as an invited paper, "In the Shadow of Memory," at the IARPP Conference, Tel Aviv, Israel, June 2009.

transformed. All this makes animation an inspired choice for a film about blocked mourning and posttraumatic stress, lived as disembodied and dissociated states.

If the animation decreases some of the libidinal or identificatory or aggressive charge in a film about war, the music is rather another story. The soundtrack makes a kind of homage and reference to other war films like *Apocalypse Now*, where the music evoked the era, creating an eerie backdrop to mayhem and damage. The phallic pleasure and potency of rock music permeates modern culture, of course. Here we see the power of rock music to tailor narcissism and phallic pleasure to warmaking. Rock music is a recruiting tool, an energizer, and an anesthetizer in the presence of violence and danger.

Only in its last 30 seconds does the film become actuality footage. This terrible moment strikes me as not just the arrival of reality but as the appearance of the Real in the Lacanian sense (Lacan, 1977). When Lacan identified the Real, in distinction from the Imaginary and the Symbolic, he meant to convey the presence of unassimilable horror, the press of trauma on memory and unconscious that floods the psyche, and yet can never be fully metabolized. We feel the effects of the Real without necessarily ever knowing it directly. The film's ending is beyond bearing and also reminiscent. Uncanny.

This film is really one of the perfect objects for a psychoanalytic inquiry. It begins with a dream then cuts to a conversation between two men about their shared history in the Lebanon War. There is a hug at the end of the talk. We might say the session ends and then there is a flashback. It's perhaps an apocryphal but certainly a useful idea that everything we need to know in an analysis is delivered in the first session, here the first 10 minutes of the film. The uncanny project of the next 90 minutes is to open the viewer and the narrator to what was always there, hidden in plain sight, waiting the capacity in viewer and narrator and the other characters to bear to know what had happened in the camps at Sabra and Shatila. Uncanny is the term I need here to convey what the film repeatedly reveals. Starting with the rather blanked out statement from two different veterans (Ari and Carmi) we hear the same remark about the massacre: "It's not in my system." And yet with each conversation, each interview, each dyadic session of two men talking, the "not in my system" turns to "knew it always." Psychoanalysis' term *uncanny* captures that mixture of surprise and inevitability in traumatic memory (Freud, 1919). It is this doubled state, this subthreshold knowledge, this experience of standing in the spaces that defines Bromberg's (1998) work on dissociation and the relational preoccupations with shifting and multiple self-states. The disavowed that is always already known sits at the heart of trauma. It is a contradiction built much more explicitly into the German word *unheimlich,* which embeds home and familiarity in this concept of strangeness, eerie, ghostly, foreignness.

This film operates in the tension of evacuated knowledge on one hand and "knew it always" on the other hand.

The structure of the film owes a lot to the pace and shape of therapeutic inquiries. Scenes of two men talking are followed by flashback, reverie, and an increasingly desperate pursuit of understanding. The filmmaker is set off on a quest by the dream that inaugurates the film. In each dyadic encounter with another vet, a different dilemma of traumatic aftermaths is illuminated: the cost of soldiering, the stress of remembering and of forgetting.

Boag, one of Ari's wartime buddies, has a dream. It is Boag's dream that makes Ari remember his own odd dreamlike image of emerging from the water on the beach, an image that, as in an analysis, is full of interpretability and multiple meanings. The full content and meaning of this image, its anchors in reality and interpersonal reality and history, remain enigmatic: rebirth, emerging naked from water, invading, arriving in Palestine from Europe with all markers of identity stripped away. Is the emergence from the water a step toward death or away from it? Both, of course.

There is an odd almost educational scene in which the filmmaker and a memory researcher talk about the nature of memory, of falsified memories, of memory as constructive and dynamic. I actually think that in this film the flow of memory is via many unconscious processes—somatic, affective, and cognitive—and via the relational matrices that are engaged and replayed throughout the film. Memory is dynamic both in a one-person and a two-person sense. Boaz's and Ari's hug may be as evocative a stir on Ari's memory as the dream. There are contagions of affect, of despair, and of fear, moving like electric circuits through all these men in the film.

In all the veterans, in different ways, minds are blown, things are "not in the system," and lives are suspended or disavowed. Ari meets up with Carmi, another veteran found, in a snow-filled scene in Europe. The snow covers the landscape of a man whose brilliant physics brain has been, for decades, addled with drugs. Snowed in. Ari and Carmi talk as they wander through the wintry scene but it is freezing, too cold to think or talk. The chill here is psychic as much as literal. I am indebted to a comment from my colleague Eyal Rozmarin, who points out that to evoke Europe in such a film is to link the war and the Holocaust, and that in such scenes, the very ambience of Europe is often disavowed. I quote from a personal communication from Eyal (May 5, 2009):

> The barren snowy fields so reminiscent of the black and white photos of the camps ... And if you recall it is in this very scene that the father points to his child-son playing war games with a gun and says something like "I did not teach him that." On the snowy fields of Europe the fate of the transgenerational relay of victimization and violence was sealed.

So Carmi, a veteran of a contemporary war, is trapped in Europe but living at a remove, depersonalized, sleepwalking.

Sense memory can be one of the powerful narrative threads in recovering from trauma. When Patchouli oil links Ari and Frenkel and their shared history, we meet another wrecked life, a man now living like a marionette, twirling and gesturing and miming jujitsu moves, a life as stalled as Carmi's but in a different way. Much later in the film, this scene with Frenkel will come back to haunt us and to be reunderstood. I would like to invoke here Freud's concept *nachtraglichkeit*, where the present unsettles the past and that now unstable past unsettles the present (Freud, 1919). We see here the nonlinear dynamic recalibration of memory. The elegant, beautiful, heroic but also grandiose, waltz through the streets of Beirut under fire, has, in Frenkel's present, become a grotesque repetition. Like Kurt Goldstein's accounts of World War I shell-shocked veterans (Shepherd, 2001) whose paralysis captured a moment just before the bad thing happened, Frenkel lives just before the entry into one of the circles, like Dante's circles of hell, surrounding Sabra and Shatila. Circles is a term used by one of the characters, Ori, to describe distinct levels of distance and responsibility for the refugee camps, but the link to Dante's idea of the circles of hell seems clear.

I want to invoke my title, Winnicott's phrase, reported by Margaret Little (1985): Dread is just memory in the future tense. Loewald (1972/1980) also has a good way of talking about this. Transference and trauma are experiences set in the near future, the about to happen, which, you come to know, has actually already happened. Uncanny. Frenkel lives suspended just before the massacre, permanently in fighting pose, hypervigilant, warding off the knowledge of the horror that has already occurred and remains to be metabolized.

This film is a living record of almost every posttraumatic stress disorder (PTSD) symptom. Fogged out or hypervigilant, rigid or floaty, minds blown by what was done, what was witnessed, and what was not done, the men, who are veterans of the 1982 Lebanon War, bear their wounds in an oddly passive, almost resigned way. Along the journey Ari makes toward the recovery of his own mind, the film touches on many ideas about militarism and masculinity, warmaking and erotics, homosexual and heterosexual. Eros and excitement lure men toward death, it is a theme that reappears several times. And subtly, men notice the transfer of warmaking from father to son.

Making the film itself, of course, we might say, is an act of resistance. And at the core crisis of the film, Ori makes a powerful, linking interpretation, tying together the narrator's personal history, his parents' experience in Auschwitz, the battle in Beirut, and the massacre in which Israelis were passive and yet conscious. These links fuel and are fueled by guilt, the veterans' most dreadful burden. Veteran guilt is perhaps the most hidden, shameful secret among the postwar residues. It is a guilt that often has and finds no witness. A number of scholars have noticed the endlessly repeated sequence in which postwar medical authorities erase or disavow the soldiers' problems at war's end. It is a phenomenon, first noted in the psychoanalytic canon, by Ferenczi (1921) in his Symposium on War held in Budapest, 1919, and only really altered in the aftermath of Vietnam in America, where PTSD came home to sit on our streets, homeless and maddened. Finally attention was paid.

In looking at the depiction of the veteran in American post-World War II films, particularly films in the late 1940s, Robert Sklar has noted that veterans do appear, but very often they are damaged, crazy, and deeply guilty. So much for the good war. There is a documentary film *Let There Be Light*, made by John Huston (1946) in an army hospital in Long Island in 1945. A voice-over narrates an intriguing fact: 20% of the U.S. war injuries were "psychoneurotic," to use a term from that period. The film shows actual veterans recovering memories of battle and death. The U.S. Army took one look at this footage and suppressed this film until the 1980s. In most circumstances, soldiers are left to deal with guilt and shock on their own and unacknowledged.

In *Trained to Kill* (2005), the late Theodore Nadelson recounted his work at the Veterans Administration in Boston. In a lifetime of work with veterans, he felt suffused in their guilt, in their shames in relation to identity and masculinity, themes we see in *Waltz with Bashir*. There is a narrative arc in Nadelson's book and it is a tragic story, made with almost perfect pitch. The basic storyline is repeated in *Waltz with Bashir*. Military training and battle are core experiences—heightened, intense, and both constitutive of and embedded in masculinity. Central chapters in the book describe, in a way that brings the reader into a countertransferential and voyeuristic relation to the experiences, the exhilaration in battle, and the arousal that comes with doing damage to another being, another body. We are left with our own moral queasiness to ponder over. Later, Nadelson gives us, in many simple but telling clinical moments, an endlessly repeating saga. Over and over, patients tell of the surviving of some hideous moment in warfare. It could be danger or violence, killing or the threat of being killed, the loss of a friend or a death the soldier himself causes, all experienced in the intense comradeship and male love that glues the military group together.

Nadelson's work and training placed him right in the nexus of psychoanalytic and psychiatric treatments of trauma. From this unique position, he understood the power of shame dynamics to enforce obedience and conformity, the potent brew of aggression and arousal in any man's encounter with militarism. And he lived in daily clinical experience the long, indeed often endless, aftermath of postwar trauma. By bringing into tension the heroics and excitement of socializing a soldier and the gruesome saga of postwar trauma, Nadelson forces us, his readers, to live in the agonized understanding he must have known very well. During and after World War II (Nadelson's first early exposure to military experiences), these two spheres were sequestered. The heroics and bravery of war trumped any deep appreciation of the long-term cost to combatants.

Guilt is clearly an element in these Lebanon War veterans' psyche. One of the characters, Rony, runs away in a moment of intense attack on a group of Israelis. He has stepped away from the demand to take over command of a group after the leader has been killed by a sniper. In a discussion of the real-life figures on whom the characters in the film were based, Ari Folman alludes to the long-term aftermath of this man's flight. In his own eyes, he had become less than human and certainly less than manly. In a dream scene, this man, perhaps tellingly, has a reverie of life at his mother's side, staying close to mother and to childhood. It has a terrible irony in the context of the other damaged men who cling to an often cartoonish masculinity and faux adulthood. Like many veterans, Rony's deeds separate him from the others he needs to contain him and be his benign witness.

Inexorably, in the film, the pieces of the puzzle surrounding the massacre begin to assemble and clarify. Yet Ari remains unclear about his dreamscape of arriving on the beach, unclear about who was where, in what circle of responsibility were he or his comrades, and Ori offers his interpretation quietly: "The murderers and the circle around them were one and the same. Maybe that's why you couldn't remember the massacre. You feel guilty. Against your will you were cast in the role of Nazi." We must be aware of the many landmines in those sentences.

We might reflect on the dream image at the beginning of the film, the pack of dogs terrorizing Boaz. These dogs constitute really quite an enigmatic image, evoking Nazis' demonic forces and the terrifying aspect of subhuman avenger. Again I quote Eyal Rozmarin: "Not only the Nazi demons but the actual dogs they used in the camps. Many survivors came back dreading dogs for life because of it, and we have all in our collective unconscious a complex relationship with German shepherds and rottweilers (May 2009)."

In Boaz's dream, the dogs then rush in from the nightmares of the parent's and grandparent's generation; guilt and dangerous victimhood are enigmatically, excessively unclear. Ferenczi's paper "The Confusion of Tongues" (1949) describes the core muddle in dyadic scenes of abuse. Through defensive identifications and affect contagion and the fluidity of inside and outside, the child being injured and attacked remains forever unsure of who did what to whom. Might we think of these processes at a group level, as processes in history of the sort Volkan (2006) has described for group identities.

I have been interested in the critical reception to this film. I saw it first at the New York Film Festival where it was enormously successful and admired. One persistent criticism has been that this film tells only the Israeli story. Where are Palestinians? I think this idea, whatever its political merits, misses the agenda and purpose of this film, and the film's place in a potential space and process of mourning and recovery from trauma. The enemy—Arab, Palestinian, and German—*is* in the film and pretty much always as Other: nonhuman, dehumanized, cried over, destroyed, and fled from. The dogs in the first dream are enigmatic, standing for every soldier, enemy, or friend, tearing you to pieces or demanding you be given up. You could be in any war on any side. And, let's notice also that the horses that destroy the peace and sanity of the photographer are described as Arabian stallions.

In this film I would say Folman makes a convincing case that recognition comes after mourning, that the *post-* in posttraumatic stress must become genuinely POST, that wars cast a long shadow of amnesia, at all levels, personal and political. Davoine and Gaudilliére, two analysts working in France, would insist that it takes a half-century to process a war. I was 4½ when my father came home to Canada at the end of the Second World War, and I was probably 64 when I really understood fully how not all right he and many of his fellow soldiers were.

I am mindful in this chapter that I have moved us around history, cultures, different historical epochs, different wars, different kinds of trauma. I am making links, but of course there are crucial differences. In many ways, this book is in part devoted to finding the nuance in how memory and trauma shadows psychic life. I am trying to link historical eras and wars and trauma carried into the present, but we must be equally mindful of differences. This is and is not everyman in every war. Some postwar casualties are obvious: in the corridors of VA and military hospitals throughout the century and in the streets of urban cities in America in the post-Vietnam era. Some markers of trauma are subtle, masked, displaced. So it is to the specific experience of the Lebanon War and its sequelae that I want to refer now.

This chapter was first prepared to be given as a presentation in Tel Aviv, in the summer of 2009. A showing of the film, a Q and A with Ari Folman, and commentaries by me and by a Palestinian Israeli Mustafa Qossoqsi were arranged for the conference of the International Association for Relational Psychoanalysis and Psychotherapy (IARPP), which was titled "In the Shadow of Memory." I was really honored and delighted when Emanuel Berman invited me to participate in the panel. As I began to work on the film and write about it, I realized it intersected with my own history, unexpectedly. I had been in Israel for the first time in 1984, making a radio documentary for the CBC on women and peace movements, and I interviewed many women active either in feminist work on in peace movements. I went back to the transcript of those meetings and was surprised at the depressed, uncertain tone of many of the women I spoke to. Did feminism have anything to do with peace? What role might women have in working for peace? Negotiating the complexity of gender roles and the peace process was a daunting, uncertain project for the Palestinian women I interviewed in Gaza and for the Israeli women who were at the intellectual and political heart of groups like Peace Now (Galia Golan, Alice Shalvi). I'll quote from one interviewee, Raya Harnick, whose son was killed in the war in Lebanon:

> If a mother who lost a son speaks out against the politics of war, the answer is always that we can't discuss or answer a mother who lost a son. I mean it's been pushed until it's an emotional side and not the rational side. In my opinion to shut up the mothers. So the answer I get is that they can't discuss it with me because I have lost a son in the war, meaning I have lost a son and my mind. ("Women and Peace," CBC Broadcast, *Ideas*, 1985)

In that moment, 1984, she could hold together the necessity of an army, the betrayal of soldiers, including her own child, the visionary hope of Israel, and her own loss. Would that ambiguity and complexity, the depressive position a psychoanalyst might say, be possible in discourse today? I don't think so. In this sense the film harkens to a period (the 1980s) of more discursive freedom.

I found myself attentive to the tone—reflective, unsettled—of many of these women I met, and hearing there a comparable affective tone in this film. This seemed so in contrast to the really intensely furious review by Gideon Levy in *Ha'artez*. I was actually shocked by that tone, by the resolute contempt for so many choices in the film that I found moving and potent. Certainly, in my own history, I have been totally enraged at the action of my government, but I am mindful that I have been working on this film review and this essay from the safety and remove of New York. But what was upsetting to me was Levy's deep contempt and distrust in regard to soldiers' trauma. Beyond aesthetic considerations, or even the political issue of the Othering and erasing of the Palestinian position, Levy is reactive to soldiers' suffering as self-involved, childish, dishonest, or overdrawn. This seems a crucial point to speak to. It is a verbal register, for me, very reminiscent of a moment in the U.S. press recently, when a military man was quoted as saying "PTSD Spells BABY." In *Waltz with Bashir*, there are ruined men: mindless, futureless, and stripped of memory and of genealogy. What would it mean to take in all the implications of these truths?

These questions about discursive space came home to me with a vengeance in the experience of reading the paper and participating in the panel on the film in Israel. I could feel a nameless anxiety as the day approached, apparently forgetting my own clinical advice carried in the title: "Dread is just memory in the future tense." People wept through the film, people walked out, people boycotted, people wanted to talk. Ari Folman began the Q and A in a sort of louche filmmaker cool style and got increasingly upset and triggered as the discussions proceeded. It certainly occurred to me and to many that there is secondary traumatization showing and commenting on this film. The film has added complexity in that the filmmaker is also a character in the animated documentary. What to ascribe to the intentions of the filmmaker, what to the unconscious, what to the character? Not always clear.

What can I say in reflection about this paper? That it is "careful," adroitly speaking truth in a small discursive space. This problem was deeply more demanding and difficult for my Palestinian colleague, Mustafa Qossoqsi. It felt to me that his sentences were like tightropes over a dangerous abyss. He was honest and respectful, and spoke to the omissions in the film. Later in the discussion, a very anguished member of the audience, Doron Levene, an Israeli living in London who has spoken movingly about the film (Levene, 2006) came on stage to speak of remorse of his personal anguish and guilt over what was done in that war. Not only speaking personally, Levene asked Folman about *his* remorse. The landmines I had been worrying about, in my commenting on how Israeli responsibility is handled in the film, seemed right then to have exploded. Folman was unresponsive to Levene's anguish in that moment. His film is more ambivalent about remorse and guilt. And perhaps he shares Levy's encomium against complaining and the admission of suffering and weakness.

Toward the end of the film as Ari begins to remember the road to the massacre and its revelations, there is an almost hallucinatory sequence in the airport. At first, he imagines he might travel anywhere, to any time zone, buy anything, acquire and move in the world. Then he sees that death, looting, and bombing are everywhere. Time has stopped with death and the restoration of temporalities requires mourning and the working through of layers of trauma. Tragically perhaps, this task of mourning, in the current situation in the Middle East, has both a terrible imperative and its own long, complex, shelf life. Time is running out and still in many, many ways, as this film so brilliantly depicts, time has stopped.

I had a new vision of this difficulty listening to a plenary talk by Neil Altman at the IARRP meeting in Tel Aviv in June 2009. He spoke of the dilemma of recovery in the context of ongoing trauma and new conflicts. There is no recovery time. And to echo another insight of Eyal Rozmarin (see Chapter 15, this volume), there is the curious problem of an orchestration of collective mourning in the context of a radical prohibition on individual registers of suffering and damage. There can be a national day set aside for the collectivity to experience mourning, but individual suffering is, and has been, a lacuna in interpersonal and individual life since the 1940s. *Waltz with Bashir*, whether intentionally or unconsciously driven, breaks with that pact, to some degree, locating problems of a deep sort at the level of the individual. Whatever was right about Gideon Levy's critique of the place of the Palestinian in this film, his derision with regard to war trauma seems more and more like a cultural/individual symptom, a refusal to see damage to the individual. The tension between collective and individual consciousness, placed in the context of tensions around temporality, are part of the constriction of discursive space. It is interesting to think that "recovery" or some repair from war trauma on the individual level, must depend, in part, on openings and containment at the social, collective level, as well as at the institutional or therapeutic sites for healing.

REFERENCES

Bromberg, P. (1998). *Standing in the spaces: Essays on clinical process, trauma, and dissociation.* Hillsdale, NJ: The Analytic Press.

Ferenczi, S. (1921). Symposium on psychoanalysis and the war neurosis held at the Fifth International Psycho-Analytical Congress, Budapest, 1919. *The International Psycho-Analytical Library, 2,* 5–21.

Ferenczi, S. (1949). Confusion of the tongues between the adults and the child: The language of tenderness and of passion. *International Journal of Psycho-Analysis, 30,* 225–230.

Freud, S. (1919). The uncanny. In J. Strachey (Ed. & Trans.), *The standard edition of the complete psychological works of Sigmund Freud* (Vol. 17, pp. 217–256). London: Hogarth Press.

Folman, A. (Writer & Director). (2008). *Waltz with Bashir* [Motion picture]. United States: Sony Pictures Classics.

Huston, J. (Writer & Director). (1946). *Let there be light* [Documentary film]. United States: U.S. Army Pictorial Services.

Lacan, J. (1977). *Ecrits.* New York: Norton.

Levene, D., (2006). Comment posted on the IARPP online conference on the concept of the "Third" in pyschoanalysis. New York, November 2006.

Little, M. I. (1985). Winnicott working in areas where psychotic anxieties predominate. *Free Associations, 1D,* 9–42.

Loewald, H. (1980). The experience of time. In *Papers on psychoanalysis* (pp. 138–147). New Haven, CT: Yale University Press. (Original work published 1972)

Nadelson, T. (2005). *Trained to kill: Soldiers at war.* Baltimore, MD: Johns Hopkins University Press.

Shepherd, B. (2001). *A war of nerves: Soldiers and psychiatrists in the twentieth century.* Cambridge, MA: Harvard University Press.

Volkan, V. (2006). *Killing in the name of identity: A study of bloody conflicts.* Charlottesville, VA: Pitchstone Publications.

Chapter 18

Resistance to resistance

Lynne Layton

Even after a life-changing analysis in which I felt I had finally attained humorous acceptance of "the bitch within," I find that I still prefer the familiarity and general recognition that comes from being a "good girl." Because of this, all sorts of political activism, especially political protest, do not come easily to me. It was easier in college, when there were large social movements against the Vietnam War, against sexism and racism—but even there I shunned the more aggressive Students for a Democratic Society (SDS) and believed that reason would eventually prevail against such perverse logic as "we had to burn the village to save it." By the time I got to graduate school in the post-Watergate 1970s, resistance movements had begun their retreat to academia, and there I found a much more comfortable home and community for my left-wing thinking and faith in reason. During those years, however, I was involved in several single-issue protests, one of which was a multiyear struggle against Washington University Medical School's program, funded by AID, to invite third world doctors to train in the latest sterilization techniques. At the time, The Agency for International Development (AID) had a published goal to sterilize a certain proportion of third world women by a certain date, and the program offered free laparoscopes to the participating doctors. I'll never forget the moment when, standing outside the medical school with a picket sign, I was accosted by an angry woman who screamed: "Your mother should have been sterilized!" I felt anger, to be sure, but I also felt shame. Her attack stirred up something old in me, a feeling that I was not good, that I was disobedient. How I wished I could revel in feeling disobedient and antiauthority, as the male leaders of the group seemed able to do. But sadly, as committed as I am to social justice, I have never been able to revel in open rebellion, and I still can't. Indeed, I've seen my past repeated in the present: At the beginning of the Iraq War, when there was a movement and a hope that U.S. belligerence might be checked, I took political action. But as the war drifted on and the movement dwindled to those few people who, every week, courageously stand outside on the same street corner with their peace signs, I found myself going indoors to pound my keyboard, not the pavement. How *do* those people sustain enough hope to stay out there week after week? Perhaps what keeps me indoors isn't just a collusion with gender norms, that is, my discomfort going against the gender stereotype that brings approval; perhaps it is something, too, about how one sustains hope.

Much as I hate approaching strangers with my political agenda, I did force myself to do it during the 2008 presidential elections. I felt it was perhaps my last chance to resist the cynicism and helplessness that had just about overtaken me. I was heartened by the fact that a movement of which I was a small part had prevailed against the American Psychological Association (APA) in winning the referendum against psychologist participation in illegal detention sites. Old hopes for change were reawakened. I felt I was again part of a movement. With a movement, I find hope and I find just enough social acceptance to enable resistance; without it I find myself stuck in a state of uncomfortable disavowal in which I experience a resistance to resistance (that is, unless you count watching Jon Stewart and Keith Olbermann as resistance). Without a movement, I find myself unable to brave the majority's indifference if not outright hostility. I cannot consistently be the "prisoner of hope" who struggles for a better world with no guarantee of success *at all* (West as cited in Deavere Smith, 1992).

How does one remain a "prisoner of hope," even when there is no movement to sustain hope (Dimen, 2006)? And what role does the need for social acceptance play in resistance to resisting oppression? In this chapter on the complexities of resistance, I focus on two sites that might help us address these questions: psychoanalysis and our contemporary sociopolitical world. First, I ask what psychoanalysis, a profession that knows quite a bit about resistance, can teach us about the obstacles to sustaining hope for a better world. Can it tell us anything about what we need to contend with in ourselves and in our relations to others to bring that world about? I shall look at a particular form of resistance that, both in the clinic and in the streets, is terribly difficult to work with: disavowal—a simultaneous knowing and refusing to know that often eventuates in sadomasochistic dynamics and forms the core of what I will call both individual and social perversion.

Second, I want to look at our specific historical moment to try to understand the immobilizing effects on the populace of what Sennett (2006) calls "the new capitalism," Bauman (2001) "the politics of disengagement," and Harvey (2005) "neoliberalism." Whatever one calls it, the policies of both government and corporate capitalism in the past 30 years have created what many social theorists refer to as a situation of "precarity" (see Beck, 1999; Butler, 2004; Sennett, 2005; H. Stein, 2000), a heightening of feelings of disposability, uncertainty, and fear of falling (Ehrenreich, 1989). The forms of vulnerability to which the politics of disengagement have given rise, and the way those forms were managed culturally in the Bush era, issued in a psychology and a politics marked by a heightened individualism and splitting into "us versus them": an inability to see the self in the other and to recognize the ways in which all of our lives are intertwined (see Layton, 2009a). A backlash against all forms of dependence and vulnerability prepared the population to ally either with a discourse that fostered war or a discourse that fostered apathy and cynicism—each of which is based in a disavowal of interdependence (Layton, 2006a). The prisoner of hope wannabe, like myself, is particularly prone to the latter, and while I do not think my own psychic issues are universalizable, I'm guessing that there is something in my own practice of disavowal, a collusion with social norms I know to be unjust (rather than a resistance against them), that might shed light on why there has been so little resistance to the travesties of the Bush regime, in particular, the Iraq War.

RESISTANCE AS PERVERSION OF TRUTH

In "Analysis Terminable and Interminable," Freud (1937) wrote about how the defenses, originally acquired to ward off threats posed to the ego by a dangerous and unpleasurable outer and inner reality can themselves become a threat to the self: "if the perception of reality involves unpleasure," Freud writes, "that perception—i.e., the truth—must be sacrificed" (p. 237). The ego, further weakened by internal dangers that threaten unpleasure, defends itself by distorting reality and, in fact, by seeking situations that justify "its maintaining its habitual modes of reaction" (pp. 238), even when unnecessary. We recognize a description of the repetition compulsion, which Freud here frames in slightly different terms than he had in previous work. It appears here as an outcome of situations in which the truth of a perception, if painful, is sacrificed in favor of a less painful lie. Tying this observation to the issue he is investigating—what makes analysis so difficult and what makes some analyses fail—Freud goes on to say that one half of the job of analysis is to lift repressions, while the other half is to recognize which defense mechanisms are exacting too high a price for the ego. These ego-modifying defense mechanisms, he argues, emerge in analysis as resistances to cure: "They are resistance not only to the making conscious of the contents of the id, but also to the analysis as a whole and, thus, to recovery" (p. 239). The outcome of an analysis "depends essentially on the strength and on the depth the root of these resistances that bring about an alteration of the ego." (p. 240). Freud suggests that resistance to cure, which appears in repetition compulsions and in character, rests not on repression but on a resistance to the analysis of resistance, a perhaps more conscious process of disavowal: When a reality is too painful to tolerate, we may both register the reality and simultaneously sacrifice what we know to be true to a more tolerable lie.

Post-Freudian writings on resistance elucidate in important ways the nature of the "new danger" to the ego, for instance, the threat analysis poses to one's loyalty to old objects, the only objects one has and thus the very core of what one defines as one's self. Or the threat that comes from the outside: the fear that the analyst will repeat the original trauma (Stolorow, 2007). But I'd like to look more closely at Freud's experience-near description of sacrificing the truth to a lie, because I think that this aspect of resistance to resistance captures something important not only about the difficulty of our work as analysts but about our experience as citizens who must "choose" either to resist against or collude with social phenomena we know to be unjust. More specifically, I want to focus on disavowal, the defense against knowing what we know to be true.

Freud's insight about lying was central to Bion and to many of his followers. Bion (1962a, 1991) asserts that the capacity to bear frustration leads to the capacity to think; when the raw emotion evoked by frustration is not adequately contained, lying, rather than thinking, may become a customary way of defending against what he calls catastrophic change. Lies may well be painful to live with, but they are less painful than the truth, which can threaten to annihilate the self. Bion (1962/1991, 1970/1993) made the important distinction between suffering pain and feeling pain: To suffer pain is to tolerate difficult to bear frustration and to mentalize it, the prerequisite for learning from experience. To feel pain is to disavow the painful truth and instead construct experiences of pleasure and pain as things outside the self that are under one's control, for example, cutting. In the perverse state of disavowal, such constructed experiences need to be ritualistically repeated precisely because they do not promote growth. There is no learning from experience (see Abel-Hirsch, 2006). Like Bion, I would argue that the inability to tolerate painful truths results from some kind of relational failure, a condition in which it is has been unsafe to depend on an other, be that other a parent or a cultural surround. Without adequate care, the vulnerable self is abandoned to overwhelming anxiety, if not fears of annihilation. A situation of relational precarity promotes the "choice" to feel pain rather than suffer pain.

Pistiner de Cortiñas (2009) connects Bion's thoughts on lying to Freud's work on disavowal as follows:

> In the personality who has the habit of lying, we find a scission that Freud described in "Fetishism" … a part which is in contact with reality and the other part that disavows it. The patient in the habit of lying usually uses ambiguity to "navigate" through this scission. This characteristic of "yes, but no," indicates the tenacity with which the scission is maintained.
>
> Conflict implies having contradictions, vertices that might be in disagreement; in ambiguous functioning the definition of vertices and the acceptance of contradictions are eluded, and what usually shows up are masks and incongruence. (p. 115)

Pistiner de Cortiñas goes on to postulate what she calls an "ambiguous position" (p. 122), and she gives a few compelling case vignettes to exemplify this particular refuge taken by patients who cannot bear to know the truth. In one case, a patient who habitually lied to himself and his analyst dreamed of being in the yellow zone of traffic lights, "avoiding the clear definition of the red and green lights" (p. 123). In another dream, the patient found trash in his home and, rather than throw it away, he sought a place in which to hide it (p. 123). This well describes the state I've often found myself in these past 10 years or more—I have neither withdrawn completely from political activity nor fully committed myself to it.

Lies come into rivalry with the truth the patient has come to analysis to rediscover; Pistiner de Cortiñas (2009) argues that, because the lying patient "sustains his independence from everything that is not his own creation" (p. 117), the central problem in the analysis "is the nature of the co-operation [with the analytic process], and not the content of the problem for which he has required analysis" (p. 117). She issues a further warning to the analyst: "lies seek collusion" (p. 125). Gerson's (1996) argument that resistance involves a mutual collusion between analyst and patient is instructive here, as is Ruth Stein's (2005) work on perverse pacts, which she defines as "a relationship between two accomplices, a mutual agreement ... that serves to cover over or turn the common and mutual gaze of the accomplices from the catastrophic biographical events that had befallen each of them" (p. 787). Again, I think this captures something important about our times and about liberal responses to them: A social reality built on cynical lies evokes a painful turn away from what we know to be true. But the disavowal works in two directions. As Howard Stein (2000) has argued, a media that turns away from painful truths, such as the body count of Iraqi citizens, not only colludes with government to trick the population into quiescence, but also colludes with the population's social defense against knowing what it knows.

What are some of the ways that the social defense of disavowal is manifest? We all need to locate our experience in existent cultural narratives or in personal adaptations of these narratives. Dominant narratives explain what is happening in terms that generally do not challenge the habitual ways of thinking that support the power status quo; as Gramsci (1971) has argued, such narratives pass as "common sense." For example, the existence of torture in illegal detention sites will be framed by government, media, and populace as the work of a few bad apples. Such decontextualizations and dehistoricizations are at the heart of the individualistic thinking proper to bourgeois ideology; to fit our experience to existing narratives, we "learn" to ignore the links that would render counternarratives meaningful, for example, narratives in which such phenomena are understood as systemic. Indeed, for many, systemic narratives simply don't make sense. For some, notably those who have something to gain from supporting the status quo, be it material reward or social acceptance, adopting these narrative frames might involve cognitive dissonance—it is not in their interest to know what they know. Others are aware that collusion with dominant narratives demands a suppression of important parts of self. But we want to trust our leaders even when we suspect they do not have our interests in mind. As Ferenczi (1933/1955) has noted, maltreatment often issues in a wish to keep the perpetrators good, a wish that makes us distrust our own senses (when a trusted leader like Colin Powell said Iraq had weapons of mass destruction, I found myself beginning to doubt what I knew).

For those who are aware of how they suffer from existing power relations, systemic narratives that run counter to the dominant frame make perfect sense and some, especially those whose life depends on it, resist. Yet many of us in this situation too often feel helpless to challenge individualistic narratives in any way than by yelling back at the television or writing articles like this one.* Finally, as Al Franken's (2003) book title avers, there are the lying liars of the past 8 years, who justified war with lies like the one that linked Iraq and al Qaeda. It is hard for anyone to say they did not know these were lies—evidence was downplayed in the press, but the evidence was there to find in popular books, documentaries, and news articles. Those who directly challenged the lies were named enemies of freedom; once the antiwar movement withered, it was easier to collude by silence. All of these forms of disavowal engender resistance to resistance and have the feel of a socially shared perversion.

Pistiner de Cortiñas (2009) offers some thoughts that connect the function of lying and the psychology of the liar to cultures that are committed to lying. I was particularly taken by the way she links lying to the inability to tolerate helplessness:

> In thoughts and in the developing of thinking, a tolerance of the emotions that stimulate feelings of helplessness has been achieved, these being uncertainty, ignorance, and the finite-infinite relationship. In lies, when it comes to facing feelings of helplessness, there is an increment in the doses of omnipotence and the need for obtaining collusions that disavow helplessness. (p. 121)

Key to the capacity for truth telling is the capacity to tolerate uncertainty, helplessness, and vulnerability rather than disavow the reality that evokes

---

* I exaggerate a bit: I did in fact write to the *Boston Globe* reporter who covered the August 16, 2008, APA rally and pointed out to her that the protest was not really about rooting out bad apples or determining whether having psychologists on site could help assure that the interrogations were ethical—the APA position. A few days later an unsigned *Globe* editorial did capture the broader concern about the detention sites being illegal in their flouting of the Geneva Conventions. I, of course, have no way of knowing if my letter had an effect.

those states. So what happens to possibilities for resistance in a culture that makes uncertainty and shame about the vulnerability it evokes a way of life?

THE SOCIAL ROOTS OF CONTEMPORARY DISAVOWAL*

As babies and children, we need containment, holding, and, more generally, a caretaking attitude toward our vulnerabilities if we are not to be overcome by anxiety. As citizens, we need the same from our social environment. Maltreatment from those on whom we depend creates the conditions in which disavowal becomes an individual and social defense. Over the past 30 years, we have been essentially abandoned by our collective containers (Peltz, 2005), and this has bred an anxiety that has, in many ways, been socially channeled toward finding individualistic rather than collective solutions to collective problems. Individualism has been a central feature of U.S. life from the country's earliest days, to be sure. But the events of our recent history have bred a heightened form of individualism that is distinct from earlier forms.

Many social theorists have ascribed the collective trauma in which we have found ourselves since the post-Vietnam '70s to the imposition of free market capitalism, a response to the economic crises of the '70s and to the demands of the left-wing social movements of the '60s for economic equality. In *A Brief History of Neoliberalism*, David Harvey (2005) suggests that the inflation, surging unemployment, and crisis in capital accumulation that occurred in the United States by the late '60s led the upper classes to panic about their loosening hold on power and wealth. That panic eventuated in the repudiation of Keynesian principles of government intervention in the economy, and led to the embrace of a neoliberal, free market ideology that ended the postwar compromise between labor and capital, broke the back of unions, deregulated public services, and generally made "big government" into a villainous term. The changes wrought by neoliberalism created a shameful gap between rich and poor. Even before the economic collapse of 2008, many of us were already living in conditions of precarity caused by shifts in labor markets that offer little to no security to the workforce (e.g., the loss of secure manufacturing jobs to low-wage service jobs with no benefits; the end of defined pension benefits; the hiring of temporary workers on short-term contracts; the shifting of jobs overseas; the decline of union power to negotiate collective contracts).

In *The Culture of the New Capitalism*, Richard Sennett (2006) writes about how changes in economic policy have affected political and social

---

* I have made the argument that follows at greater length in Layton (2009b).

institutions. Sennett argues that political institutions have modeled themselves on "cutting edge" corporate cultures, a key feature of which is to be less and less accountable for the negative effects their policies have on workers/citizens. These new cultures, based on a consultant model that discourages long-term attachments and rewards risk taking and shaking things up, cultivate an idealized self that "publicly eschews long-term dependency on others" (p. 177). The "politics of disengagement" that characterize the postwelfare state make it difficult both to locate a clear collective opponent and clear collectives with whose grievances one might ally. Individual competition replaces collective struggle (Bauman, 2001). Even if collectives could be clearly located, employers' structural lack of loyalty to the workforce makes it dangerous for workers to resist. Where we once had collective struggle in the workplace, we now have the occasional, highly courageous whistleblower. The post-1960s despair about resisting capitalism has even found its way into what does remain of resistance to social injustice: as many critics have noted, there has been a decoupling of social movements aimed at recognition of difference (e.g., women's movements, gay and lesbian liberation movements) from social movements aimed at a more equitable redistribution of wealth (Bauman, 2001; see also Fraser, 1997; Rorty, 1998).

An indifferent institutional order breeds "low levels of informal trust and high levels of anxiety about uselessness" (Sennett, 2006, p. 181). In our increasingly individualist meritocracy, a few people are recognized as truly talented, and the rest are relegated to the nonspecial status of a disposable mass. The untalented masses come to feel that they have only themselves to blame for being not special. Along with the self-esteem and harsh superego issues that this obviously would produce—it is no surprise perhaps that self-psychology, with its focus on self-esteem regulation, develops during this period—Sennett finds that an important consequence for individual psychology is that people feel anxious not so much about failure as about being found useless or redundant. People who have been deemed redundant or whose jobs are outsourced are, indeed, disposed of in very traumatizing ways (H. Stein, 2000). The threat heightens the sense of precarity for all of us.

During the Bush years, tax cuts for the wealthy, corporate welfare, and the costs of a terribly unpopular war decimated public services, which left most of us on our own to fend against very real anxieties that we will end up without health care, without pensions, without social security. Harvey (2005) writes:

> As the state withdraws from welfare provision and diminishes its role in arenas such as health care, public education, and social services, which were once so fundamental to embedded liberalism, it leaves larger and larger segments of the population exposed to impoverishment. The

social safety net is reduced to a bare minimum in favour of a system that emphasizes personal responsibility. Personal failure is generally attributed to personal failings, and the victim is all too often blamed. (p. 76)

At the same time, vulnerable and dependent feelings have become increasingly shameful states since the Reagan revolution. And they have become all the more shameful since the events of 9/11 and the U.S. response to those events: during the Bush years, vulnerability and fear were simultaneously stoked, by both media and government, and treated as shameful (see Hollander & Gutwill, 2006).

Peltz (2005) traces some of the psychic effects of public indifference on the professional middle-class patients she treats. Like the social theorists cited earlier, she, too, argues that when the government abdicates responsibility for containing anxiety and for "holding" the vulnerable and the needy, dependency becomes more and more shameful. She suggests that those who have been lucky enough not to fall through the now huge holes in the social safety net have taken refuge in a manic defense against need. Left on our own by government and corporate cultures to sink or swim, unable or unwilling to make the links necessary to understand our predicament and rebel against it, many of those fortunate enough to be able to do so have defended against feelings of abandonment, helplessness, and vulnerability with the "lie" of self-sufficiency. As Peltz elaborates, the professional middle-class has accepted as normal a state in which we daily run ourselves ragged; we feel virtuous when we can fit 100 activities in the shortest amount of time and we feel like lazy slobs when we cannot. Like the patients Peltz describes, most of us in the professional middle class have dutifully shaped our subjectivities in accord with dominant individualistic norms that, even more so than in past eras, unlink the social from the individual. And so we consistently rail against ourselves when, for example, our small businesses fail or when we are unable to balance career and child care. We imagine that there are stronger, special others who can do it all and that if only we weren't weak, inferior beings, we, too, would succeed. Adapting to the "politics of disengagement," we have successfully made ourselves solely responsible for our so-called failures.

In so doing, many of us have anxiously retreated to the private or professional sphere, disavowing our connections to the goings on in the wider world. Sociologist John Rodger (2003) has a term for the disengaged response to the disengagement of the post-welfare state: *amoral familism*. The ethic of social solidarity that characterized the welfare state, he argues, has been replaced by a tendency to limit concern only to the self and to those in one's intimate circle. Taking together Peltz's and Rodger's observations, we might guess that the lack of resistance to the Iraq War has something to do with the fact that most middle-class families do not feel directly affected by the tremendous loss of human life: Iraqis are "them," and even the volunteer poor who make up most of our fighting forces are "them."

Resistance is hard to sustain in such conditions of anxious vulnerability. As analysts are well aware, two common defenses of disavowed vulnerability are retaliation and withdrawal, and I think we have seen the U.S. population take refuge in one or both of these defenses during the Bush years (Layton, 2006a). Retaliation and withdrawal are at the very heart of sadomasochistic object relations, two breakdown products of an inability to suffer painful truths. We have seen retaliation in the hateful politics of homophobes, racists, radical antiabortionists, and government leaders who locate all good in themselves and all evil outside. We have seen withdrawal and cynical apathy in those who feel helpless to change things. Most of us on both sides have opted to save our own skins, but it is not hard to see why. Phenomena such as the aforementioned decoupling within social movements of a politics of recognition (where the focus is on respect for difference and equality for marginalized groups) from a politics whose focus is redressing economic inequality (with a redistribution of wealth); the state of amoral familism; and the manic defense against vulnerability are not just about greed and self-interest, and not a product of an exploitive human nature. Rather, they are collective and individual responses to profound relational failures: the indifference, lack of accountability, and outright hostility that have brought about a loss of hope and a loss of trust in those who are supposed to be concerned with our welfare. If the social surround makes you feel ashamed of needing anything from the outside, if your leaders take little to no responsibility for the welfare of the citizenry, there are few options left besides feeling that you can only rely on yourself—even as you know you cannot. Yet that feeling is perhaps at the heart of the fetish structure of social perversion.

THE FETISH STRUCTURE OF SOCIAL PERVERSION

In an attempt to link disavowal to the foregoing description of the social world in which we have found ourselves these past years, I would like to explore in more detail the psychological structure that supports the state of resistance to resistance. Nick Totton (2006) has proposed that rather than look at civilization as neurotic, we think of it as perverse, "structured as a response to massive collective trauma" (p. 144). I have described the trauma; let us now look at the nature of perversion. For the late Freud (1927), perversion rests on the disavowal particular to the fetish structure. Although Freud customarily roots the fetish structure in castration anxiety, he in fact describes the first instance of disavowal as a resistance to acknowledging the reality of dependence on others. The hungry baby, who has no control over the appearance or disappearance of the mother, hallucinates the breast, thus finding an omnipotent solution to a painful reality: I don't need you; I'm self-sufficient. The capacity to hallucinate a way out of painful tension can be a source of creativity, to be sure. But when that capacity becomes a regularly practiced disavowal of dependence, interdependence, and need for an other, we have the makings of a perverse situation.

For Freud (1927), the etiology of fetishism lies in the boy's inability to tolerate the fact that girls are castrated boys. In the fetish structure, a boy disavows the awareness that girls do not have a penis and that he, too, can be castrated. The fetishist oscillates between acknowledging that there are castrated beings and denying it. But, as Bass (2000) has pointed out, Freud's description of the fetish is itself based in disavowal: girls are not castrated boys; they are different from boys. In the fetish structure, the inability to fully acknowledge difference leads to an oscillation between a fantasized presence of a penis and an equally fantasized absence of a penis. Thus, in Freud's version of the fetish, there is no difference, only the sameness of masculinity (see Chasseguet-Smirgel, 1986; Irigaray, 1985): if boys are A, girls are –A. True difference, on the contrary, would take in the reality that boys are A and girls B.

Bass (2000) argues that the trauma that gives rise to disavowal is the trauma of the encounter with difference. But my sense is that the difficulty acknowledging difference is the result of a prior trauma. To understand this further, I think we need to look at Freud's rather consistent disavowal of the significance of helplessness and dependency. Often in his writings, he mentions annihilation or separation anxiety only to insist, with little evidence, on the primacy of castration anxiety. At times, his leap from dependency to castration has the feel of a non sequitur. Freud ends the paper "Analysis Terminable and Interminable" (1937), for example, with the assertion that the bedrock beyond which analysis may not be able to go is the contempt for femininity that marks both sexes. The argument appears to come out of nowhere. Feminist scholarship has suggested that cultural contempt for femininity has little to do with having or not having a penis, but rather with the repudiation of the human attributes that have traditionally defined femininity: emotionality rather than reason, dependence rather than autonomy, vulnerability rather than stoic strength. The contempt for femininity is thus indeed at the core of the fetish structure and of perversion, but the fetish structure that does not tolerate difference arises from a denial of dependency, not castration anxiety.

In the fetish structure, we find an oscillation between fantasies of omnipotent self-sufficiency and fantasies of "perfect love," of being totally taken care of. This perverse condition most likely emerges when the other on whom we need to rely is unavailable, incapable of or unwilling to nurture, to allow dependence, and to contain vulnerability. The subject poorly cared for disavows that painful truth and seeks solace alternately in self-sufficiency and/or undifferentiated merger, pure difference or no difference.

For those caught in such an oscillation, there is little room for the kind of resistance that redresses social inequality in a progressive way, one that counters splitting, projection, and the repudiation of otherness.

To be sure, there are omnipotent forms of resistance that attract submissive adherents, but such forms, marked as they are by unmourned trauma and disavowal, generally give rise to solutions that look a lot like the old problems. Hope requires something that would facilitate an exit from the familiar oscillation between the split complementarities of pure submission and pure domination (Benjamin, 1988, 2004). In recent years, our cultural parents have utterly failed at this task; on the contrary, they have done everything possible to fortify the polarized "us versus them" discourse that has characterized the United States since the "Reagan revolution" legitimized a retreat from demands for redistribution of wealth and encouraged a backlash against social movements for equal rights. The hope offered by President Obama in the 2008 election rests in large part in his resistance to the powerful narratives that sustain this profound and traumatizing relational failure. But we cannot succumb to the fantasy that Obama will do all the work of making change, for it is already clear that on several fronts Obama represents the interests of capital and militarism and that those interests are in many cases inimical to social change. The resistance movement that formed in opposition to Bush's social and foreign policies will need to stay mobilized.

Perversity is a lying relation to reality, turning a blind eye to truth. I have argued that at the heart of the perverse relation to reality is an inability to bear self-annihilating feelings that arise from not being able to rely on another for recognition, containment, and care. My definition of social perversion centers on the self-estrangement we practice to live in a precarious state that systematically makes dependency shameful, that unlinks the individual and the social in a way that makes it possible to disavow the ways in which we are all interconnected, and therefore encourages the collapse of difference into hierarchies of superior–inferior. Such disavowals underlie the social perversions of racism, homophobia, sexism, and the destruction of the environment. And they underlie as well our failure to resist the Iraq War. Paul Haggis' film, *In the Valley of Elah* (2007), offers a compelling "case study" of the way the structure of social perversions operates in the resistance to war resistance. In the film, culturally tabooed connections are made between the Vietnam War and the Iraq War, connections that resist the norm to unlink the social and the psychic (Layton, 2006b). The film offers thoughts about how resistance as disavowal gets in the way of political resistance to the painful truths of war—and it underscores how the abandonment by social containers connects with the so-called bedrock contempt for femininity, that is, the culturewide inability to tolerate vulnerability, uncertainty, dependence, and interdependence.

IN THE VALLEY OF ELAH

Sergeant Hank Deerfield (Tommy Lee Jones) fought in Vietnam. As the film begins, Hank finds out that his son, Specialist Mike Deerfield, who recently returned from combat in Iraq, has gone AWOL. Hank sets off on a journey to find his son. We soon discover that Mike was brutally murdered, dismembered, and abandoned by his own troop mates when a fight over something insignificant spiraled out of control. The film's "message" is that unspeakable things have happened in Iraq and that no one comes back without severe psychic trauma. Specifically, those who return have a propensity to violence, often perpetrated against the people they most love. That message echoes even in a small subplot in which a distraught young woman turns for help to the police after her soldier husband tortures and kills their dog in front of her and her son. The police dismiss her worry over her husband's mental state; indeed, they mock what they consider the triviality of her concern. But when the husband later drowns his wife in the same tub in which he killed the dog, the police are mute with horror and shame.

Like the woman in the subplot, Hank Deerfield, too, is frustrated by both military and civilian police, and he undertakes his own investigation into what happened to his son. The heart of the film is Hank's psychic journey, a journey in which the perverse disavowals of everyday life give way to unbearable truths. Arrogant, certain, always sure he knows best, trusting no one, looking everywhere for the devil but within, Hank, a good soldier (Grand, 2007), sets out to find his son. On his way to Mike's barracks, he notices that the flag in front of the high school is flying upside down. He gets out of his car and sternly teaches the immigrant worker how to put the flag on correctly, enlightening him on what an upside down flag signifies: a national call of distress.

Hank's first act as investigator is to steal his son's cell phone and hire someone to unscramble the videos Mike shot in Iraq. As each video is unscrambled, the audience and Hank come closer to knowing what Mike experienced in Iraq; the more we know, the more we can imagine what happened to him and his mates when they came home. But from the first moment that Hank learns his son is gone, he is haunted by a recurring dream that takes the form of snippets of a conversation between him and his son. At the outset, all we hear of it is "Dad, Dad …" Later we hear "something happened Dad" and "please get me out of here."

The film turns out to be about the violence perpetrated on the soldiers by their individual and cultural fathers—and how that violence is passed on. Hank will have to bear many painful truths before he is able to face the entirety of that crucial conversation with his son: He will have to confront his ugly racism; bear the unsettling of some of the very certainties that ground him, such as "the people you fought next to would never hurt you"; and face a devastated wife who accuses him of killing both her soldier sons because of the macho code to which he made them adhere. He will have to watch a video that shows his son laughing as he tortures Iraqi prisoners. In the end, Hank will turn to Ortiz, the Mexican platoon mate who Hank was at one point certain had killed Mike and whom Hank subjected to a venomous attack of racial hatred. It is Ortiz who gives Hank the key to understanding what happened to Mike: in their first week in Iraq, Ortiz recounts, Mike was ordered to drive into an Iraqi child in the street and keep driving (Ortiz describes these orders as having come from above). Mike did not want to do it, but he obeyed orders. Against orders, however, he got out of the car to return to the child; he was ordered back to the car by the very platoon mate who later stabbed him to death. A video Mike made tells the true story. But even though they know they hit a child, the group agrees on the lie that they hit a dog. Such disavowals become their reality and lead to the ultimate self-estrangement that destroys their humanity.

After he hears Ortiz's story, Hank is able to remember the full conversation with his son, which, in fact, occurred just after the incident Ortiz describes. With Hank, the audience sees Mike call home from Iraq and ask his father to get him out of there. Hank responds, "That's just nerves talkin'." Mike cries, and Hank says: "For chrissake." Hank asks if anyone is there who might see Mike cry. Mike: "No." Hank: "That's good." Mike: "Bye, dad."

Hank begins the mourning process that can only occur when the unbearable truth replaces the lie, when he lets himself know what he knows. But Hank's mourning involves much more than recognizing how his love for Mike was contingent on Mike conforming to a cultural stereotype of masculinity, much more even than his emotional abandonment of Mike in Mike's time of need. Because the audience knows that the kind of atrocities that occur in Iraq also occurred in Vietnam, we must imagine that Hank is for the first time facing the truth of what happened to him in Vietnam. His disavowal of that truth pervaded his home and created sons who, in his wife's words, could not feel they were men had they not gone to war. Hank could not tolerate vulnerability in himself, his wife, or his sons. His collusion with the norms of patriarchal masculinity, fortified by the code of soldiering (Grand, 2007), cost him his family, and Hank's obsessionally ordered life, to which we are privy from the outset, is completely undone by the end. When Hank returns home, he finds a package from Mike that contains the flag Hank had given him before he went to Iraq. In the package is also a note that says, "For Dad." Hank takes the flag to the high school and runs it up the flagpole—upside down. Linking personal to national distress, Hank repudiates what the official discourses, by which he has lived his life, perversely sever.

In The Valley of Elah clearly suggests a connection between the soldierly world Mike inherits from Hank and sexism and racism; it links hatred of the vulnerable other to hatred for the vulnerable self and shows the destruction such hatred unleashes (note that Haggis is also the director of *Crash*, 2005, a film that portrays the ways that racial, sexual, class, and gender hatreds defend against knowing the multiple ways that we are implicated in each other's lives and in each other's suffering). To link the film to the biographical fragment with which I began this essay, I would note that both Mike and I encounter resistance to resisting social injustice when such resistance requires breaking with the perverse gender stereotypes that have brought us love and social acceptance (see Kaplan, 1991, who roots perversion in conformity to gender stereotypes)—perverse because they require that we disavow parts of the self that are essential to being human. For Mike, being a good boy means sacrificing the truth for a love that requires him to disavow vulnerability; for me, being a good girl entails sacrificing the truth for a love that requires that I disavow assertiveness and avoid interpersonal conflict. But since norms for what it means to be the ideal human are generally consonant with norms for what it means to be the ideal male—that is, the contempt for femininity common to both sexes—both Mike and I enact what we might think of as the bedrock of resistance: Each of us is dependent on love and caretaking, and neither of us can bear the vulnerable state in which we would find ourselves were we to lose love. Thus, we find ourselves stuck in the fetish structure of social perversion, the escape from freedom (Fromm, 1941) that takes refuge from a frightening state of isolation in an oscillation between fantasies of self-sufficiency and wishes for perfect love. These fantasies have been generally imagined to be split between the sexes (Benjamin, 1988), but perhaps they are not, at bottom, as split as they seem. Neither Mike nor I can bear the loss of love and social acceptance that we would face were we to demand recognition of all the parts of ourselves.

For those stuck in the fetish structure, as I have said, there are few possibilities of true resistance, and I have argued that cultural conditions of precarity support the dominance of the social perversion. So where might we look to find a way out? In the sense in which I have defined perversion, it would follow that every repetition compulsion has a perverse core. As Freud (1937) suggests in "Analysis Terminable and Interminable," a key frustration for analysts is the patient's resistance to giving up the habitual ways of defending against danger—danger that the patient might even court so as to feel justified in not giving up the habitual defenses. What Freud does not discuss is why this should be so, and his concluding contention that analytic bedrock is fear of losing the penis (passivity) or penis envy is, as I have said, quite unconvincing. Freud's (1920) earlier suggested cause, the death drive, is equally unconvincing. Resistance to giving up habitual defenses has rather to do with there being a painful place in all of us where we have felt unloved or not well-loved, where our vulnerability has been maltreated. Unable to mourn this painful loss, most of us disavow it to keep ourselves connected to the very people who have caused the pain. The disavowal forges a bond of love that is also a bond of hate—and guilt about hate (of all Freud's theories of causation of the repetition compulsion, it is perhaps the primal horde theory that comes closest to what I am arguing, e.g., Freud, 1921). *In the Valley of Elah* directly connects Mike's inability to resist what he clearly knew was destroying him in Iraq to his inability to counter his cultural and personal fathers (his mother lashes out at Hank: "Living in this house he never could have felt like a man if he hadn't gone [to war]"). We consciously and unconsciously collude with social norms, even when such collusion causes pain, because we long for acceptance and for caretaking responses to our vulnerability from both our loved ones and our cultural surround.

There is some reason to think that resistance to resistance is less deep-rooted in the social case than in the individual case, that is, careful tending from the social surround might more quickly counter the defense of disavowal than does the analyst's careful tending. A series of studies by Gaertner and Dovidio (2005) on what they call *aversive racism* provides a case in point. The authors define aversive racism as a characteristic of those who profess to be nonracist but whose racism emerges in subtle unconscious ways (ways that minorities are easily able to register). In some studies, the authors found, for example, that environmental factors determine whether a White aversive racist will help a Black person in trouble. In conditions where expectations are ambiguous or where there is another White person who could help, the aversive racist will not help, but if he is the only person available, he will. It seems that some conditions favor disavowal and others discourage it. The authors' common ingroup identity model constructs real-world possibilities of enlarging the definition of ingroups, without denying difference, and in these conditions aversive racism diminishes or disappears. Studies such as these suggest that our theories that posit a natural need for ingroups and outgroups (Dalal, 2006), enemies and allies (Volkan, 1988), may be mistaking for human nature what is actually the result of socialization processes.

The way a culture manages vulnerability matters, and a culture whose norms for proper human being do not rest on repudiation of vulnerability and dependence—the contempt for femininity—may well change the nature of resistance. Thus far we have not seemed to find a way out culturally from the oscillation between the wish for perfect love and the wish to be self-sufficient, but perhaps President Obama's communitarian ethic, his demand for accountability, and his awareness of the way we are all implicated in each other's lives will set a different course and establish conditions in which resistance to war and to the inequities that exist in contemporary capitalism again feel possible.

REFERENCES

Abel-Hirsch, N. (2006). The perversion of pain, pleasure and thought: On the difference between "suffering" an experience and the construction of a thing to be used. In D. Nobus & L. Downing (Eds.), *Perversion: Psychoanalytic perspectives/perspectives on psychoanalysis* (pp. 99–107). London: Karnac.

Bass, A. (2000). *Difference and disavowal: The trauma of Eros*. Stanford, CA: Stanford University Press.
Bauman, Z. (2001). *Community: Seeking safety in an insecure world*. Cambridge: Blackwell.
Beck, U. (1999). *World risk society*. London: Polity Press.
Benjamin, J. (1988). *The bonds of love*. New York: Pantheon.
Benjamin, J. (2004). Beyond doer and done-to: An intersubjective view of thirdness. *Psychoanalytic Quarterly*, 98(1), 5–46.
Bion, W. R. (1991). *Learning from experience*. Northvale, NJ: Jason Aronson. (Original work published 1962)
Bion, W. R. (1993). *Attention and interpretation*. London: Karnac. (Original work published 1970)
Butler, J. (2004). *Precarious life: The powers of mourning and violence*. London: Verso.
Chasseguet-Smirgel, J. (1986). *Sexuality and mind: The role of the father and the mother in the psyche*. New York: New York University Press.
Dalal, F. (2006). Racism: Processes of detachment, humanization, and hatred. *Psychoanalytic Quarterly*, 75(1), 131–161.
Deavere Smith, A. (2003). *Twilight: Los Angeles, 1992*. New York: Dramatists Play Service.
Dimen, M. (2006). Something's gone missing. In L. Layton, N. C. Hollander, & S. Gutwill (Eds.), *Psychoanalysis, class and politics: Encounters in the clinical setting* (pp. 195–201). New York: Routledge.
Ehrenreich, B. (1989). *Fear of falling: The inner life of the middle class*. New York: Pantheon Books.
Ferenczi, S. (1955). Confusion of tongues between adults and children. In *Final contributions to the problems and methods of psychoanalysis* (pp. 156–167). New York: Basic Books. (Original work published 1933)
Franken, A. (2003). *Lies and the lying liars who tell them: A fair and balanced look at the right*. New York: Dutton.
Fraser, N. (1997). *Justice interruptus*. New York: Routledge.
Freud, S. (1920). Beyond the pleasure principle. In J. Strachey (Ed. & Trans.), *The standard edition of the complete psychological works of Sigmund Freud* (Vol. 18, pp. 1–61). London: Hogarth Press.
Freud, S. (1921). Group psychology and the analysis of the ego. In J. Strachey (Ed. & Trans.), *The standard edition of the complete psychological works of Sigmund Freud* (Vol. 18, pp. 65–144). London: Hogarth Press.
Freud, S. (1927). Fetishism. In J. Strachey (Ed. & Trans.), *The standard edition of the complete psychological works of Sigmund Freud* (Vol. 21, pp. 152–159). London: Hogarth Press.
Freud, S. (1937). Analysis terminable and interminable. In J. Strachey (Ed. & Trans.), *The standard edition of the complete psychological works of Sigmund Freud* (Vol. 23, pp. 209–253). London: Hogarth Press.
Fromm, E. (1941). *Escape from freedom*. New York: Farrar and Reinhart.
Gaertner, S. L., & Dovidio, J. F. (2005). Understanding and addressing contemporary racism: From aversive racism to the common ingroup identity model. *Journal of Social Issues*, 61(3), 615–639.
Gerson, S. (1996). Neutrality, resistance, and self-disclosure in an intersubjective psychoanalysis. *Psychoanalytic Dialogues*, 6, 623–645.

Grand, S. (2007). Maternal surveillance: Disrupting the rhetoric of war. *Psychoanalysis, Culture & Society, 12*(4), 305–322.

Gramsci, A. (1971). *Selections from the prison notebooks* (Q. Hoare & G. N. Smith, Eds.). New York: International Publishers.

Haggis, P. (Director). (2005). *Crash* [Motion picture]. United States: Lions Gate Entertainment.

Haggis, P. (Director). (2007). *In the valley of Elah* [Motion picture]. United States: Warner Independent Pictures.

Harvey, D. (2005). *A brief history of neoliberalism*. Oxford, UK: Oxford University Press.

Hollander, N. C., & Gutwill, S. (2006). Despair and hope in a culture of denial. In L. Layton, N. C. Hollander, & S. Gutwill (Eds.), *Psychoanalysis, class and politics: Encounters in the clinical setting* (pp. 81–91). New York: Routledge.

Irigaray, L. (1985). The blind spot of an old dream of symmetry. In G. Gill (Trans.), *Speculum of the other woman* (pp. 13–129). Ithaca, NY: Cornell University Press.

Kaplan, L. J. (1991). *Female perversions*. New York: Doubleday.

Layton, L. (2006a). Retaliatory discourse: The politics of attack and withdrawal. *International Journal of Applied Psychoanalytic Studies, 3*(2), 143–155.

Layton, L. (2006b). Attacks on linking: the unconscious pull to dissociate individuals from their social context. In L. Layton, N. C. Hollander, & S. Gutwill (Eds.), *Psychoanalysis, class and politics: Encounters in the clinical setting* (pp. 107–117). New York: Routledge.

Layton, L. (2009b). Who's responsible? Our mutual implication in each other's suffering. *Psychoanalytic Dialogues, 19*(2), 105–120.

Layton, L. (Ed.). (2009a). Us versus them [Special issue]. *Psychoanalysis, Culture & Society, 14*(1).

Peltz, R. (2005). The manic society. *Psychoanalytic Dialogues, 15*(3), 347–366.

Pistiner de Cortiñas, L. (2009). *The aesthetic dimension of the mind: Variations on a theme of Bion*. London: Karnac.

Rodger, J. (2003). Social solidarity, welfare and post-emotionalism. *Journal of Social Policy, 32*(3), 403–421.

Rorty, R. (1998). *Achieving our country*. Cambridge, MA: Harvard University Press.

Sennett, R. (2006). *The culture of the new capitalism*. New Haven: Yale University Press.

Stein, H. (2000). Disposable youth: The 1999 Columbine high school massacre as American metaphor. *Journal for the Psychoanalysis of Culture & Society, 5*(2), 217–236.

Stein, R. (2005). Why perversion? "False love" and the perverse pact. *International Journal of Psychoanalysis, 86*(3), 775–799.

Stolorow, R. D. (2007). *Trauma and human existence: Autobiographical, psychoanalytic, and philosophical reflections*. New York: Analytic Press.

Totton, N. (2006). Birth, death, orgasm, and perversion: A Reichian view. In D. Nobus & L. Downing (Eds.), *Perversion: Psychoanalytic perspectives/perspectives on psychoanalysis* (pp. 127–146). London: Karnac.

Volkan, V. (1988). *The need to have enemies and allies*. Northvale, NJ: Jason Aronson.

Index

A

Abu Ghraib prison, 4, 74, 143, 155
ACLU, *see* American Civil Liberties Union
Activism, *see* Psychoanalytic activism
Adams, John Quincy, 157
Adult onset trauma, 30, 40
Afghanistan veterans, *see* Posttraumatic stress disorder
Agency for International Development (AID), 359
AID, *see* Agency for International Development
al-Qaeda, 153, 169
Ambiguous position, 363
American Civil Liberties Union (ACLU), 88
American exceptionalism, 156
American foreign policy, violence in, 153–174
    Abu Ghraib prison, 155
    al-Qaeda, 169
    American exceptionalism, 156
    American self-preservation, 156–158
    Christians, 162
    city on the hill, mission of, 163
    country vulnerable to humiliation, 170
    disintegration anxiety, 167
    emotional state of the nation, 154
    enhanced interrogation techniques, 155
    God's American Israel, 156
    grandiose self-image, 170
    grandiosity and American identity, 158–163
    greed vs. grandiosity, 163–164
    images of national superiority, 159
    informal empire, 159
    invisible empire, 159
    Iraq War, 154
    lessons learned, 165–166
    myth of America, 157
    narcissism, 168, 171
    national dialogue, 171
    national propensity for violence, 166–169
    national self-representation, 159
    omnipotence, self-concept of, 160
    Puritan belief, 162
    reality-based community, 163
    repair of American grandiosity, 170–172
    required leadership, 171
    responses to helplessness, 154–156
    Second Coming, 156
    self-determination, 158
    self-interest vs. shortsightedness, 164–165
    shameful degradation, 156
    silent empire, 164
    virtue over vice, 157
    World Trade Center attack, 153
American Psychological Association (APA), 67, *see also* Torture, American Psychological Association and
    Division of Psychoanalysis, 115
    ethics code, 83, 144
    Ethics Committee, 87
    Ethics Office, 123
    future, 85

interrogations at Guantánamo, 75
issues supported by, 143
links between military-intelligence establishment and, 84
lobbyists, 75
national security policy, 138
PENS Task Force, 117
policy, opposition to, 79–80
policy of engagement, 89
referendum, 82
Americas Watch, 282
Amnesty hearing, 52
Amoral familism, 367
Antiapartheid activist, 49
Anti-past, 19
Antique imaginings, 238
Antiwar films, 349–358
  animation, 349
  Apocalypse Now, 350
  Auschwitz, 352
  circles, 352
  conundrum of, 349
  depersonalization, 349
  educational scene, 351
  eros, 352
  hallucinatory sequence, 357
  landmines, 357
  Lebanon War, 350, 354
  Let There Be Light, 353
  markers of identity, 351
  nachtraglichkeit, 352
  narcissism, warmaking and, 350
  Nazis, 354
  New York Film Festival, 354
  Peace Now, 355
  postwar medical authority, 352
  psychoneurotic war injuries, 353
  PTSD symptoms, 352
  shifting self-states, 350
  subthreshold knowledge, 350
  uncanny project, 350
  unheimlich, 350
  veteran guilt, 352, 354
  Waltz with Bashir, 352, 356
Antiwar work, psychoanalysis and, 1–2
APA, *see* American Psychological Association
Apocalypse Now, 350
Argentine National Commission of the Disappeared, 286
Aversive racism, 374

B

Balkan war, 46
Barbarism, document of, 235, 320
Battlemind, 35
Behavioral Science Consultation Team (BSCT), 72, 114
Bêta elements, 214
Beyond the Pleasure Principle, 182
Binding legacy, 307
Black sites, 68
Black Skin, White Masks, 336
Blindness of the seeing eye, 225
Blue and Brown Book, The, 217
Body receptors, 204
Bolshevik Revolution, 179
Bosnian war
  refugees, 45
  survivors of, 57
  war crimes tribunal, 54
Bounded subject, 198
Bourgeois ideology, 363
Brainwashing, 252, 256, 267
Bridge world, 214
Brief History of Neoliberalism, A, 365
British Psychoanalytic Society, 16
Brunswick memo, 110
BSCT, *see* Behavioral Science Consultation Team
Buddhist values, 262
Bush, George W., 67, 162
Bush Doctrine, 161

C

Carter, Jimmy, 160, 163, 171
Casus belli, 201–221
  Aby Warburg's delusional research, 211–213
  Aby Warburg's madness, 210–211
  addiction to generalities, 217
  All Quiet on the Western Front, 203
  bearing witness, 204
  Bellevue clinic, 210
  bêta elements, 214
  Beverly Hills (July 1983), 208–210
  biological diagnosis, 205
  bloody rough way, 216
  body receptors, 204
  bridge world, 214
  casus belli, 202

cathartic techniques, 202
changes of aspect, 218
child neuropsychiatry, founder of, 205
children, potential for rehabilitation, 205
collaborationist regime, 201
condemnation of apocalypse, 203
danced causality, 214
definition of, 175
diaries, 203
do not harm, 217–219
Esquirol Clinic, 219
ethnic cleansing, 217
family resemblances, 214
faradization, 206
feedback loop, 202
ferocious psychosis, 207
field of catastrophes, 201–204
god Python, 208
Greek tragedies, 202
historical materialism, 205
Jewish Laws, 201
keeper of secrets, 206
living archaeology, 213
Middle Ages, 215
Mnemosyne, 212
mystification, 219
nameless dread, 214
nameless science, 214
Native American rituals, 213
Nazi-inflected psychology, 205
neuronal man, 206
objectification, 206, 207
petrifaction, 203
prophetic rantings, 211
Psychoanalytical Institute of Berlin, hijacking of, 204
rhythm of work, 202
science of tyranny, 201
second philosophy, 216
seismograph of my soul, 212
self-censoring, 203
snake dance ritual, 213–215
storytelling, 204, 207–208
surviving images, 212
symbolism, 214
thoughts without a thinker, 218
today, 206–207
torture (Berlin 1937), 204–206
undoing of character, 201
World War I, veteran of, 215–217

Catastrophe, psychoanalytic politics of, 29–43
adult onset trauma, 30, 40
battlemind, 35
children's objects, 29
combat troops and returning veterans, military policy on, 34–37
consequences of violence, denying of, 31–32
contextualized trauma , 40
deterrent to seeking treatment, 36
exposure therapy protocols, 39
fascination of the abomination, 31
feedback loop, 38
Green Beret, 37
Holocaust, 33, 40
labeling, 33
massive trauma, 39
measure of dissociation, 31
mental health claims, disability payments for, 37
Oklahoma City bombing, 31
paradox, 32
pornography of violence, 30–31
posttraumatic stress disorder
  contradictory history of, 32–34
  in psychoanalytic practice, 37–41
powerless disgust, 31
psychic trauma, 30
Rand Foundation, 34
shame, 39
tension between individual and societal needs, 29–30
toxic narratives, 32
underserved populations, 37
World War II, 34
Catastrophic change, 362
Catharsis, effect of, 53
Central Intelligence Agency (CIA), 67, 69, 108
Changes of aspect, 218
Cheney, Dick, 161
Child neuropsychiatry, founder of, 205
Children, potential for rehabilitation, 205
Churchill, Winston, 185
CIA, see Central Intelligence Agency
CISD model, see Critical incident stress debriefing model

City on the hill, mission of, 163
Clinton, Bill, 54, 161
Coalition for an Ethical Psychology, 109, 127
Coercive persuasion, 252
Collateral damage, 230
Collective identity, 305–326
   alienation, 311
   ambition, 324
   American immigration officers, 321
   binding legacy, 307
   caregiving, 311
   collective's truth, 325
   collective unconscious, 310
   commands of god, 307
   compulsive funk, 315
   document of barbarism, 320
   epiphany, 316
   family problems, 311
   family relationships, 306
   first responsibility, 325
   frayed conscience, 319
   Gaza offensive, 305
   hegemonic social order, 310
   Holocaust
      alphabet game, 306
      jokes, 307
      memory, 322
      nightmares, 313
   identity, 325
   impotence, 317
   Israeli social contract, 313
   Jewish fundamentalism, protection of, 305
   kibbutz, 312, 316, 318
   moment of recognition, 315
   Nazis, 321
   Oedipal complex, 308
   organization of society, 309
   propaganda industry, 321
   psychoanalysis, 308
   rationale for sacrifice, 307
   ruling reason, working against, 324
   Six-Day War, 323
   subjectivity, 311
   thirdness, 317
   transference, 311
   uber-Jews of Zion, 318
   unconscious, 310
   will for self-realization, 322
Colonialism, 328
Combat, 223–241
   American invasion of Iraq, 225
   antique imaginings, 238
   atrocity-producing situations, 237
   autobiographical subjectivity, 226
   blindness of the seeing eye, 225
   chivalrous passage of arms, 235
   collateral damage, 230
   cultural blindness, 224–226
   document of barbarism, 235
   dread, maternal surveillance, and remorse, 239
   familial communication, 228
   feminine nonviolence, 235
   fixed identities, 224
   grief, 231, 233
   heroic phantasm, 224
   insurgent bombings, 235
   I–Thou relation, 223
   just cause, 226
   just war, 238
   malignant dissociative contagion, 225
   maternal accusation, 227
   maternal admonishment, conjuring of, 235–237
   maternal phantasm, 235
   maternal surveillance
      grief restoring absent images, 233–235
      return of the repressed, 232–233
   moral third, 235
   mourning play, 237
   omniscient Mother, 236, 238
   peace advocate, 237
   personal unconscious, 233
   political debate, 234
   protest theatre, 237–238
   sex, 231
   shame, 235
   state of rupture, 223
   transference representation, 236
   Vietnam, 223
   war photography, 224
   warrior mythology, 233
   warrior-sacrifice, 236
   war zone, 226–232
Commune, 310
Communist doctrine, 267, 271
Complex posttraumatic stress disorder, 284
Concentration camp survivors, 46, 284
Confounding identity, 329, 334

Congregation of Christian Mothers for Peace and Justice, 299
Consciousness-raising technologies, 257
Craddock Four, 52
Crash, 372
Critical incident stress debriefing (CISD) model, 7
Cult(s)
  adherents, 262
  soft induction into, 262
  survivors, 257
  true purpose of, 262
Culture of the New Capitalism, The, 365

D

Danced causality, 214
Debs, Eugene, 346
Deep throat, 109
Defeating process, 15
Descartes, 216
Diagnostic and Statistical Manual, 33
Dialectics of hope, 10
Dirty War, 49, 298
Disavowed mourning, 195
Discipline and Punish: The Birth of the Prison, 260
Disintegration anxiety, 167
Document of barbarism, 235, 320
"Doer–done to" dynamic, 52, 55
"Do not harm" standard, 76, 217
Doublethinking, 265
Dread, *see* Antiwar films
Dreyfus Affair, 310

E

Ego-modifying defense mechanisms, 361
Empathy, 274
Enhanced interrogation techniques, 125, 152, 155
Enlightenment, 198
Ethnic cleansing, 2, 217
Externalized psychosis, 259
Extraction of the Stone of Madness, The, 206
Extreme influence, 273
  power of, 262
  processes of, 257

F

False self, 273
Familial communication, 228
Family problems, 311
FAQ document, *see* Frequently asked questions document
Faradization, 206
Fatherless Society, The, 184
Feedback loop, 38, 202
Fetishism, 362, 369
Filling up the soul, 262
Films, *see* Antiwar films
Foreclosure of alterity, 194
Foreign policy, *see* American foreign policy, violence in
Forward psychiatry, principles of, 15
Four Freedoms speech, 185
Franken, Al, 364
Free market capitalism, 365
Frequent-flyer program, 69
Frequently asked questions (FAQ) document, 79
Freud, Sigmund, 16, 177, 216, 328
From the Unconscious Life of Our School Youth, 184
Frontier, 16

G

Gaslighting, 259–261
Gaza offensive, 305
Gendering of human rights, *see* Women, Latin American terrorist state and
Geneva Conventions, 109, 113, 129
  defendants, 339
  detainees currently denied protections under, 129
  forces not abiding by, 71, 109
  German officer, 246
  Globe editorial, 364
  illegal enemy combatants, 143
  internalized, 68
  interrogations violating, 132
  psychologist participation in interrogations, 144
  psychologist role in national security activity, 137
Globalization, ethical responsibility entailed by, 196
Global war on terror, 10

despair generated by, 91
   major theaters of, 86
God's American Israel, 156
Grave reservation, 8
Greed neurosis, 180
Guantánamo Bay, 118, 143
   ethical standards of health
      professionals, 118
   GTMO-ization program, 73
   SERE standard operating
      procedure, 110
   standard operating procedure, 118
Guilt
   collective, victims of doctrine of, 285
   cosmic, 269
   Lebanon War veterans, 354
   maternal, 295
   mind control and 258
   persecutory, personification of, 235
   as preferred forms of
      communication, 270
   reintroduced, 137
   veteran, 352, 354
   victim, 343
   violations of human rights, 145
   violence and, 30
   White liberal guilt expungement, 151

H

Hallucinations, 269, 357, 368
Healing truth, marker of, 58
Heart of Darkness, 30
Hippocratic Oath, 119
Historical borders, 16
Historical materialism, 205
History, learning from, 15–27
   Algerian War, 22
   anti-past, 19
   black holes of patients, 18
   borders, protection of, 16
   British Psychoanalytic Society, 16
   death area, 18
   defeating process, 15
   effort to drive the other crazy, 18
   forward psychiatry, principles
      of, 15
   French Army, 23
   Freud, 19
   frontier, 16
   historical truth, 19
   Hitler, 24

International Symposium
   for Psychotherapy of
   Schizophrenia, 18
interpretation of violence, 23
judicial institution, 26
laissez faire attitude, 25
natural borders, 16
Ottoman Empire, 24
patient resistance, 15
transitional subject, 20
World War II, 17, 20
History Beyond Trauma, 215, 345
History versus subjective
   experience, 58
Holmes, Oliver Wendell, 182
Holocaust
   alphabet game, 306
   jokes, 307
   metaphor for, 243
   nascent Jewish hegemony, 313
   nightmares, 313
   oral testimonies, 40
   survivors, 33, 40, 121
   traumatic preoccupation with, 344
Homosexual phantasies, 189
Human Rights Violations hearings, 52
Human vulnerability, forms of, 197
Hunger, other war of, 292
Hussein, Saddam, 153
Hypnotic living death, 259

I

IARPP, *see* International Association
   for Relational Psychoanalysis
   and Psychotherapy
ICRC, *see* International Committee of
   the Red Cross
ICTY, *see* International Criminal
   Tribunal for the former
   Yugoslavia
Identification, politics of, *see*
   Palestine, resistance to Israeli
   occupation of
Identification Papers, 327
Identity, *see also* Collective identity
   American, 158
   confounding, 329, 334
   denial of, 268
   -diffusion, 265
   false, 228
   fixed, 224

former sense of, 196
fragmented, 286
gender, 281
ingroup, 374
investments, 335
Jewish, 328, 342, 343
Latin American woman, 292
markers of, 351
national, 136, 169
negative, 269, 271
professional, 90, 268
sexual, 230
shared, 322
victim, 344
war hero, 223
Ideological illumination, 255
IEDs, see Improvised explosive devices
Iliad or Poem of Force, The, 34
Illegal enemy combatants, 143
Impersonal intimacy, 262
Implantation, 272
Improvised explosive devices (IEDs), 6, 12
Incest taboo, 188
Informal empire, 159
Ingroup identity model, 374
Inner captor, 275
In the Penal Colony, 261
Insidious trauma, 283
Intelligence, exploitation of assets, 83
Internalized shame, 270
Internal object relations, 198
International Association for Relational Psychoanalysis and Psychotherapy (IARPP), 355
International Committee of the Red Cross (ICRC), 114
International Criminal Tribunal for the former Yugoslavia (ICTY), 45, 47
International Rehabilitation Council for Torture Victims (IRCT), 86
International Symposium for Psychotherapy of Schizophrenia (ISPS), 18
Interpretation of Dreams, The, 177
Introjective processes, 254
Intromission, 272
In the Valley of Elah, 371–374
Iraq veterans, 11, see also Posttraumatic stress disorder
Iraq War, 154, 244, 359, 370

IRCT, see International Rehabilitation Council for Torture Victims
ISPS, see International Symposium for Psychotherapy of Schizophrenia
Israel–Palestine conflict, see Palestine, resistance to Israeli occupation of
I–Thou relation, 223

J

Jewish fundamentalism, protection of, 305
Jewish identity, 342, 343
Jewish Laws, 201
Johnson, Lyndon, 165
Jointness, 111
Joint Personnel Recovery Administration (JPRA), 71
Jones, Jim, 262, 264
JPRA, see Joint Personnel Recovery Administration
Just war, 238

K

Keeper of secrets, 206
Kennedy, John F., 159, 233
Kid therapist, 4
Kissinger, Henry, 161
Know Hope, 91
Korean War, 166, 209

L

Latin American terrorist state, see Women, Latin American terrorist state and
Lebanon War, 349, 350, 355
Leniency, 267, 269
Let There Be Light, 353
Linguistic borders, 16
Looping of consciousness, 254
Love-bombing, 262

M

Malignant dissociative contagion, 225
Marital problems, 6
Maternal accusation, 227
Maternal phantasm, 235

Maternal surveillance
  grief restoring absent images, 233–235
  protest theatre, 237–238
  remorse and, 239
  return of the repressed, 232–233
McCain, John, 159
McGovern, George, 160, 163, 171
Memory
  combat, see Combat
  Holocaust, 322
  uncanny, 350
Militarism deconstructed, 175–176
Military psychiatry, ethic, 6
Mind control, 251–277
  abandonment of self, 272
  badness of self, 263
  black hole, 262
  brainwashing, 252, 256
  Buddhist values, 262
  chaotic-change moment, 255
  childrearing, 258
  coercive demands, 261
  coercive persuasion, 252
  cognitive organization, changes in, 257
  Communist doctrine, 267, 271
  confession, 268
  consciousness-raising technologies, 257
  crimes against the soul, 260
  cult adherents, 262
  cult survivors, 257
  cultural revolution, 272
  culture revolution in China, 268
  definition of, 252
  denial of identity, 268
  doublethinking, 265
  EEG pattern, 255
  empathy, 274
  externalized psychosis, 259
  extreme influence, 273
    power of, 262
    processes of, 257
  false love, 266
  false self, 273
  filling up the soul, 262
  gaslighting, 259–261
  hallucinations, 269
  hate turning into "loving kindness," 261
  Holy Spirit, 257
  humiliation, 265
  hypnotic living death, 259
  idealization of analyst, 273
  ideological illumination, 255
  impermeability, 273
  impersonal intimacy, 262
  implantation, 272
  incarceration, 260
  indoctrination, 258
  infantilization of candidates, 258
  inner captor, 275
  institutional effects, 273
  intimate engagement, 265
  introjective processes, 254
  intromission, 272
  learning sessions, 263
  leniency, 267, 269
  looping of consciousness, 254
  love
    hostility turned into, 266–270
    transmutation into coercion, 262–264
  love-bombing, 262
  masochism, 261
  modern soul, 260
  negative identity, 269
  organized sadism, 260
  People's Temple, 264–266
  physiological processes, 255–257
  politics of experience, 258
  power of affects, 258
  prison, 260
  psychic equivalence, 263
  psychological factors, 257–258
  psychotherapy, 272–275
  psychotic insight, 267
  redemption seeker, 261
  reeducation programs, 252
  religious brainwashing routines, 267
  revolutionary universities, 267
  self-abhorrence, corrected, 271
  self-condemnation, 268, 272
  self-loathing, 270–272
  self-sacrifice, 264
  self-strengthening tracks, 252
  sensory assaults, 255
  sexual abuse, 258
  shamanistic healing, 255
  shame, 270
  signal-shame, 270
  silencing of human consciousness, 253

snapping, 254–255, 256
soul murder, 258
Spiritual Inner Awareness, 262
stereotypy, 255
superego complex, 273
symmetrical contrasts, 251
technical transformation of individuals, 260
technologies of care, negative, 260
technologies of experience, 257
terms, 252
thought reforms, 252
trance states, 254, 255
types of processes, 261
unspeakably abused youth, 260
victim softening, 261
Voodoo, 252
Mind and Matter, 206
Misogyny, *see* Women, Latin American terrorist state and
Mizrahi Jews, 338
Model
 BSCT, 126
 consultant, 366
 critical incident stress debriefing, 7
 feminine, 294
 ingroup identity, 374
 sociopathic, 137
 sovietic pedagogy, 205
Modern soul, 260
Moral third, 235
Moses and Monotheism, 216
Mothers and the Grandmothers of the Plaza de Mayo, 298
Mourning play, 237
Movimiento de la Izquierda Revolucionaria, 297
Munich complex, 188
Mystification, 219
Myth of America, 157

N

Nameless dread, 214
Narcissism
 cultural shift toward, 136
 gratification, 171
 warmaking and, 350
National security lobby, 149
Native American rituals, 213
Natural borders, 16
Navajo soldiers, 248

Nazi-inflected psychology, 205
Negative identity, 269
Neoliberalism, 360
Neuronal man, 206
New capitalism, 360
NGOs, *see* Nongovernmental organizations
Nixon, Richard, 161
Nixon tapes, 125
Nongovernmental organizations (NGOs), 81
Nuremberg defense, 90

O

Obama, Barack, 69, 81, 191, 250, 370
Object-relational tradition of psychoanalysis, 192
Oedipal complex, 308
Oklahoma City bombing, 31
Omnipotence, self-concept of, 160
Omniscient maternal eye, 238
Operation Defensive Shield, 339
Organized sadism, 260
Orwell, George, 190, 265, 271

P

Palestine, resistance to Israeli occupation of, 327–347
 American socialist, 346
 anti-identitarian position, 329
 colonialism, 328
 colonizer's prescribed identifications, 332
 compassionate radical, 340
 confounding identity, 329
 diary, 335
 furious provocative radical, 340
 Green Line, 344
 guilt, 341
 hawiyyeh, 330, 331
 hysteria, 332, 333
 identification with Palestinians, 341
 impersonation, 332
 intifada, 343
 invocation of Biblical precepts, 342
 Israeli soldiers refusing to serve, 337–346
 Israel–Palestine conflict, 329
 Jewish identity, 342, 344
 mimicry, 333

Mizrahi Jews, 338
Operation Defensive Shield, 339
Other, 336
otherness, 336
Palestinian professor and Israeli interrogator, 330–337
Parents Circle, 335
patriotism, 341
personhood, 336
political possibilities, 329
power hierarchies, 335
psychic damage, 345
racism, ACT UP, 328
radical leftists, 340
rape, 343
refugees, 342
refuseniks, 338
revolutionist agenda, 339
settler colonialism, 329
subjectivity, 336
symbolic violence, 328
terrorist population, 333
victim position, 344
voluntarist notion of identification, 328
we-ness versus underdog schemas, 339, 341
West Bank
  invasion of, 335
  settlements, 337
Zionist position, 342
Peace Now, 355
Peace Psychology resolution, 144
Pearl Harbor, 192
PENS, see Psychological Ethics and National Security
Pension-struggle neurosis, 180
People's Temple, 264–266
Perpetrator-friendliness, 55
Perpetratorhood, 61
Physicians for Human Rights, 84
Physicians for Social Responsibility, 86
PNAC, see Project for the New American Century
Politics of disengagement, 360, 366, 367
Politics of experience, 258
Posttraumatic stress disorder (PTSD), 1, 3–14
  Abu Ghraib prison, 4
  adaptive symptoms, 3
  arguments about symptoms, 36
  chronic symptoms, 39

cliché of, 33
combat kills, 3
complex, 284
contradictory history of, 32–34
critical incident stress debriefing model, 7
detainees, 86
diagnosis, 1, 32
dialectics of hope, 10
film, 352
global war on terror, 10
grave reservation, 8
kid therapist, 4
marital problems, 6
military psychiatry, ethic, 6
Osama Bin Laden, 9
in psychoanalytic practice, 37–41
Readjustment Counseling Center, 10
redeployment, 8
retraumatization, prevention of, 5
symptoms, 34, 36
traumatic stressors, 283
understanding of, 201
Vietnam veterans, 4
war within, 4
Powell, Colin, 364
Power status quo, 363
Precarity, situation of, 360
Prescribed identifications, 332
Prisoner of hope, 360
Project for the New American Century (PNAC), 161
Psychoanalysis
  antiwar work, 1–2
  collective identity, 308
  final stage of, 138
  intimate desire vs. personal identity, 311
  object-relational tradition, 192
Psychoanalysis, vulnerability, and war, 177–200
  abstinence, 181
  acceptance of loss, 196
  antiwar sentiment, 184
  Bolshevik Revolution, 179
  bounded subject, 198
  casus belli, definition of, 175
  classical liberalism, collapse of, 179–185
  closure, 193
  depressive position, 189

disavowed mourning, 195
economy of grieving, 195
Enlightenment, 198
feminine alternative to Freud, 188
foreclosure of alterity, 194
Four Freedoms speech, 185
globalization, ethical responsibility entailed by, 196
global justice, dawn of, 191–197
greed neurosis, 180
homosexual phantasies, 189
humanity, 198
human vulnerability, forms of, 197
incest taboo, 188
internal object relations, 198
linguistic links, 193
mass event, personalized, 193
mercenary army, 192
Munich complex, 188
object-relational tradition of psychoanalysis, 192
pacifism, 178
Pearl Harbor, 192
pension-struggle neurosis, 180
politics of vulnerability, 197–198
problem of motivation in democratic wars, 177–179
riddle of the neurosis, 198
sexuality, 183
shell shock, 181, 198
social democracy, triumph of, 185–191
socialism, 190
vulnerability, 192
war on terror, 191–197
World Trade Center attack, 191
World War I, 179–185
World War II, 185–191
Psychoanalytic activism, 107–141
 action, 138
 amendment to reaffirmation, 131–133
 American denial of responsibility, 114
 beginnings, 112–114
 brainstorming, 108
 Brunswick memo, 110
 BSCTs and PENS, 114–117
 clandestine agencies, 135
 clinical listening, 107
 Coalition for an Ethical Psychology, 109, 127
 coalition formation, 126–127
 complacency, 116
 conditions of confinement, 113
 covert torture program, 107
 deep throat, 109
 dirty wars, 116
 enhanced interrogation program, 125
 friendly amendment, 128
 future operations, 137–138
 Geneva conventions, 109, 113, 129
 government abuses of detainees, 107
 Guantánamo Bay, 110, 118
 Hippocratic Oath, 119
 Hollywood moment, 109–112
 Holocaust survivors, 121
 implementation policy, 133
 inoculation program, 109
 interrogator attitude, 113
 jointness, 111
 mind-altering drugs, 121
 "moratorium" amendment, 130
 narcissism, cultural shift toward, 136
 Nixon tapes, 125
 outside doctors, 114
 PENS confidentiality, breaking of, 120–125
 PENS listserv, 124
 PENS process, 125–126
 preexisting condition, 114
 press reports, 119
 rape, 130
 Resolution Against Torture, 127–128
 Resolution Against Torture, reaffirmation of, 128–131
 resolution to exit the procedure, 129
 risk risking, 109, 135
 sleep deprivation, 117
 sociopathic model, 137
 tacit political unconscious, 135
 threat to well-being, 136
 torture, 125
 town meeting, 131
 turning point, 117–120
 United Nations
  Convention Against Torture, 128
  Special Rapporteur on Torture, 134
 "war on terror" reporters, 112

waterboarding, 130
WithholdAPAdues, 132
World Trade Center attack, 112
Yoo–Bybee torture memos, 128, 130
Psychological Ethics and National Security (PENS), 143
Psychological fallout, 47
Psychologists for Social Responsibility (PsySR), 86
Psychology's militarism, paradox of, 65–66
Psychoneurotic war injuries, 353
Psychotic insight, 267
PsySR, *see* Psychologists for Social Responsibility
PTSD, *see* Posttraumatic stress disorder

R

Racism
  ACT UP, 328
  aversive, 374
  confronting, 371
  disavowals, 370
  social movements against, 359
Rand Foundation, 34
Rape and Sexual Abuse: Torture and Ill Treatment of Women in Detention, 290
Readjustment Counseling Center, 10
Reagan, Ronald, 159, 160
Reagan revolution, 370
Redemption seeker, 261
Regeneration, 8
Relational failure, 362
Religious brainwashing routines, 267
Remarks on Frazer's Golden Bough, 214
Repetition compulsion, 373
Resistance, 303–304, *see also* Palestine, resistance to Israeli occupation of; Psychoanalytic activism
Resistance, resistance to, 359–376
  Agency for International Development, 359
  ambiguous position, 363
  amoral familism, 367
  apathy, 361
  aversive racism, 374

bedrock contempt for femininity, 370
bedrock of resistance, 373
"big government," 365
bitch within, 359
bourgeois ideology, 363
capacity for truth telling, 364
catastrophic biographical events, 363
catastrophic change, 362
contempt for femininity, 373
"cutting edge" corporate cultures, 366
disavowed helplessness, 364
dominant narratives, 363
ego, new danger to, 361
ego-modifying defense mechanisms, 361
fantasies of "perfect love," 369
fetishism, 362, 369
fetish structure of social perversion, 368–370
finite-infinite relationship, 364
free market capitalism, 365
government intervention, 365
guilt, 373
habitual defenses, resistance to giving up, 373
ingroup identity model, 374
In the Valley of Elah, 371–374
"lie" of self-sufficiency, 367
lying, 362
mourning process, 372
mutual collusion, 363
national call of distress, 371
neoliberalism, 360
new capitalism, 360
norms, 373
Obama's communitarian ethic, 374
pain, choice to feel, 362
penis envy, 373
perception of reality, 361
perversion of truth, 361–365
perversity, 370
political agenda, 360
politics of disengagement, 360, 366, 367
power status quo, 363
primal horde theory, 373
prisoner of hope, 360
Reagan revolution, 370
relational failure, 362
repetition compulsion, 373
sadomasochistic dynamics, 360
self-esteem regulation, 366

shameful states, 367
situation of precarity, 360
socially shared perversion, 364
social perversion, 360, 373
social roots of contemporary disavowal, 365–368
Students for a Democratic Society, 359
tax cuts, 366
traumatizing relational failure, 370
U.S. individualism, 365
vulnerability, cultural management of, 374
welfare state, social solidarity, 367
Resolution Against Torture, 127–131
Revolutionary universities, 267
Riddle of the neurosis, 198
Risk risking, 109, 135
Rogers, John, 262
Roosevelt, Franklin, 185
Roosevelt, Theodore, 157

S

Sadomasochistic dynamics, 360
Salvadoran Revolution, 292
Science versus theatre, 58
SDS, see Students for a Democratic Society
Second Coming, 156
Seductive Poison, 264
Self-condemnation, 268, 272
Self-esteem regulation, 366
SERE techniques, see Survival, Evasion, Resistance, and Escape techniques
Settler colonialism, 329
Seventh Million, The, 344
Sexual abuse, 258
Shamanistic healing, 255
Shameful degradation, 156
Sheehan, Cindy, 237, 238
Shell shock, 181, 198
Signal-shame, 270
Silence rule, 298
Six-Day War, 323
Sleep manipulation, 74, 79
Snake dance ritual (1923), 213–215
Snapping, 256
Societal repair, see Testimony, inevitable tensions in
Sociopathic model, 137

Soul murder, 258
Spanish Civil war, 310
Spiritual Inner Awareness, 262
State of rupture, 223
Stereotypy, 255
Storytelling, 204
Students for a Democratic Society (SDS), 359
Subthreshold knowledge, 350
Superego complex, 273
Survival, Evasion, Resistance, and Escape (SERE) techniques, 110
Survivor(s)
    adult onset trauma, 40
    antiapartheid violence, 60
    appearance before TRC and ICTY, 58
    Balkan war, 46
    Bosnian war, 57
    concentration camp, 46, 284
    cult, 257
    expectations, 30
    Holocaust, 33, 40, 121
    indifference to survival, 38
    measure of dissociation, 31
    need, 29
    Oklahoma City bombing, 31
    –perpetrator interactions, 54
    postconflict countries, 51
    potential, 5
    psychodynamic treatment, 41
    psychological fallout, 47
    reassurance to, 5
    search for accountability and justice, 2
    shame, 39
    symptoms not officially recognized, 34
    torture, psychological disorders, 289
    toxic narratives, 32
    victimization of, 60

T

TASSC, see Torture Abolition and Survivors Support Coalition International
Technologies of experience, 257
Telescoping of Generations, The, 345
Testimony, inevitable tensions in, 45–63
    acknowledgment, 54, 60

amnesty hearing, 52
amnesty for perpetrators, 52
antiapartheid activist, 49
apartheid-era violence, 52
Balkan war, survivors of, 46
Bosnian war, survivors of, 57
"breaking out" of silence, 46
bureaucratic production of truth, 60
catharsis, effect of, 53
concentration camp survivors, 46
Craddock Four, 52
Dirty War, 49
dissatisfaction with justice rendered, 59
"doer–done to" dynamic, 52, 55
financial resources, 54
healing truth, marker of, 58
history versus subjective experience, 58
human rights violation hearings, 52, 56
International Criminal Tribunal for the Former Yugoslavia, 450
"liminal" state of not knowing, 52
perpetrator-friendliness, 55
perpetratorhood, 61
perpetual tyranny, 46
politically selected victims, 58
protected witness, 51
psychological fallout, 47–49
public hearings, 50
public/private matters, 58–61
recognition, characterization of, 60
science versus theatre, 58
social violence, impact of, 60
societal repair, 47
South African TRC, origins of, 49–54
survivor, victimization of, 60
testifying under protection, 51
truth commission, 55
types of truth, 58
uncanny, 53
unmediated trauma, 53
uses of truth, 54–58
victory over abuse, 57
Thirdness, 317
Torture, American Psychological Association and, 143–152
Abu Ghraib prison, 143
casebook, 144
conspirators, 146
dialogue with myself, 145
economic privilege, 151
enhanced interrogation techniques, 152
Guantánamo Bay detention center, 143, 147
illegal enemy combatants, 143
national security lobby, 149
Peace Psychology resolution, 144
Psychological Ethics and National Security, 143
uncanny, 145
UN World Congress Against Racism, 146, 149
White liberal guilt expungement, 151
Torture, psychologists defying, 67–105
accountability, state licensing boards and, 88
American Civil Liberties Union, 88
APA Ethics Committee, 87
APA referendum, 82
Behavioral Science Consultation Teams, 72–74
black sites, 68
detainee abuse, 67
detention settings, 82
"do not harm" standard, 76
ethical discussion, 89–91
exploitation of assets, 83
fear up, 82
frequent-flyer program, 69
frequently asked questions document, 79
Geneva Conventions, 68, 71
global war on terror, 86
despair generated by, 91
major theaters of, 86
government-sponsored research, 90
GTMO-ization program, 73
innocent civilians, 77
interrogation settings, typology of, 68–70
Know Hope, 91
laudable goals, 80
legality of techniques, 70
military JAGs, 78
nongovernmental organizations, 81
Nuremberg defense, 90

opposition to APA policy, 79–80
organizational reform, 84–85
personal accountability, 87–89
potential insurgents, 68
professional association response, 74–79
psychologists, 70–72
public reckoning, 85–87
referendum implementation, 81–82
referendum passage, 80
roles, 82–83
sleep manipulation, 71, 74, 79
systematized torture regime, 68
task of dismantling abuse, 91–92
tasks ahead, 80–81
truth commission, 86
U.S. interrogation operations, 68
U.S. program of torture, 85
Torture Abolition and Survivors Support Coalition International (TASSC), 86
Toxic narratives, 32
Tractatus Philosophicus, 215
Trained to Kill, 353
Trance states, 254, 255
Transitional subject, 20
Trauma
adult onset, 30, 40
categories of symptoms, 283
contextualized, 40
energy stemming from, 204
insidious, 283
psychic, 30
psychotherapy of, 207
repeated, 287
self-destructive reaction to, 192
unmediated, 53
Trauma and Recovery, 281
TRC, *see* Truth and Reconciliation Commission
Truth, *see* Testimony, inevitable tensions in
Truth and Reconciliation Commission (TRC), 47, 49
Tutu, Desmond, 49

U

Übertragung, 218
Uncanny project, 350
UNCAT, *see* United Nations Convention Against Torture
unheimlich, 350
United Nations
Convention Against Torture (UNCAT), 128
domestic violence reports, 282
Principles of Medical Ethics, 118
Special Rapporteur on Torture, 134
World Congress Against Racism, 146, 149

V

Veteran(s)
Afghanistan, 11
guilt, 352
Iraqi, 11
Lebanon War, 354
Vietnam, 4, 13, 224
World War I, 215–217
Victim(s)
collective guilt, 285
collective identity, 344
coping mechanisms of, 282
identification with, 250
politically selected, 58
position, Palestine, 344
softening, 261
Vietnam War, 166, 223, 310
Violence, *see* American foreign policy, violence in
Volunteer army, 192
Voodoo, 252

W

Waltz with Bashir, 349
War imaginary, 175
War and militarism deconstructed, 175–176
War of Nerves, A, 24
War of Nerves: Soldiers and Psychiatrists in the 20th Century, A, 205
War photography, 224
War stories, 243–250
assassination, 250
"eliminate difference," 248
entrance wound, 245–246
excitement, 247

"first, do no harm" admonition, 247
Geneva rights, 246
harm engendered, 247
harm transfigured, 250
helpless disidentification, 244
infiltration, 246–249
inventory, 245
metaphor for the Holocaust, 243
Navajo soldiers, 248
Nazis, 246
point of attachment, 249
prayer, 244
primal words, 249
secret foreknowledge, 248
self-surgery, 246
sniper, 249–250
surgical textbook, 244
victim, identification with, 250
Wayward Youth, 184
Welfare state
 creation of during World War II, 177
 social solidarity, 367
We-ness versus underdog schemas, 339, 341
Wesley, John, 266
What Maisie Knew, 243
Wilson, Woodrow, 157, 158
WithholdAPAdues, 132
Women, Latin American terrorist state and, 279–301
 Americas Watch, 282
 Argentine National Commission of the Disappeared, 286
 body autonomy, assault on, 288
 coercive violence, 281
 complex posttraumatic stress disorder, 284
 concentration camp survivors, 284
 Congregation of Christian Mothers for Peace and Justice, 299
 depersonalization, 290
 dichotomized object, 282
 Dirty War, 284, 298
 doctrine of collective guilt, 285
 ego ideal system, 299
 family, 291
 feminine model, 294
 gender difference
  exacerbation of, 289–296
  homogenization of, 284–289
 glorification of women, 282
 Guatemalan Mutual Support Group for the Appearance of Our Relatives Alive, 291
 hunger, 292
 identificatory subject of desire, 300
 identity of the Latin American woman, 292
 ideological cleaning of body politic, 279
 insidious trauma, 283
 internalization of aggression, 289
 international Human Rights organizations, 296
 media, 295
 middle class women, 294
 militarization of social life, 286
 misogynist ideology, 281
 misogynist military forces, 279
 Mothers and the Grandmothers of the Plaza de Mayo, 298
 Movimiento de la Izquierda Revolucionaria, 297
 opposition to terrorist state, 298
 paradoxical problem, 280–282
 pathological effects of state-induced trauma, 299
 posttraumatic stress disorder, 283
 pregnancy, 296
 psychic integrity, 289
 psychic numbing, 283
 psychological response to state terror, 297–298
 rape by army personnel, 290
 reengulfment, male dread of, 282
 repeated trauma, 287
 resistance to torture, 297
 Salvadoran Revolution, 292
 "secret" concentration camps, 285
 self-blame, 294
 self-esteem, 288
 sensory deprivation, 287
 sexual assault, 283, 290, 295
 sexual torture for men, 288
 silence rule, 298
 society as war zone, 285
 state ideology, 294
 state terror, psychological impact of, 279
 targets of military repression, 281
 torture techniques, 295
 traumatic stressors, 283

victims, coping mechanisms of, 282
violence against women, 282–284
wife battering, 291
Women's Rights Project, 282
World Trade Center attack, 112, 153, 191, 245

World War I, 15, 210, 215–217
World War II, 17, 34, 209

Y

Yoo–Bybee torture memos, 128, 130